Mathematica Beyond Mathematics

Although many books have been written about Mathematica, very few of them cover the new functionality added to the most recent versions of the program. This thoroughly revised second edition of *Mathematica Beyond Mathematics: The Wolfram Language in the Real World* introduces the new features using real-world examples based on the experience of the author as a consultant and Wolfram certified instructor. The examples strike a balance between relevance and difficulty in terms of Mathematica syntax, allowing readers to incrementally build up their Mathematica skills as they go through the chapters While reading this book, you will also learn more about the Wolfram Language and how to use it to solve a wide variety of problems.

The author raises questions from a wide range of topics and answers them by taking full advantage of Mathematica's latest features. For example: What sources of energy does the world really use? Are our cities getting warmer? Is the novel El Quixote written in Pi? Is it possible to reliably date the Earth using radioactive isotopes? How can we find planets outside our solar system? How can we model epidemics, earthquakes and other natural phenomena? What is the best way to compare organisms genetically?

This new edition introduces the new capabilities added to the latest version of Mathematica (version 13), and discusses new topics related to machine learning, big data, finance economics, and physics.

New to the Second Edition

- Separate sections containing carefully selected additional resources that can be accessed from either Mathematica or online
- Online Supplementary materials including code snippets used in the book and additional examples.
- Updated commands to take full advantage of Mathematica 13.

José Guillermo Sánchez León (http://diarium.usal.es/guillermo) is an engineer, physicist and mathematics PhD holder. He is currently a mathematics and statistical consultant and teaches Mathematical modeling at the University of Salamanca. He has worked in the energy industry and conducted research in a variety of fields: Modeling, optimization, medical physics, astronomy, finance and others. In 1999, he was awarded a research grant at Wolfram Research Inc. headquarters in Champaign (Illinois, USA) after his statistical applications with Mathematica project won a competition sponsored by the company. Since then, he has been an active Mathematica and webMathematica alpha and beta tester.

He is also a Wolfram certified instructor and has extensive experience in teaching and developing programs with both applications. Among his more than 100 articles, there are several where *Mathematica* and web *Mathematica* have been used extensively.

Mathematica Beyond Mathematics

The Wolfram Language in the Real World
Second Edition

José Guillermo Sánchez León
University of Salamanca, Spain

Translated by
Rubén García Berasategui

CRC Press
Taylor & Francis Group
Boca Raton London New York

CRC Press is an imprint of the
Taylor & Francis Group, an **informa** business

A CHAPMAN & HALL BOOK

Second edition published 2023

by CRC Press
6000 Broken Sound Parkway NW, Suite 300, Boca Raton, FL 33487-2742

and by CRC Press
4 Park Square, Milton Park, Abingdon, Oxon, OX14 4RN

Library of Congress Cataloging-in-Publication Data

Names: Sánchez León, José Guillermo, author.
Title: Mathematica beyond mathematics : the Wolfram language in the real world / José Guillermo Sánchez León, University of Salamanca, Spain ; translated by Rubén García Berasategui.
Description: Second edition. | Boca Raton, FL : Chapman & Hall/CRC Press, 2023. | Includes bibliographical references and index.
Identifiers: LCCN 2022030387 (print) | LCCN 2022030388 (ebook) | ISBN 9781032004839 (hbk) | ISBN 9781032010236 (pbk) | ISBN 9781003176800 (ebk)
Subjects: LCSH: Mathematica (Computer file) | Mathematics--Data processing.
Classification: LCC QA76.95 .S26 2023 (print) | LCC QA76.95 (ebook) | DDC 510.285/536--dc23/eng/20221104
LC record available at https://lccn.loc.gov/2022030387
LC ebook record available at https://lccn.loc.gov/2022030388

ISBN: 978-1-032-00483-9 (hbk)
ISBN: 978-1-032-01023-6 (pbk)
ISBN: 978-1-003-17680-0 (ebk)

DOI: 10.1201/ 9781003176800

Typeset in Times New Roman
by KnowledgeWorks Global Ltd.

Publisher's note: This book has been prepared from camera-ready copy provided by the authors.

Contents

Preface

Mathematica and the Wolfram Language

If you have used *Mathematica* occasionally or heard of it, you may have the false impression that it is a program for performing complicated calculations, usually for academic purposes. However, this idea is far from the truth. Actually, *Mathematica* is much more than that. By putting together the computational power and ease of use of the Wolfram Language, *Mathematica*'s high-level general-purpose programming language, the program can be used in any scientific or technical field: aerospace engineering, environmental sciences, financial risk management, medical imaging and many others.

Mathematica can be considered a tool that empowers non-professional programmers to develop applications, although if you are a professional programmer, you will see that the software provides a development environment similar to the ones available for C++ or Python. You can even use the program to control external devices.

About the Book

Although many books have been written about *Mathematica,* very few of them cover the new functionality added to the most recent versions of the program. This text introduces the new features using real-world examples, based on the experience of the author as a consultant. In the process, you will also learn more about the Wolfram Language and how you can use it to solve a wide variety of problems. Both are the most important objectives of the book. To accomplish that, the author raises questions from a wide range of topics and answers them by taking full advantage of *Mathematica's* latest features. For example: What is the hidden image in *"The Ambassadors"* painting by Holbein? What sources of energy does the world really use? How can we calculate tolerance limits in manufacturing processes? Are our cities getting warmer? Is the novel *"El Quijote"* written in Pi? How do we know how old our planet is? How can we find planets outside our solar system? How big is our galaxy? And the universe? How do we know it? How can we model the distribution of radioactive isotopes in the human body? And a tsunami? What are and how can we create Mandelbrot fractals? How can we measure the genetic distance between species? How can we perform financial calculations in real time? How do we value financial derivatives? How can we make entertaining simulations for teaching mathematics, physics, statistics, ... ? Why are there no free quarks?

The answers to the previous questions will not only help you master *Mathematica*, but also make you more familiar with the corresponding topics themselves.

Book Objectives

This book will not only be useful to newcomers but also to those familiar with the program and interested in learning more about the new functionality included in the latest versions.

Those readers with minimal or no knowledge of *Mathematica* are strongly advised to read Chapter 1 along with Stephen Wolfram's book *An Elementary Introduction to the Wolfram Language* available from within the program documentation under the resources section: **Help ▸ Wolfram Documentation ▸ Intro Book & Course ≫**

This text will also make it easy to start programming using the Wolfram Language and to learn how to take full advantage of its capabilities.

The final objective of *Mathematica Beyond Mathematics* is to help you avoid feeling overwhelmed by the software's vast capabilities. The author has explored a significant part of them choosing the most relevant parts and illustrating them with examples from many different sources including the program documentation. Links to additional resources are also provided. The main aim of all this is to reduce significantly the amount of time required to master the tool. The *Mathematica* functions used will be explained using short sentences and simple examples.

Use of the Book

Although this book is not a manual, inexperienced readers should at least read the first four chapters in consecutive order to gain a solid understanding of how *Mathematica* works. Regarding the rest of the chapters, you should keep in mind the following: a) Chapter 5 deals with relatively advanced probability and statistics topics; c) Chapter 6 is a short introduction to machine learning and neural networks; d) Chapters 7 to 11 explore topics related to a single area of knowledge (mathematics, astronomy, nuclear physics, modeling, and economics and finance). You can read them according to your preferences; and d) Chapter 12 is for those readers facing problems requiring big computational resources (parallel computing, grid-enabled calculations, etc).

The principal theme of each chapter is used as the motivation to illustrate certain features of the program that you may find useful for solving a great variety of problems. For example: Chapter 8, related to astronomy, explains how to create dynamic images; Chapter 10, covering the modeling of biological systems, discusses the resolution of differential equations in *Mathematica* using concrete examples.

Second Edition and the Supplementary Materials

Mathematica Beyond Mathematics 2nd Edition improves and updates the first edition. In this endeavor, Ruben Garcia Berasategui, a lecturer at Jakarta International College, has played a fundamental role, not only by translating the text but also by making insightful comments and suggestions.

Everything shown in the text has been done using *Mathematica*, including the access to information sources.

At the end of each chapter, there is a section containing additional resources carefully selected that can be accessed most of the time from either *Mathematica* or online.

In the book's website: http://www.mathematicabeyondmathematics.com/ you can find supplementary materials including code and files used in the examples.

Although most of the contents of the first edition are still relevant, this second edition also covers functionality added in versions 12 and 13 and includes many more examples. To keep the number of pages similar to the first edition, some of the previous contents have been moved online (on the book's website).

This edition has been written and edited using *Mathematica* 13, although it should still be useful with future releases of the program as well. At least until 2025, if there are any changes to the program's functionality as described in the book, they will be covered as supplementary material in the website.

The author would appreciate any comments or suggestions. Please send them to: guillermo2046@gmail.com, with the subject: '*Mathematica* Beyond Mathematics'. If any errors are found, an errata list will be created and made available online (on the book's website).

Salamanca (Spain), November 2022

Author

J. Guillermo Sánchez León (http://diarium.usal.es/guillermo) is an engineer, physicist, and mathematics PhD holder. He is currently a mathematics and statistical consultant and teaches mathematical modeling at the University of Salamanca. He has worked in the energy industry and conducted research in a variety of fields: Modeling, optimization, medical physics, astronomy, finance, and others. In 1999, he was awarded a research grant at Wolfram Research Inc. headquarters in Champaign, Illinois, US after his statistical applications with *Mathematica* project won a competition sponsored by the company. Since then, he has been an active *Mathematica* and web*Mathematica* alpha and beta tester. He is also a Wolfram-certified instructor and has extensive experience in teaching and developing programs with both applications. Among his more than 100 articles, there are several where *Mathematica* and web*Mathematica* have been used extensively.

Reviewer

Rubén García Berasategui is a math, finance, and statistics lecturer at Jakarta International College. He holds a bachelor's degree in business administration and an MBA. He's been using *Mathematica* since 1997 and in 2012 became the first Wolfram certified-instructor in Southeast Asia. He has been training hundreds of newcomers to *Mathematica* and sharing his passion for the program with them ever since.

Acknowledgments

The author would like to express his gratitude to Ruben Garcia Berasategui for his help in writing both editions. Most likely, I would not have written the book without his participation.

The author would also like to thank Callum Fraser (Editor of CRC Press, Taylor & Francis Group), for his vision in realizing the need in the market for a book about Mathematica like this one, and Mansi Kabra (Senior Editorial Assistant of CRC Press , Taylor & Francis Group) for his encouragement and support during the production process.

The author would also like to say thank you to Addlink Software Científico (http://www.addlink.es), sponsor of the first Spanish edition of this book; Antonio Molina and Juan Antonio Rubio for their editorial guidance; and J. M. Cascón (USAL), J. López-Fidalgo (UCLM), S. Miranda (Solventis), R. Pappalardo (US), J. M. Rodríguez (USAL) and C. Tejero (USAL), for their comments and corrections to the first edition.

1

Getting Started

This chapter introduces Wolfram Mathematica, a software program that goes beyond numeric and symbolic calculations. It is a modern technical computing system based on a high-level general-purpose programming language, the Wolfram Language (WL). You will be able to start using the application after a few minutes with it but, if you want to take full advantage of its capabilities, you will have to spend time to master its language although probably less than with other programming languages such as C++ or Java, given its gentler learning curve. The sections in this chapter provide an overview of the program's capabilities, show you how to get started, interact with Mathematica's front end and explain the main ideas behind WL through examples ranging from basic plotting to image manipulation and machine learning.

1.1 *Mathematica,* an Integrated Technical Computing System

Mathematica took a revolutionary step forward with the introduction of the **free-form linguistic input**, which consists of writing in plain English a request that the program will try to answer. Using this format ensures that newcomers can get started very quickly. For experienced users, it means help in finding the most appropriate functions to perform a desired calculation. To get an idea of *Mathematica*'s capabilities before beginning to use the program, let's take a look at the following examples (for now, you just need to read what follows, later we will show you how to write inputs and run commands):

- We can even ask very specific questions: To learn more about the temperature in a location, just write down the name of the place followed by "temperature." The most important advantage of this approach is that the information from the answer can be used inside the *Mathematica* environment.

In[ ]:=

Out[ ]= 28. °C

The `AirTemperatureData` function is shown in the output. It is one of the WL functions for accessing scientific and technical databases from many different fields over the Internet. The data have been already formatted by Wolfram Research for their direct manipulation inside the application.

- Here is another example of this kind of functions. We can choose a chemical compound and get many of its properties. In this case, we get the plot of the caffeine molecule.

In[ ]:=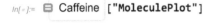

Out[ ]=

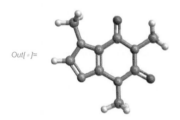

With very brief syntax we can build functions that in other programming languages would require several lines of code.

- In less than half a line we can write the necessary instructions to make an interactive model of the Julia fractal set.

```
In[ ]:= Manipulate[JuliaSetPlot[0.365 - k i, PlotLegends → Automatic],
          {{k, 0.385}, 0.3, 0.7}]
```

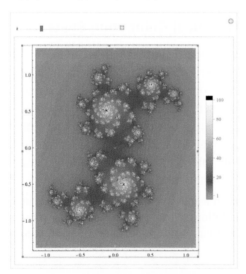

- The code below, generates a map with a 10-km-radius around where you are (this is usually based on your IP address):

```
In[ ]:= GeoGraphics[GeoDisk[Here, Quantity[10, "km"]]]
```

Apart from the examples given above showcasing some of *Mathematica*'s functionality, it is also worth mentioning:

(i) The *Wolfram Predictive Interface*, which makes suggestions about how to proceed after executing a command, (ii) The context-sensitive input assistant to help us choose the correct function, (iii) The possibility of running the program from the browser (WolframCloud.com or https://www.wolfram.com/wolfram-one/), including specific cloud-computing-related commands, (iv) The machine learning capabilities that, for example, enable us to analyze a handwritten text and keep on improving its translation through repetition, (v) Very powerful functions for graphics and image processing. We will refer to these and other new capabilities later on.

Visit https://www.wolfram.com/language/gallery/, the Wolfram Language Code Gallery, to find additional examples covering a wide variety of fields.

1.2 First Steps

1.2.1 Starting

The program can be run locally with **Mathematica** or **Wolfram|One** or over the Internet using **Wolfram Cloud** or **Mathematica Online**.

Mathematica: https://www.wolfram.com/mathematica
Wolfram|One: https://www.wolfram.com/wolfram-one
Wolfram Cloud: https://www.wolframcloud.com
Mathematica Online: https://www.wolfram.com/mathematica/online

From here on, unless mentioned otherwise, we will assume that you have installed **Mathematica** locally and activated its license. However, if you are using any of the other two alternatives, the process is quite similar. When installing it, an assistant will guide you through the activation process. If this is the first time, it will take you to the user portal (https://user.wolfram.com), where you will have to sign up. The website will ask you to register using an email address and create a Wolfram ID and a password. It will also automatically generate an activation key. Remember your Wolfram ID and password because they will be useful later on if, for example, you would like to install the program in a different computer or have an expired license. In both cases, you will need a new activation key. The Wolfram ID and password will also be necessary to access the Wolfram Cloud or Wolfram|One, and we will give a specific example at the end of the chapter.

To start the program in Windows, under the programs menu in your system, click on the **Wolfram Mathematica** icon.

Under OS X, the program is located in the applications folder, and you can launch it either locating it with Finder, using the Dock or from the launchpad (OS X Lion or more recent versions).

By default, a welcome screen (Figure 1.1) similar to this one will appear:

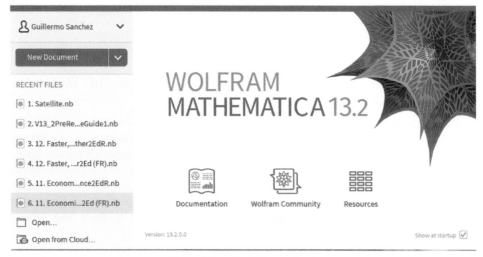

Figure 1.1   The Welcome Screen

This screen contains several external links (hyperlinks). For example, by clicking on **Resources**, you will be able to access a broad collection of reference materials and tutorials. In the upper left-hand corner you will see this icon: ![icon] . If you are a registered user, chances are you already have a Wolfram ID. In that case, click on the icon and sign in. Once the process has been completed, the icon will be replaced with the name that you used when you signed up, as shown in the previous image. The advantage of this is to gain access from within *Mathematica* to your files in the cloud (WolframCloud) but for now, you can skip this step if desired. If you click on New Document, below the icon shown at the beginning of the paragraph, a new blank *Notebook* will be created. Notebooks are the fundamental way to interact with *Mathematica*. You can also create them by clicking on **File ▸ New ▸ Notebook**.

A *notebook* is initially a blank page. We can write on it the text and instructions that we want. It is always a good idea to save the file before we start working on it; as in most programs, this can be done in the menu bar: **File ▸ Save As ▸** "Name of the File" (*Mathematica* by default will assign it the extension .nb).

It is usually convenient to have access to the **Formatting** toolbar, select in the menu bar **Window** and check **Toolbar ▸ Formatting**. Please note the drop-down box to the left of the toolbar (Figure 1.2), that we'll refer to as: *style box*. It will be useful later on.

Figure 1.2   *Mathematica*'s Formatting and Default Toolbar.

Once you begin to type, a new cell will be created. It will be identified by a blue right-bracket square   (]) ("cell marker") that appears on the right side of the *notebook*. Each cell constitutes a unit with its own properties. To see the cell type, place the cursor inside or on the cell marker. The type will appear in the *style box* (Figure, Input, Text, Title, Program...). We can also change the format directly inside the box. Type in a text cell and try to change its style.

```
This is a cell formatted using the Program style.
```

When a blank *notebook* is created, it has a style (cell types, fonts, etc.) that assigns to each type of cell certain properties (evaluatable, editable, etc.). The first time a *notebook* is created, *Mathematica* assigns a style named *Default* (by default). Later on we will see how to choose different styles. In the *Default* style, when a new cell is created, it will be of the *Input* type, which is the one used normally for calculations. When we evaluate an *Input* type cell, "In[n]:= *our calculation*" will be shown and a new cell of type *Output* will be generated where you will see the calculation result in the form of "Out[n]:= *result*" (n showing the evaluation order). However, in this book we sometimes use an option to omit the symbols "In[n]:=" and "Out[n]:=". These and other options to customize the program can be accessed through **Edit ▸ Preferences ...**.

To execute an *Input* cell select the cell (placing the cursor on the cell marker) and press ⎇SHIFT+⎇ENTER   or ⎇ENTER on the numeric keypad. You can also access the contextual menu by right-clicking with the mouse and selecting "Evaluate Cell".

In[ ]:= **2 + 2**

Out[ ]= **4**

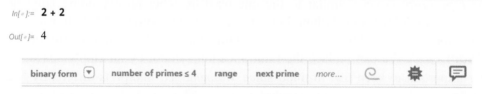

Figure 1.3   The Suggestions Bar.

When a cell is executed, a toolbar appears below its output as shown above (Figure 1.3). This bar, named the **Suggestions Bar**, provides immediate access to possible next steps optimized for your results. For example, If you click on "**range**" you will see that a new input is being generated and a new suggestions bar will appear right below the result:

In[ ]:= **Range[4]**

Out[ ]= {1, 2, 3, 4}

The **Suggestions Bar** is part of the predictive interface. It tries to guide you with suggestions to simplify entries and ideas for further calculations. It is one more step to reduce the time it takes to learn how to use the program.

All written instructions inside a cell will be executed sequentially.

To create a new cell we just need to place the cursor below the cell where we are and then start writing. We can also create a new cell where the cursor is located by pressing [ALT]+[ENTER].

To facilitate writing, *Mathematica* provides **Palettes**. These can be loaded by clicking on **Palettes** in the menu bar. There are several palettes to make it easier to type mathematical symbols such as the **Basic Math Assistant** or the **Other ▸ Basic Math Input** palettes. It would be useful from now on to keep one of them open.

- Using a palette or the keyboard shortcuts write and execute the following expression. Please remember: an empty space is the equivalent of a multiplication symbol: that is, instead of 4×5 you can write 4 5.

$$In[\cdot]:= \quad \frac{4 \times 5}{5}\, a \,+\, \sqrt{4}\; b \,-\, 3^2\, b$$

$$Out[\cdot]= \quad 4\,a - 7\,b$$

Alternatively, you can use keyboard shortcuts to write special symbols, subscripts, superscripts, etc. The most useful ones are: subscript [CTRL] + _ , superscript [CTRL] + ^ or [CTRL] + 6, fraction [CTRL] + / and the square root symbol [CTRL] + 2. Depending on your keyboard configuration, to use some of the previous symbols you may have to press first + [SHIFT] (key CAPS) for example: superscript [CTRL] + [SHIFT] + ^.

The previous *Input* cell was written in the standard format (*StandardForm*). It is the one used by default in *Input* cells. To check it, position the cursor on the cell marker and in the menu bar choose: **Cell ▸ Convert To ▸**. *InputForm* will appear with a check mark in front of it.

$$In[\cdot]:= \quad ((4*5)/5)*a + Sqrt[4]*b - 3^2*b;$$

The *InputForm* is very similar to the one used by other programming languages such as C, FORTRAN or Python. For instance: multiplication = " * ", division = " / ", addition = " + ", subtraction = " - " , exponentiation = " ^ ", and square root of a = "Sqrt[a]". We can replace the multiplication symbol by an empty space. It is very important to keep that in mind since *Mathematica* interprets differently **"a2"** and **"a 2"**: **"a2"** is a unique symbol and **"a 2"** is **a*2**. We have also added semicolon ";", thus the instruction will be executed, but its output will not be shown. This is useful to hide intermediate results.

Initially, it was its symbolic calculation capabilities that made *Mathematica* popular in the academic world.

- As an example, write and execute the following in an input cell (use a palette to help you enter the content).

$In[\circ]:= \partial_{x,y} \sin\left[x \sqrt{y}\right]$

$Out[\circ]= \dfrac{\cos(x \sqrt{y})}{2 \sqrt{y}} - \dfrac{1}{2} x \sin(x \sqrt{y})$

A customized style, to which we will refer later, has been defined in this notebook so that the outputs are shown in a format that corresponds to traditional mathematical notation. In *Mathematica* it is called *TraditionalForm*. We could have defined a style so that the inputs also use the traditional notation as well, but in practice it is not recommended. However, you can convert a cell to the *TraditionalForm* style anytime you want.

- Copy the *input* cell from above to a new cell and in the menu bar select **Cell ▶ Convert To ▶ TraditionalForm**. Notice that the cell bracket appears with a small vertical dashed line. You will be able to use the same menu to convert to other formats.

$In[\circ]:= \dfrac{\partial^2 \sin(x \sqrt{y})}{\partial x\, \partial y}$

$Out[\circ]= \dfrac{\cos(x \sqrt{y})}{2 \sqrt{y}} - \dfrac{1}{2} x \sin(x \sqrt{y})$

The **natural language** or **free-form** format enables us to write in plain English the operation we want to execute or the problem we want to solve. To start using it, press = at the beginning of the cell (input type) and you will notice that ▤ ("Free-form Input") will appear. After that, write your desired operation. Your instruction will automatically be converted into standard *Mathematica* notation, and then the answer will be displayed. Let's take a look at some examples.

- To add the numbers from 1 to 10 you can write "sum 1 to 10" or other similar English expressions such as "sum from 1 to 10".

$In[\circ]:=$ ▤ **sum 1 to 10** » ☐

`Sum[n, {n, 1, 10}]`

$Out[\circ]= 55$

If you follow the previous steps, the entry proposed may be different from the one shown here. The reason behind this is that when using the free-form format, the program interprets your input by connecting to Wolfram|Alpha online, and the knowledge engine is constantly evolving.

Move the cursor over `Sum[i,{i, 1, 10}]` and the message "Replace cell with this input" will appear. If you click on the symbol ⊞ located in the upper right-hand corner, additional information about the executed operation will be shown.

If you have received any message related to connection problems, it is because the free-form format and the use of data collections that we will refer to later on, requires that your computer is connected to the Internet. If it is and you still have connectivity problems it may be related to the *firewall* configuration in your network. In that case you may have to configure the *proxy* or ask your network administrator for assistance.

How do I test internet connectivity?: http://support.wolfram.com/kb/12420

- To solve an equation such as $3x^2 + 2x - 4 = 0$, you can literally type what you want *Mathematica* to do. The output shows the correct *Mathematica* input syntax and the answer, which in this case includes both, the exact solution and the decimal approximation.

In[ ]:=

> **solve 3 x^2 + 2 x – 4 = 0**
>
> **Solve[3*x^2 + 2*x – 4 == 0, x]**

Out[ ]= $\left\{\left\{x \to \frac{1}{3}\left(-1 - \sqrt{13}\right)\right\}, \left\{x \to \frac{1}{3}\left(\sqrt{13} - 1\right)\right\}\right\}$

> When you don't know the correct function syntax, you can use the free-form format to write what you want to do. *Mathematica* probably will show you the correct syntax immediately; click on it and only the input in the correct notation will remain with the free-form format entry disappearing.

The free-form input format generally works well when it's dealing with brief instructions in English. In practice it is very useful when one doesn't know the specific syntax to perform certain operations in *Mathematica*.

It's also very useful to use **Inline Free-form Input**.

- In this example, we assume we don't know the syntax to define an integral. In an input cell we click on **Insert ▸ Inline Free-form Input** and will appear. Inside this box we type **integrate x^2**.

In[ ]:= Integrate x^2

Out[ ]= $\dfrac{x^3}{3}$

- The entry is automatically converted to standard WL (Wolfram Language) or *Mathematica* syntax or by clicking over the symbol, the cell will go back to its free-form format.

In[ ]:= **Integrate[x^2, x]** ✓

Out[ ]= $\dfrac{x^3}{3}$

- By clicking in ✓ the input is automatically converted to the standard syntax in *Mathematica*. It can be used as a template.

In[ ]:= **Integrate[x^2, x];**

- You can ask practically anything that can be computed. The below value is changing. The distance from the Earth to the Moon varies depending on the position of the Moon in its orbit around the Earth.

In[ ]:=

Out[ ]= 370 017. km

■ Here we combine the **Inline Free-form Input** with `QuantityMagnitude` , a *Mathematica* function that gives the magnitude value of a quantity:

In[•]:= **QuantityMagnitude** $\left[\text{⊟ earth gravity} \right]$

Out[•]= 9.80

The *Mathematica* free-form input is integrated with Wolfram|Alpha, http://www.wolframalpha.com, a type of search engine that defines itself as a computational knowledge engine and that instead of returning links like Google or Bing, provides an answer to a given question in English. If you want, you can access Wolfram|Alpha directly from within *Mathematica* defining a cell in the NaturalLanguageInput style (or Shift +Alt+ 9) or writing " ==" (🌟) at the beginning of the cell (input type).

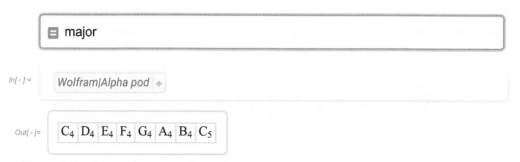

From the output, not shown, we selected the part that we were interested in ("Music notation") by clicking on the symbol ⊕ and choosing the desired option ("Formatted pod"). A new entry was generated displaying only the musical scale.

```
WolframAlpha["Do major",
  IncludePods → "MusicNotation", AppearanceElements → {"Pods"},
  TimeConstraint → {30, Automatic, Automatic, Automatic}]
```

If you click on **Play sound** you will hear the sound.

> The most important difference between using the free-form input notation and the Wolfram|Alpha one, is that in the first case the entry that we write is sent by the program to a Wolfram server that will return the syntax in *Mathematica* and it will be executed locally, while by using he the Wolfram|Alpha notation, all the processing is done in the server, and our computer only displays the output. If we have the program locally installed, it is better to use the free-form notation to get the *Mathematica* syntax. In this way, we will be able to use the output easily in subsequent calculations.

Wolfram|Alpha is specially useful when utilizing a browser since we don't need to have any software installed. Additionally, a new program named Wolfram|Alpha Notebook Edition that uses the NaturalLanguageInput style by default, is available: https://www.wolfram.com/wolfram-alpha-notebook-edition. Both applications can be used in many kind of devices (desktops, tablets, mobiles, and more.)

Besides all the previously described ways to define the cell type, there is another one that consists of placing the cursor immediately next to the cell to which you want to add another one. You will see "+" below the cell to its left. Click on it and choose the cell type that you want to define: *Input*, ▤ , ✸ (Alpha Input), Text, etc.

1.2.2 The Help System

If you have doubts about what function to employ, the best approach in many cases is to use the free-form input format and see the proposed syntax. However, if you would like to deepen your knowledge about the program you will need to use its help system.

The help system consists of different modules or guides (**Help ▸Wolfram Documentation**) from where you can access an specific topic with its corresponding instructions (**guide/...**). Additionally, you will find tutorials and external links. Clicking on any topic will show its contents.

Wolfram Language & System	Documentation Center	
Core Language & Structure	Data Manipulation & Analysis	Visualization & Graphics
Machine Learning	Symbolic & Numeric Computation	Higher Mathematical Computation
Strings & Text	Graphs & Networks	Images
Geometry	Sound & Video	Knowledge Representation & Natural Language

Figure 1.4   The Wolfram Language Documentation Center (partial screen).

In the screen-shot shown above (Figure 1.4), after clicking on **Data Manipulation & Analysis ▸ Importing & Exporting**, *Mathematica* displays the window below (Figure 1.5). Notice that besides giving us the functions related to the chosen topic, we can also access tutorials and even external resources such as video presentations.

Importing and Exporting

The Wolfram Language automatically handles hundreds of data formats and subformats—all coherently integrated through the Wolfram Language's uniform use of symbolic expressions. For each particular format, the correspondence between representations inside and outside the Wolfram Language can be specified at any level of detail using the Wolfram Language's general data elements mechanism.

Figure 1.5   How to import and export using *Mathematica*.

If we don't have a clear idea of the function we want to use or we remember it vaguely, we can use the search toolbar (Figure 1.6) with two or three words related to what we are looking for.

Figure 1.6   The Search Toolbar.

Another type of help is the one given when a mistake is made. In this case we should pay attention to the text or sound messages that we have received. For example, if we try to evaluate the following text cell, we will receive a beep.

"This is a text entry"

With **Help ▸ Why the Beep?...** we will find out the reason behind the beep.

An extra source of help is the Input Assistant with its context-sensitive autocompletion and function templates features: when you start writing a function, *Mathematica* helps you automatically complete the code (similar to the system used by some mobile phones for writing messages). For example: If you start typing **Pl** *Mathematica* will display all the functions that start with those two letters right below them (Figure 1.7). Once you find the appropriate entry, click on it and a template will be added.

Figure 1.7   The Input Assistant in action.

Sometimes, a down arrow will be shown at the end of the function offering us several templates (Figure 1.8). Choose the desired one.

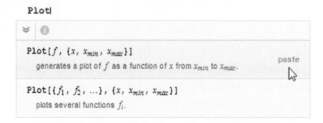

Figure 1.8   Function templates for the **Plot** function.

Additionally, a color code enables syntax error detection. If after typing a function its color is blue, probably it has not been entered correctly. All *Mathematica* and user-defined functions will be shown in black. For example: If you write in an input cell **Nmaximize**, it will be displayed with a blue font. However, once you change it to `NMaximize`, the function is now shown in black and a template is displayed.

1.3 Editing Notebooks

We've seen that the interaction with *Mathematica* is done through the Front End (what we usually see when using *Mathematica*), and this generates files named "*Mathematica notebooks*". What you are currently reading is one such *notebook*.

Each *Mathematica notebook* (*.nb) is a complete and interactive document containing text, tables, graphics, calculations, and other elements. The program also enables high-quality editing so notebooks can be ready for publication. You can save notebooks in many different formats (*.pdf, *.tex, *.ps, *.cdf, ...).

It is important to keep in mind the task at hand when using Mathematica. For example, you may want to prepare a report, create a tutorial or write a book chapter (this book has been written and edited in Mathematica and saved as a PDF file). Therefore, the first thing to do before starting using notebooks is to choose the appropriate style although you can always change it afterwards.

1.3.1 Notebook Structure

Each cell has a style. For example, the previous cell has the style **Subsection** associated to it. This was done by selecting its cell marker (]) and choosing Subsection in the *style box*.

Notebooks are organized automatically in a hierarchy of different cell types (title, section, input, output, figures, etc). If you click on the exterior cell marker, the cells will be grouped. For example, if you click on a section, all the cells associated to it will collapse and only the section title will be visible. Click once more to see the entire section again.

This is an ordinary text with the color formatted Format ▶ Text Color.

Click on a text cell and modify its format.

With Insert ▶ Hyperlink we can create hyperlinks to jump from one location to another inside the same *notebook,* to another *notebook* or to an external link.

The **Format** menu contains options to change the font, **face**, size, color, etc., of the contents of the cells. We can even use special characters such as Å.

This *notebook*, like any other, has a predefined style. This style tells *Mathematica* how to display its contents. There are many such styles. To choose one you can use **Format ▶ Stylesheet**. To define the style of a particular cell use **Format ▶ Style**. Additionally, inside a style, we can choose the appropriate working environment: **Format ▶ Screen Environment**. Try changing the appearance of a *notebook* by modifying the working environment.

1.3.2 Choosing Your Style

To create or edit an article, book, or manual, the **Palettes ▸ Writing Assistant** palette is of great help. Load it. Note that is divided into three sections: *Writing and Formatting*, *Typesetting*, and *Help and Settings*. See the help included in the palette.

If you click inside the palette on the button **Stylesheet Chooser**... (alternatively you can go to **Format ▸ Stylesheet ▸ Stylesheet Chooser**...) a separate window with icons representing different styles will appear (Figure 1.9). If you press on the upper part of any of the icons (**New**) a new sample *notebook* formatted according to that style will be generated. You will be able to use it and modify its contents to suit your needs. If you click on the bottom part of the icons (**Apply**), the style will be applied to the currently active *notebook*. For example, in the next screenshot (Figure 1.9), by clicking on the upper part of the Textbook icon, a *notebook* with a Textbook template style will be opened.

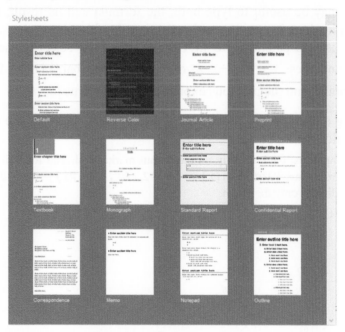

Figure 1.9 The Stylesheet Chooser.

When a new cell is created, depending on the stylesheet, by default it will be of a certain format type, for example with the *Default* stylesheet new cells are of type *Input*.

> When typing in a new *notebook* choose the stylesheet from the beginning with **Format ▸ Stylesheet ▸ Stylesheet Chooser...** . After clicking on the upper side of the icon representing the desired style (New), you will be able to use and modify the template according to your needs.

1.3.3 Typesetting Advice

If you are interested in emphasizing text containing formulas using classical mathematical notation, begin by typing in a text cell. Select the formula and define it as Inline Math Cell by pressing CTRL + 9 or, in the Writing Assistant palette, choose **Math Cells ▶ Inline Math Cell**. Finally, in the same palette in Cell Properties click on Frame and select the appropriate one . That's how the following example was done:

> ¿What is the domain of the function $f(x) = \sqrt{x^2 - 1}$?

If you'd like to type a sequence of formulas and align them at the equal signs, in Writing Assistant, choose **Math Cells ▶ Equal Symbol Aligned Math Cell**.

$$a + b = c$$
$$a = c - b$$

Figure 1.10   Aligning equations at the equal sign.

Sometimes it may be useful to type both, the formula and its result, as part of the text. This can be done by writing and executing the function in the same text cell. For example: using a palette write $\int x^3\, dx = \int x^3\, dx$. Now select the second term of the equation and press SHIFT+CTRL+ENTER, or, in the menu bar, choose **Evaluation ▶ Evaluate in place**, and the previous expression will become $\int x^3\, dx = \frac{x^4}{4}$.

When creating a document, you may be interested in seeing only the result. In that case you can hide the Input cell.

■ To hide the input just click directly on the output marker cell. The following example was created by executing **Plot[Sin[x], {x,-2 Pi, 2 Pi}]** and then hiding the command to display only the graph.

In[ ]:= **Plot[Sin[x], {x, -2 Pi, 2 Pi}]**

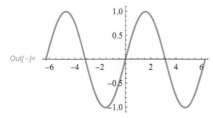

All *Mathematica* functions are in English, but the menus and palettes can also be displayed in any of the additional 12 languages offered since version 11. The alternatives include Chinese, French, German, Italian, Japanese, Korean and Spanish among others. Users can take advantage of this functionality by going to **Edit ▶ Preferences ▶ Interfaces**. Once an option is selected, every function will be automatically annotated with a "code caption" in the chosen language. Since *Mathematica* 11, the program includes a real-time spell checker that works with any of the supported languages. Alternatively, users of previous versions can review the English contents of a notebook with **Edit ▶ Check Spelling...** .

With **Edit ▶ Preferences** and with **Format ▶ Option Inspector** (see tutorial/OptionInspector) you will be able to customize many features of *Mathematica* that can be applied globally, to a *notebook* or to a particular cell. For example: with **Format ▸ Option Inspector ▸ Global Options ▸ File Locations** you can change *Mathematica* default directories.

1.3.4 The Traditional Style

We've seen that *Mathematica* by default uses the style: StandardForm in input cells (*Input*). However, if we'd like to create professional looking documents, it's recommended to use the TraditionalForm style that is similar to the one used in classical mathematical notation.

If the main purpose of a document is to be read by others, it may be convenient to present the inputs and outputs using the traditional notation. It's likely that the reader will not realize that it was created with *Mathematica*. This is specially true when the document is saved using the cdf format to which we will refer in the following section.

> You can write your inputs in **StandardForm**. Once you verify that they work, you can convert them to the traditional style.

■ Example: Let's type:

$$In[\circ]:= \texttt{Limit}\left[\frac{(x + \Delta x)^2 - x^2}{\Delta x}, \Delta x \to 0\right]$$

$$Out[\circ]= 2\,x$$

■ We convert the cell to the traditional form by selecting it and clicking on: **Cell ▸ Convert to ▸ TraditionalForm**. Then the cell above will be shown as follows:

$$In[\circ]:= \lim_{\Delta x \to 0} \frac{(x + \Delta x)^2 - x^2}{\Delta x}$$

$$Out[\circ]= 2\,x$$

> If an entry is in the traditional style, to see how it was originally typed convert it to the standard style by selecting the cell and choosing from the menu bar **Cell ▸ Convert to ▸ StandardForm**.

We can make an output to be shown in the traditional form by adding **//TraditionalForm** at the end of each *input*. There are several ways to apply this style to all the outputs in a notebook.

■ Go to **Format ▸ Option Inspector▸** (by category) **Cell Options ▸ New Cell Defaults ▸ CommonDefaultFormatTypes** and specify for "Output" TraditionalForm instead of StandardForm. The form for "Input" can also changed to TraditionalForm, but it is not usually recommendable.

■ Another easy way is to type the following:

```
In[◦]:= SetOptions[EvaluationNotebook[],
        CommonDefaultFormatTypes → {"Output" → TraditionalForm}]
```

It's recommended to include the function above in a cell at the end of the notebook and define it as an initialization cell (click on the cell marker and check **Cell ▸ Cell Properties ▸ Initialization Cell**). This way, the cell will be the first one to be executed every time we open the notebook.

In many of the styles available in **Stylesheet Chooser...** outputs are displayed using the standard form. In this chapter, we have used a customized stylesheet that shows outputs in the traditional format. In the rest of the book, we will almost always use the standard form since what we are trying to highlight is how to create functions with *Mathematica* and for that purpose, the classical format is not adequate.

1.3.5 Automatic Numbering of Equations and Reference Creation

Textbooks and scientific articles frequently use numbered equations. This can be done as follows:

- First open a new notebook (**File ▸ New ▸ Notebook**) and select one of the stylesheets available in the format menu. For instance **Textbook**: **Format ▸ Stylesheet ▸ Book ▸ Textbook**.

- Then create a new cell and choose **EquationNumbered** as its style (**Format ▸ Style ▸ EquationNumbered**). The cell will be automatically numbered, then you can write the equation, for example: $a + b = 1$.

$a + b = 1$

$$(1.1)$$

If you have opened a new notebook using the Textbook style, you will probably see (0.1) instead of (1.1). It's possible to modify the numbering scheme, but it is beyond the scope of this chapter. When you need to number equations, sections, and so on, remember that it is a good idea to select the stylesheet directly from the Writing Assistant palette: **Palettes ▸ Writing Assistant ▸ Stylesheet Chooser...** . After selecting **Textbook (New)**, a new notebook will be generated with many different numbered options such as equations, sections, etc. You can then use that notebook as a template.

If you wish to insert an automatic reference to the equation proceed as follows:

- Write an equation in a cell with the **EquationNumbered** style. Then go to **Cell ▸ Cell Tags ▸ Add/Remove...** and add a tag to the formula. Let's give it the tag: "par" (we write "par", from parabola, although we could have used any other name). A good idea would be also to go to **Cell ▸ Cell Tags** and check **▸ Show Cell Tags**, keeping it checked while creating the document, and only unchecking it after we are done. This will enable us to see the cell tags at all times. The first equation done this way is written below:

par

$x^2 = 1$

$$(1.2)$$

- If you wish to refer to this equation, type the following: (1.), put the cursor after the dot and go to **Insert ▸ Automatic Numbering**. In the opening dialog (Figure 1.11) scroll down to **EquationNumbered**, choose the tag of the equation, in our case: "par" and then press OK (see the screen below). After this has been done, the number of the equation appears between the parentheses: (1.2). If you click on the "2" the cursor will go immediately to the actual equation.

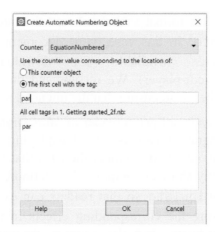

Figure 1.11   Creating an automatically numbered object.

1.3.6 Cell Labels Display

As we have mentioned previously, any time we execute an input cell, a label in the form In[n] is shown. However, we can choose not to display it. This can be done from **Edit ▸ Preferences** or typing **SetOptions[SelectedNotebook[], ShowCellLabel → False]**. You can also reset the counter (set In[n] to 1) in the middle of a session by typing the following in an input cell: $Line=0.

1.3.7 Save as...

Documents created with *Mathematica* are saved by default with extension "nb". If you share any of these documents with someone, that person will require access to the program.

However, *Mathematica* can also save documents in other formats. To do that, in the menu bar, choose **File ▸ Save As ▸**. Then, from the list of options, select the desired one.

Very often, we may be interested in saving a document in *pdf* format to reach a broad audience. We can also use *TeX* (widely used in professional journals in mathematics) or *HTML/XML*, if we want to use it in a website. Probably, it would be better to save the web files in *XML* since everything can be included in a single file. *HTML*, in contrast, contains a main file and numerous additional ones associated to it. However, if we'd like to keep *Mathematica's* interactive capabilities, the most appropriate format would be *cdf* (Computable Document Format).

1.3.8 The Computable Document Format (*.cdf)

As previously discussed, if you'd like other people without access to *Mathematica* to read your documents without losing their interactivity, you should save them in the cdf format: **File ▸ CDF Export ▸ Standalone** or **File ▸ Save As ▸ Computable Document** (**.cdf*).

The main advantage of this format is that it keeps the interactivity of the objects created with `Manipulate`. In this case, your documents can be read with the Wolfram *CDF Player* (available for download for free from http://www.wolfram.com/cdf-player). We recommend the inclusion of the player web link when sending the file to ensure that the intended reader can open it .

The Wolfram *CDF Player* is not just a reader of *Mathematica* documents for users without access to the program. It can also avoid displaying the functions used in the inputs, showing only the outputs (remember the trick that we have seen previously to hide the inputs and only show the outputs). Furthermore, it can include dynamic objects created with `Manipulate` so that readers can experiment with changing parameters and visualizing the corresponding effects in real time. Additionally, readers will not even know that the original document was generated in *Mathematica.* To see all these capabilities in action take a look at the following cdf document:

http://www.wolfram.com/cdf/uses-examples/BriggsCochraneCalculus/BriggsCochraneCalculus.cdf

Many of the Wolfram Research websites, such as Wolfram|Alpha (http://www.wolframalpha.com), allow visitors to generate and download documents in *cdf format*. The same happens with the Wolfram Demonstrations Project (http://demonstrations.wolfram.com) where the files are available in cdf format and you can download and run them locally.

1.3.9 The Wolfram Cloud

As stated at the beginning of the chapter, *Mathematica* can also be ran in the cloud, in a special online platform named Wolfram Cloud: http://www.wolframcloud.com/. This platform can also be used to store files remotely and, as a matter of fact, all the files generated locally are interchangeable with the cloud-generated ones. If you don't have access to *Mathematica* locally but you would like to start familiarizing yourself with the program, you can do so directly using Wolfram Cloud's free basic plan. If you are planning to use this option heavily, the cost will depend on your requirements and whether you already own a *Mathematica* license or not. In any case, you must register as a user using a Wolfram ID.

After signing up, you can access it by visiting http://www.wolframcloud.com. If you already registered in the user portal, you can use the same Wolfram ID.

Once you enter your Wolfram ID and password, the screenshot shown in Figure 1.12 will appear although you might see an initial screen that is somewhat different since the online platform is constantly evolving.

WOLFRAM CLOUD

Integrated Access to Computational Intelligence

The Wolfram Cloud combines a state-of-the-art notebook interface
with the world's most productive programming language—scalable for
programs from tiny to huge, with immediate access to a vast depth of
built-in algorithms and knowledge. Learn more »

*Wolfram Cloud technology powers Wolfram|One, Wolfram Programming Lab,
Wolfram Mathematica, Wolfram Enterprise Private Cloud and Wolfram|Alpha, as
well as Wolfram instant APIs, instant Web Apps and more...*

BASIC PLAN   HAVE AN ACCOUNT?

Figure 1.12   Welcome Screen in the Wolfram Development Platform.

You can now create a new notebook by clicking on: **Create a New Notebook**, and
reproduce most of the examples discussed in the chapter. You can also save the
notebook online and when accessing the file again, no matter from what location,
continue working on it. Additionally, you have the possibility of accessing cloud
files from a local *Mathematica* installation: **File ▶ Open from Cloud...** or **File ▶
Save to Cloud....** *Mathematica* has specific functions related to the Wolfram
Cloud. In later chapters, we will refer to them.

1.4 Basic Ideas

Users with prior knowledge of other programming languages very often try to
apply the same style (usually called procedural) when using *Mathematica.* The
program itself makes it easy to do so since it has many similar instructions such as:
For, Do, or Print. Although this programming style works, we should try to avoid it
if we want to take full advantage of the Wolfram Language's capabilities. For that
purpose we should use a *functional* style, which basically consists of building
operations in the same way as classical mathematics.

This section is for both, those unfamiliar with the WL (or *Mathematica*) and users
that still use procedural language routines. Because of that, if you have ever
programmed in other languages please try not to use the same programming
approach.

Although we use the formal Wolfram Language syntax in this section, many of the
examples shown below can be replicated using natural language.

In this brief summary, we will refer to some of the most frequently used functions
and ideas. We will often not explain the syntax or we'll do it in a very concise way.
For additional details, just type the function, select it with the cursor, and press
<F1>. If you don't understand everything that you read, don't worry, you will be
given more information in later chapters.

> When you don't know the syntax of a function, type it, select it with the cursor, and press the <F1> key.

1.4.1 Some Initial Concepts

- Let's begin by writing the expression below using the **Basic Math Assistant** palette;

$In[\circ]:=$ **expr1 = 5 !** $\dfrac{\text{Cos}[3\,\pi]}{\sqrt[3]{2}}\, x^2 + \pi\, \dfrac{e^{-7}\,\text{Sin}[1]}{y}$

$Out[\circ]=$ $\dfrac{\pi \sin(1)}{e^7\,y} - 60 \times 2^{2/3}\,x^2$

This notebook uses a style where the outputs are shown in the traditional form.

We have assigned it the name **expr1**, to be able to use the result later on. To write the transcendental numbers π and e, we have taken advantage of special symbols (e is not the same as e; e is a letter without any other meaning) located in the **Basic Math Assistant** palette. Alternatively, we could have used the keyboard and write Pi for π and E^ or Exp[] for e. Notice the use of a single equal sign "=" to indicate sameness or equivalence in equations. For comparisons, we will type a double equal sign "==".

- The same operations can be written using the typical *InputForm* notation:

$In[\circ]:=$ **5 ! * (Cos[3*Pi]/2^(1/3))*x^2 + Pi*(Sin[1]/(E^7*y))**

$Out[\circ]=$ $\dfrac{\pi \sin(1)}{e^7\,y} - 60 \times 2^{2/3}\,x^2$

- Here x is replaced with 3 and y with 5. To do that, you apply the following syntax: "*expr /. rule*" that will replace *expr* with the contents of *rule*.

$In[\circ]:=$ **expr1 /. {x → 3, y → 5}**

$Out[\circ]=$ $\dfrac{\pi \sin(1)}{5\,e^7} - 540 \times 2^{2/3}$

There is something that should draw our attention in this output. There are terms that have not been evaluated: π, $\sin(1)$, and e^7. This is because *Mathematica* does not simplify or make approximations if that means losing precision. It literally works with infinite precision. Nevertheless, we can force it to compute the decimal approximation, the usual approach in other programs. To do that, we will use the function N as shown in the next example. A similar result can be also obtained by including decimals in some of the numbers.

- In this example we use the % symbol to recall the last output. //N or N[*expr*] use machine precision in their calculations even though the output may show fewer decimals. By default, 5 decimals are shown, but with N[*expr*, n] we can use a precision n as large as desired.

$In[\circ]:=$ **% // N**

$Out[\circ]=$ -857.196

If you apply several replacement rules to the same expression in succession, the replacement takes places consecutively:

■ We first apply (/.) the rule (**a→b**) to the expression $a\,x + b\,y$ and then the second rule (**b→c**) is applied to the previous result.

In[ ]:= **a x + b y /. a → b /. b → c**

Out[ ]= $c\,x + c\,y$

When an assignment is no longer going to be used, it may be convenient to delete it using **Clear**.

■ The following input removes the assignment previously associated to **sol**.

In[ ]:= **Clear[expr1]**

■ Now, when we type **sol** we see that it has no assignment.

In[ ]:= **expr1**

Out[ ]= expr1

1.4.2 Functions

Mathematica contains thousands of prebuilt functions (Functions) that are always capitalized and whose arguments are always written inside square brackets [arg]. It's important to keep in mind that the program differentiates between lower and upper cases. For example: **NMinimize** is not the same as Nminimize. However, users can define a function using lower case.

Lists are a fundamental *Mathematica* concept, frequently used in many contexts. In a list, elements are inside curly brackets {}. Later on we will refer to them.

Remember these five rules: All functions in the WL are capitalized: **Log**; arguments are written inside square brackets: **Sin[2 x + 1]**; list elements are inside curly brackets: {a_1, ..., a_n}; ranges are written as lists: **Plot[x, {x, 0, 2 Pi}]**; and parentheses are used to group operations: **(Log[a x] + Sin[b x])/(a - b)**.

Use *Mathematica*'s help files (**Help ▶ Wolfram Documentation**) to investigate functions and their arguments. Remember that you can select a function with the cursor and press <F1> to get information about it. Besides its syntax, examples, tutorials, and other related functions are displayed.

■ Write **Pl** in an *input* cell. The context-sensitive input assistant will show you all the functions that start with Pl. Select Plot. Next to Plot a new down arrow will appear. Click on it and a template like the one shown below will be created.

$$\text{Plot}\left[\,f\,,\,\{\,x\,,\,x_{min}\,,\,x_{max}\,\}\,\right]$$

Figure 1.13   Template for the **Plot** function.

Figure 1.13 shows an *input* type cell. If you'd like to avoid its evaluation, go to: **Cell ▶ Cell Properties** and uncheck Evaluable. That is what we did in this case in the original document since the idea is to see the cell content but not to execute it.

■ In the previous template, all the basic function arguments are displayed. Now you can fill them in using the options as the next example shows. Execute the *Mathematica* expression below.

In[ ]:= `Plot[Cos[x], {x, 0, 30}]`

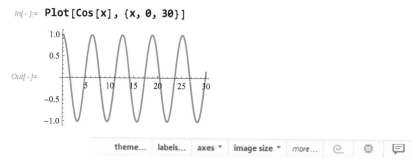

Figure 1.14   The Suggestions Bar for plots.

- The resulting output is a plot with the suggestion bar (Figure 1.14) appearing right below it. Thanks to the WolframPredictiveInterface, the Image Assistant and Drawing Tools provide point-and-click image processing and graphics editing. Click on **theme...** and a menu with several choices to customize the plot will unfold (Figure 1.15).

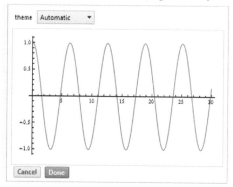

Figure 1.15   Customizing a plot using the "**theme...**" option in the Suggestions Bar.

Besides arguments, WL functions also have options. One way of seeing those options directly is Options[*func*]:

- Solve options:

In[ ]:= `Options[Solve]`

Out[ ]= {Assumptions :→ $Assumptions, Cubics → Automatic, GeneratedParameters → C, InverseFunctions → Automatic, MaxExtraConditions → 0, Method → Automatic, Modulus → 0, Quartics → Automatic, VerifySolutions → Automatic, WorkingPrecision → ∞}

Next, we'll show some examples. Replicate them using the **Basic Math Assistant** palette that includes templates for hundreds of commands.

- Definite integral example:

In[ ]:= $\int Sin[x]^4 Cos[x]^3 \, dx$

Out[ ]= $\dfrac{3\sin(x)}{64} - \dfrac{1}{64}\sin(3x) - \dfrac{1}{320}\sin(5x) + \dfrac{1}{448}\sin(7x)$

- Use `Simplify` anytime you need to simplify a complex expression.

In[ ]:= $Simplify\left[x^8 - 4x^6 y^2 + 6x^4 y^4 - 4x^2 y^6 + y^8\right]$

Out[ ]= $\left(x^2 - y^2\right)^4$

There are also other functions to manipulate expressions. You can find them in the palette **Other ▶ Algebraic Manipulation**. Note that you can manipulate an entire expression or just part of it.

- Try to use it with this example simplifying the Sin and Cos arguments separately.

$$\text{Sin}\left[x^8 - 4\,x^6\,y^2 + 6\,x^4\,y^4 - 4\,x^2\,y^6 + y^8\right] / \text{Cos}\left[x^4 + 2\,x^2\,y^2 + y^4\right]$$

- To get:

$$\text{Sin}\left[\left(\left(x^2 - y^2\right)^4\right)\right] / \text{Cos}\left[\left(x^2 + y^2\right)^2\right]$$

- In this example a random distribution of spheres in a three-dimensional space is shown.

In[ ]:= `Graphics3D[{Yellow, Sphere[RandomInteger[{-5, 5}, {10, 3}]]}]`

Out[ ]=

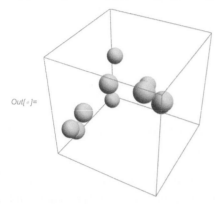

You can rotate the image to see it from different angles by clicking with the mouse cursor inside of it.

If you haven't used *Mathematica* previously, the last command may seem strange. Let's analyze it step by step:

- Highlight `RandomInteger` and press <F1>. The help page for the command will be displayed. In this case with **`RandomInteger[{-5, 5}, {10, 3}]`** we are generating 10 sublists with 3 elements, each element being a random number between −5 and 5. We are going to use them to simulate the coordinates $\{x, y, z\}$.

- The previous function is inside `Sphere` (do the same, highlight `Sphere` and press <F1>) a command to define a sphere or a collection of spheres with the coordinates $[\{\{x_1, y_1, z_1\}, \{x_2, y_2, z_2\}, \ldots\}, r]$, where r is the radii of the spheres. If omitted, as in this example, it's assumed that $r = 1$.

In[ ]:= `Sphere[RandomInteger[{-5, 5}, {10, 3}]]`

Out[ ]= $\text{Sphere}\left[\begin{pmatrix} -4 & -1 & 4 \\ -2 & 3 & -2 \\ 4 & 4 & -1 \\ 1 & -4 & -4 \\ -2 & 1 & 0 \\ -5 & -4 & -2 \\ 2 & 5 & 5 \\ 1 & 5 & 2 \\ 4 & -5 & -4 \\ 2 & 3 & -1 \end{pmatrix}\right]$

Note: In this particular notebook, list outputs are sometimes shown as matrices.

We use the command Graphics3D to display the graphs. We have added Yellow to indicate that the graphs, the spheres in this case, should be yellow and the option ImageSize to fix the graph size and make its proportions more adequate for this example. Finally we arrive at the instruction:

In[ ]:= Graphics3D[{Yellow, Sphere[RandomInteger[...] +]}]

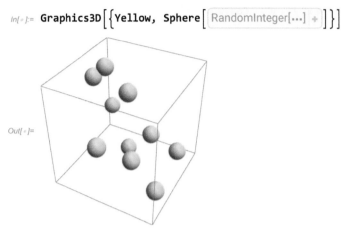

Out[ ]=

We have iconized the argument inside Sphere (highlight the part of expression to be iconized, right-click on it and then select **Un/Iconize selection**).

> When you don't understand an instruction consisting of several different commands, decompose it into its individual instructions, and analyze each one separately using the help system.

- In the next cell, we use Rotate to show a rotated output, in this case by 1 radian, compared to the usual display.

In[ ]:= Rotate[1/Sqrt[1 + x], 1]
          (*Type inside the output and you'll see that it retains its properties.*)

Out[ ]= $\frac{1}{\sqrt{x+1}}$

In the above evaluation cell (*Input*) we can include a comment in the following format: (* comment *). Nevertheless, it is recommended to include the comments in separate cells in text format. Remember that you can do that by choosing *Text* or other format in the *style box*.

- Practically anything can be used as a symbol. For example, let's first generate a small red sphere:

In[ ]:= Graphics3D[{Red, Sphere[]}, ImageSize → 30]

Out[ ]=

- Then, substitute the symbol thus obtained, , and paste it below replacing b. Finally, evaluate the cell.

In[ ]:= Expand[(1 + b)³]

Out[ ]= $b^3 + 3 b^2 + 3 b + 1$

$In[\circ]:=$ **Expand$\left[\left(1 + \ \fbox{$\bullet$}\ \right)^3\right]$**

$Out[\circ]=$ $\fbox{$\bullet$}^3 + 3\ \fbox{$\bullet$}^2 + 3\ \fbox{$\bullet$} + 1$

1.4.3 List and Association

Lists (List) are a fundamental concept in the WL. Lists or ranges are written inside curly brackets. The list elements are separated by commas: {a, b, ...}. For a description of list operations see: guide/ListManipulation. We'll describe them in more detail in later chapters.

■ Let's create a lists of lists of random numbers.

$In[\circ]:=$ **mat = RandomInteger[10, {5, 5}];**

■ We add **InputForm** to see the output displayed in the internal format used by the WL. Matrices are represented as lists of lists.

$In[\circ]:=$ **mat // InputForm**

$Out[\circ]//InputForm=$

   {{9, 5, 3, 7, 3}, {6, 1, 10, 10, 10}, {6, 8, 8, 8, 6}, {9, 7, 8, 0, 3}, {10

■ If desired, matrices can also be displayed in standard mathematical notation.

$In[\circ]:=$ **MatrixForm[mat]**

$Out[\circ]//MatrixForm=$

$$\begin{pmatrix} 9 & 5 & 3 & 7 & 3 \\ 6 & 1 & 10 & 10 & 10 \\ 6 & 8 & 8 & 8 & 6 \\ 9 & 7 & 8 & 0 & 3 \\ 10 & 4 & 7 & 7 & 6 \end{pmatrix}$$

■ Since they are lists, matrices can be handled as such. For example, here we extract [[...]] (or Part[...]) the 4th element of the 2nd sublist (2nd row, 4th column):

$In[\circ]:=$ **mat[[2, 4]]**

$Out[\circ]=$ 10

■ In the expression below we assign the symbol **sol** to the solution of a $x^2 + b x + c == 0$.

$In[\circ]:=$ **sol = Solve$\left[a x^2 + b x + c == 0, x\right]$;**

Remember that for assignments, a single equal sign "=" is used, while to indicate an equality in an equation or a comparison, we will use a double equal sign "==" that when typed will be automatically converted to "==".

■ The output is a list.

$In[\circ]:=$ **sol**

$Out[\circ]=$ $\left\{\left\{x \rightarrow \dfrac{-\sqrt{b^2 - 4 a c} - b}{2 a}\right\}, \left\{x \rightarrow \dfrac{\sqrt{b^2 - 4 a c} - b}{2 a}\right\}\right\}$

■ Next, we extract the second solution (note that both solutions are inside a list):

In[ ]:= **sol[[2]]**

$$Out[ ]= \left\{x \to \frac{\sqrt{b^2 - 4 a c} - b}{2 a}\right\}$$

- To verify that the previous result is correct, we use "*expr* **/.** *rule*", replacing the x in the equation with its values from the solution.

In[ ]:= **a x² + b x + c == 0 /. sol**

$$Out[ ]= \left\{\frac{\left(-\sqrt{b^2 - 4 a c} - b\right)^2}{4 a} + \frac{b\left(-\sqrt{b^2 - 4 a c} - b\right)}{2 a} + c = 0,\right.$$
$$\left.\frac{\left(\sqrt{b^2 - 4 a c} - b\right)^2}{4 a} + \frac{b\left(\sqrt{b^2 - 4 a c} - b\right)}{2 a} + c = 0\right\}$$

Although we can see that the substitution has taken place, it's often convenient to simplify it. For that we use Simplify[**%**], where % enables us to call the result (*Out*) of the last entry and evaluate the cell. This method is very useful in complex situations where we want to check that *Mathematica* actually returns the correct answer.

- We can check that the equality holds and therefore the solution is correct.

In[ ]:= **Simplify[%]**

Out[ ]= {True, True}

For further details about matrix operations you can consult the documentation: guide/MatrixOperations.

Association[$key_1 \to val_1$, $key_2 \to val_2$, ...] is another important structure in the WL. It represents an association between keys and values.

In[ ]:= **assoc = <|"Height" → 175, "Weight" → 75, "Sex" → "Female"|>**

Out[ ]= <|Height → 175, Weight → 75, Sex → Female|>

- Keys extracts keys from an association and Values extracts values:

In[ ]:= **Keys[assoc]**

Out[ ]= {Height, Weight, Sex}

In[ ]:= **Values[assoc]**

Out[ ]= {175, 75, Female}

- We can query the value associated with the key "Weight":

In[ ]:= **assoc[["Weight"]]**

Out[ ]= 75

1.4.4 Defining Your Own Functions

Mathematica makes easy define functions in a similar way to mathematics. The usual syntax in the case of a function of one variable is: **f[x_] := expr[x]**. The *blank* ("_") next to the variable is used to specify the independent function variable.

- Let's see an example:

In[ ]:= **f[x_] := 0.2 Cos[0.3 x²]**

- Now we can assign a value to the independent variable and we'll get the function value.

In[ ]:= **f[3]**

Out[ ]= −0.180814

- We could have also typed the previous two operations in the same cell (don't forget in this case to enter ";" at the end of each function). However, it's better to use one cell for each function until you have enough practice. It will help you find mistakes.

In[ ]:= **f[x_] := 0.2 Cos[0.3 x²];**
 f[3]

Out[ ]= −0.180814

- Now we are ready to visualize the Derivative (') of **f[x]** in a specific interval.

In[ ]:= **Plot[f'[x], {x, 0, 10}]**

Out[ ]=

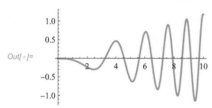

- The same approach can be extended to several variables. In this case, we use the previously defined *f (x)* function.

In[ ]:= **g[x_, y_] := f[x] 2.3 Exp[- 0.3 y²]**

- Now we can present the result in a 3D graph with **Plot3D**. We use the option **PlotRange->All** to display all the points in the range for which the function is calculated. Remove the option to see what happens.

In[ ]:= **Plot3D[g[x, y], {x, -2 π, 2 π}, {y, -2 π, 2 π}, PlotRange → All]**

Out[ ]=

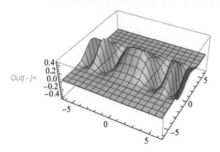

1.4.5 Pure Functions

Pure functions are often used when programming in the WL, and they will be described in detail in Chapter 2. To create them, we can either use Function or a combination of & and # symbols.

- Here is an example of a pure function ($f(x) = x^2 + 3$):

In[•]:= `f = (#^2 + 3) &`

Out[•]= $\#1^2 + 3 \,\&$

In[•]:= `f[3]`

Out[•]= 12

1.4.6 Example Data

The installation of the program includes a set of files with data from different fields that can be used to test certain functions. We will use some of the data in this book. To download them, use `ExampleData`.

To find out all the areas for which there are examples available, just type ExampleData[].

- The command below downloads the original text of Charles Darwin's book *On the Origin of Species*:

In[•]:= `txt = ExampleData[{"Text", "OriginOfSpecies"}];`

- To find out how many times the word "evolution" appears compared to the word "selection" we use the function `StringCount`, specifying the chosen word or text.

In[•]:= `StringCount[txt, "evolution"]`

Out[•]= 4

In[•]:= `StringCount[txt, "selection"]`

Out[•]= 351

A fundamental feature of *Mathematica* is its capability for importing and exporting files in different formats. The functions that we need for that are `Import` and `Export`. They enable us to import or export many different types: XLS, CSV, TSV, DBF, MDB, ...

- In this case we import the first sheet of an XLSX file from the ExampleData folder:

In[•]:= `Import["ExampleData/cities.xlsx", {"Data", 1}]`

Out[•]= $\begin{pmatrix} \text{City} & \text{Country} & \text{Population} \\ \text{Tokyo} & \text{Japan} & 8.3 \times 10^6 \\ \text{Chicago} & \text{United States} & 2.8 \times 10^6 \\ \text{London} & \text{United Kingdom} & 7.4 \times 10^6 \\ \text{Berlin} & \text{Germany} & 3.4 \times 10^6 \end{pmatrix}$

1.4.7 Additional Functions and Packages

The latest releases of Mathematica contain more than 7,000 built-in functions. However, it's always possible to add extra functionality by creating your own.

Apart from that, you can also use functions created by other Mathematica users located in the Wolfram Function Repository:

https://resources.wolframcloud.com/FunctionRepository. They can be accessed using ResourceFunction:

- We look for the functions related to astronomy (the output is not shown).

In[ ]:= **ResourceSearch["Astronomy"]**

- "AstroDistance" computes the current distance between Mars and Jupiter:

In[ ]:= **ResourceFunction["AstroDistance"][**
  Entity["Planet", "Mars"], Entity["Planet", "Jupiter"]]

Out[ ]= 3.91542 au

Several functions can be combined to develop applications for specific purposes (*packages*). The packages included in *Mathematica* can be seen in **Documentation Center ▸ Add-ons and Packages**.

Additional packages (add-ons) developed by users for specific purposes are available in http://packagedata.net/.

Normally, all these packages will be located in Addons/Applications, a subfolder in the *Mathematica* installation directory.

To load a package you can type Needs["package name`"] or <<package name`. For example, the following package includes various functions related to the properties of the radiation emitted by a black body.

In[ ]:= **Needs["BlackBodyRadiation`"]**

In[ ]:= **?"BlackBodyRadiation`*"**

Out[ ]=

> ❯ BlackBodyRadiation`
>
> **BlackBodyProfile**   **MaxPower**   **PeakWavelength**   **TotalPower**

- The following package function calculates the peak wavelength of a black body for a given temperature.

In[ ]:= **PeakWavelength[5000 Kelvin]**

Out[ ]= 5.79554×10^{-7} Meter

Chapter 2 provides a brief introduction to package development with actual examples interspersed throughout the rest of the book.

1.5 From Graphics to Machine Learning

1.5.1 Graphics

One of the most remarkable aspects of *Mathematica* is its graphical capabilities. The program's tool for 2D graphics can be accessed by pressing CTRL+D or by selecting in the menu bar **Graphics ▸ Drawing Tool**.

The most commonly used functions to represent graphics are: Plot, Plot3D, and

`ListPlot`. However, there are many more.

- When typing **Plot** the context-sensitive input assistant will show you all the words that include the word Plot, browse them. Alternatively type the following command and execute it (its output is omitted):

  ? *Plot*

The names that are shown can give you a hint about the type of plot they generate. Click on the chosen name to get detailed information. In some cases, you will be referred to the plot options.

- The graphical functions don't end there. The ones that include the word **Chart** are usually related to statistical graphs. Similarly to the previous case, you can use the expression:

  ? *Chart*

- Here we use a figure known as the "Utah Teapot", included in ExampleData, to make a 3D bar chart.

In[ ]:= **g = ExampleData[{"Geometry3D", "UtahTeapot"}];**

In[ ]:= **BarChart3D[{1, 2, 3}, ChartElements → g, BoxRatios → {4, 1, 1}]**

Out[ ]=

- In the following example we use the options **AxesLabel** to add labels to the axes and PlotLegends → **"Expressions"** to show the represented functions:

In[ ]:= **Plot[{Sin[x], Cos[x], Tan[x]}, {x, 0, 2 Pi},**
  AxesLabel → {"x", "f(x)"}, PlotLegends → "Expressions"]

Out[ ]=
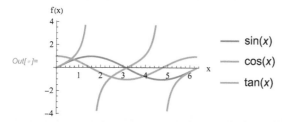

- Since Mathematica 13.0, we can create plots with multiple axes:

In[ ]:= `ListLinePlot[Table[Table[x^n, {x, 40}], {n, 3}], MultiaxisArrangement → All]`

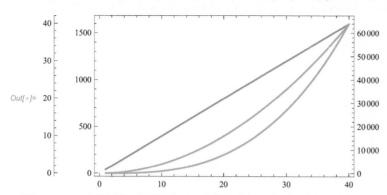

Out[ ]=

- We can even find the intersection points of a circumference and a parabola and graphically represent them. To write special symbols such as "∧" , the best thing to do is to use *Mathematica*'s palettes, which we will discuss later on.

In[ ]:= $\mathbf{pts = Solve\left[x^2 + y^2 = 1 \wedge y - 2\,x^2 + \dfrac{3}{2} = 0, \{x, y\}\right]}$

Out[ ]= $\left\{\left\{x \to -\dfrac{1}{2}\sqrt{\dfrac{1}{2}(5-\sqrt{5})},\ y \to \dfrac{1}{4}(-1-\sqrt{5})\right\}, \left\{x \to \dfrac{1}{2}\sqrt{\dfrac{1}{2}(5-\sqrt{5})},\ y \to \dfrac{1}{4}(-1-\sqrt{5})\right\},\right.$

$\left.\left\{x \to -\dfrac{1}{2}\sqrt{\dfrac{1}{2}(5+\sqrt{5})},\ y \to \dfrac{1}{4}(\sqrt{5}-1)\right\}, \left\{x \to \dfrac{1}{2}\sqrt{\dfrac{1}{2}(5+\sqrt{5})},\ y \to \dfrac{1}{4}(\sqrt{5}-1)\right\}\right\}$

```
Show[
   {ContourPlot[{x² + y² == 1, y - 2 x² + 3/2 == 0}, {x, -1.5, 1.5}, {y, -1.5, 1.5}],
    Graphics[{Red, PointSize[Medium], Point[{x, y} /. pts]}]}]
```

Out[ ]=

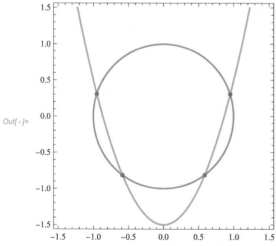

- Graphics-related functions include numerous options (output omitted):

    `Options[Plot3D]`

- The next cell displays the surface $\cos(x)\,\sin(y)$ bounded by the ring $1 \le x^2 + y^2 \le 4$ (To write "≤" you can type "<="). Note that to set the region where the function exists, we use the RegionFunction option.

In[ ]:= `Plot3D[Cos[x] × Sin[y], {x, -2, 2}, {y, -2, 2},`
         `RegionFunction → Function[{x, y, z}, 1 ≤ x^2 + y^2 ≤ 4]]`

Out[ ]=
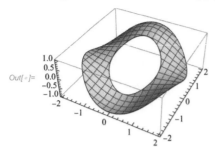

- RegionPlot3D enables us to plot multiple regions at a time:

In[ ]:= `RegionPlot3D[{1 ≤ x^2 + y^2 ≤ 2, 1 ≤ x^2 + z^2 ≤ 2, 1 ≤ y^2 + z^2 ≤ 2},`
         `{x, -2, 2}, {y, -2, 2}, {z, -2, 2}, Mesh → None]`

Out[ ]=
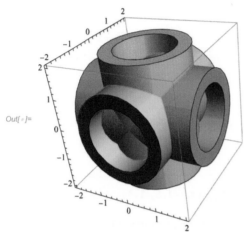

You can interact with 3D graphics: Click on the graph to rotate it, with [CTRL] + click
(or ⌘-click) you can zoom, and with [SHIFT]+click you can move it around.

- The code below generates the famous Lissajous curves. Using `Grid` we can show a two-
  dimensional mesh of objects, in this case graphs. We also use the `Tooltip` function to
  display labels related to objects (not only graphics) that will be shown when the mouse
  pointer hovers over the objects. Here, when the mouse pointer is over the chosen curve,
  you'll see the function that generated it.

```
In[ ]:= Grid[Table[Tooltip[ParametricPlot[{Sin[n t], Sin[m t]}, {t, 0, 2 Pi},
        ImageSize → 70, Frame → True, FrameTicks → None, Axes → False],
      {Sin[n t], Sin[m t]}], {m, 3}, {n, 3}]]
```

Out[ ]=

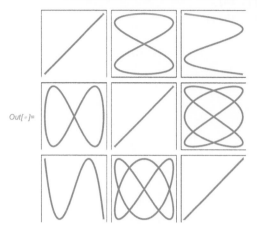

Other functions that do not include **Plot** or **Chart** but that are also useful for creating graphs are Graphics, Graphics3D, and ColorData, as the following example shows.

- Representation of the water molecule.

```
In[ ]:= Graphics3D[{Specularity[White, 50], ColorData["Atoms", "H"],
        Sphere[{0, 0, 0}, .7], Sphere[{1.4, 0, 0}, .7],
        ColorData["Atoms", "O"], Sphere[{.7, 0, .7}]}, Lighting → "Neutral"]
```

Out[ ]=

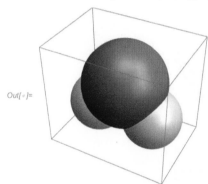

- Using Canvas, we can annotate existing graphs (click inside):

```
In[ ]:= Canvas[Plot[Sin[x], {x, 0, 2 π}]]
```

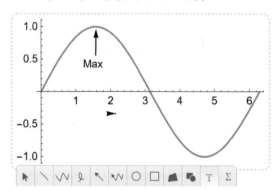

- CSGRegion a new function very useful for CAD Modeling

```
In[ ]:= CSGRegion["Union", {Cuboid[{-30, -30, -1}, {30, 30, 1}],
       Cylinder[{{-30, 0, 0}, {30, 0, 0}}, 10],
       Cylinder[{{0, -30, 0}, {0, 30, 0}}, 10]}]
```

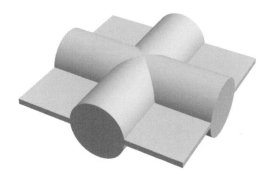

If you want to learn more about graphics, you can select the relevant text with the cursor and press <F1>. For example, type: tutorial/GraphicsAndSoundOverview, select it and press <F1>.

1.5.2 Images

Image processing has experienced a significant improvement. When clicking on an image you will see a menu with numerous options (Figure 1.16).

- The following command loads an image.

```
ExampleData[{"TestImage", "Lena"}]
```

Figure 1.16   The Image Assistant.

- The Image Assistant appears below the image. Click on any of available options. For example, after clicking on **more...** we see the following menu (Figure 1.17):

Figure 1.17   Additional capabilities available from the "**more... option**" in the Image Assistant.

- After selecting one of the **Actions & Tools**, a miniature version of the image appears where you can check the effect on the original image. For example: When selecting **invert colors**, you'll see Figure 1.18.

Figure 1.18   Inverting the colors using the Image Assistant.

After getting the desired effect, press **Apply** to modify the image. An alternative way to do that is to select the function directly from the menu: **ColorNegate**[*image*]. Copy it to an *input* cell and execute it. This last procedure is the appropriate one if you want to create a program or explain how the image has been modified. We will be using this method when we refer to image processing in later chapters.

1.5.3 Manipulate

Mathematica has functionality to enable dynamic and interactive operations. One of the most significant examples of this is the `Manipulate` function.

- Type the *Mathematica* expression shown below. Notice how sliders are created for each of the parameters being defined. Click on the ⊞ symbol included in the output to show the buttons created by Manipulate.

```
In[ ]:= Manipulate[Plot[Sin[frequency x + alpha], {x, -2 π, 2 π}],
        {frequency, 1, 5}, {alpha, 0, Pi / 2}]
```

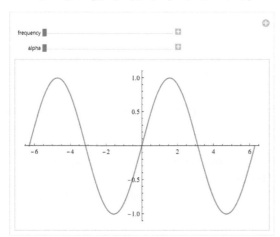

In http://demonstrations.wolfram.com, you can find thousands of small interactive applications of great educational value. Throughout this book, we'll refer to them using the word "demonstrations". If you don't know how to build them just download and open the files; later on you'll be able to modify their code to suit your needs or even make new ones. In Chapter 4, we'll learn more details.

- This example explores *Coulomb's law*. This law describes the interaction between two charges (based on '*Potential Field of Two Charges*' created by Herbert W. Franke: http://demonstrations.wolfram.com/PotentialFieldOfTwoCharges/).

```
In[ ]:= Manipulate[ContourPlot[q1 / Norm[{x, y} - p[[1]]] + q2 / Norm[{x, y} - p[[2]]],
        {x, 2, -2}, {y, 2, -2}, Contours → 10],
        {{q1, -1}, -3, 3}, {{q2, 1}, 3, -3},
        {{p, {{-1, 0}, {1, 0}}}, {-1, -1}, {1, 1}, Locator}, Deployed → True]
```

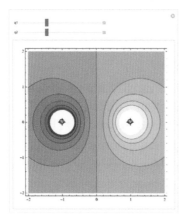

Move the locators around with the mouse to see how the potential fields change.

1.5.4 Gauges

Mathematica includes different types of customizable gauges that can behave dynamically:

```
In[ ]:= Through[{AngularGauge, VerticalGauge, ThermometerGauge, HorizontalGauge}[
        42, {0, 100}, ImageSize → Tiny]]
```

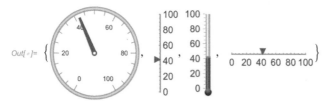

- With just a few lines of code, we can generate two dynamic clocks synchronized with both, our local time and a different time zone.

```
In[ ]:= Dynamic[Refresh[Row[
        {ClockGauge[AbsoluteTime[], PlotLabel → Style["Local", Large, Bold]],
         ClockGauge[AbsoluteTime[TimeZone → +9],
          PlotLabel → Style["Tokyo", Large, Bold]]}], (UpdateInterval → 1)]]
```

Local   **Tokyo**

If you find the previous example difficult to understand, remember that you can always access the documentation to familiarize yourself with the functions used to generate the clocks.

1.5.5 Things to Remember

It's important to keep in mind that inputs are executed based on the order of evaluation and not the order in which they are shown on the screen. This means that if a function calls previously defined functions, those functions should have been evaluated in advance. It's not enough to see them already typed. For example, if you modify an assignment that is used later on, you will have to execute again all the related cells.

- Evaluate sequentially the following three cells:

In[ ]:= **a = 3 (*first*)**

Out[ ]= 3

In[ ]:= **b = x^a (*second*)**

Out[ ]= x^3

In[ ]:= **a b (*third*)**

Out[ ]= $3 x^3$

Now assign a the value of 2 in the first cell. What happens when you re-execute the third instruction (**a b**)?

Mathematica **retains information about all the variables defined in a session, even after starting a new** *notebook*. Because of that, when starting a new section that will not be using the variables defined previously, it is a good idea to clear them first.

- One way to remove all the values and definitions is as follows:

In[ ]:= **ClearAll["Global`*"]**

There's another way to limit the use of variables to a specific context:

Define this line as a subsubsection.

- Select all the cells corresponding to this subsubsection and in the main menu choose: **Evaluation ▸ Notebook's Default Context ▸ Unique to Each Cell Group**.
- Define the function below:

In[ ]:= **f[x_] := 3 x**

In[ ]:= **f[3]**

Out[ ]= 9

Define this line as a new subsubsection.

- If you execute the same function f, you'll see that is not defined.

In[ ]:= **f[3]**

Out[ ]= $f(3)$

If you really want to remove all the information and not only the variables, the best way is to quit the session. This requires that you exit *Mathematica* (actually it requires that you exit the program *Kernel*). You can do that in several ways: With **File ▸ Exit** or if you don't want to exit the notebook use either **Evaluation ▸ Quit Kernel** or in an input cell type **Quit[]**. Once you have exited, the program removes from memory all the assignments and definitions. When you execute the next command the kernel will be loaded again.

Although *Mathematica* tries to maintain compatibility with *notebooks* created in previous versions, it is not always complete so you may have to make some modifications. When you open for the first time a notebook created in a previous *Mathematica* version, the program will offer you the possibility of checking the file automatically; Accept it and read the comments that you may receive carefully. In many cases, *Mathematica* will modify expressions directly and let you know about the changes made giving you the option to accept or reject them. In other cases it will offer suggestions.

> Remember: If you don't understand a function use the Documentation Center.

1.5.6 Entities

The Wolfram Language (WL) contains real-world data accessible through the entity framework, describe in detail in Chapter 4.

- For example, to find out the population of Japan, we would just ask *Mathematica* for the population value corresponding to the "Japan" entity. Using ▣ (**Insert ▸ Inline Free-form Input**) we write "Japan" and "Population" inside EntityValue:

EntityValue[▣ Japan , ▣ Population]

In[ ]:= **EntityValue[[Japan COUNTRY] , [population]]**

Out[ ]= $126\,476\,458$ people

1.5.7 Wikipedia

- We can even obtain information from Wikipedia using the function WikipediaData.

In[ ]:= **WikipediaData["Alhambra", "ImageList"][[3]]**

Out[ ]=

1.5.8 Machine Learning

Mathematica also includes machine learning functions. Let's see an example.

- Handwrite different numbers on a white piece of paper, scan them or type them in a touch screen (in MS Windows you can use the **Crop** application) and label them:

In[ ]:= **labeledData =** $\{2 \to 2, 5 \to 5, 4 \to 8, 0 \to 0, 2 \to 2, 7 \to 7, 5 \to 5, 1 \to 1, 3 \to 3,$
$0 \to 0, 3 \to 3, 9 \to 9, 6 \to 6, 2 \to 2, 8 \to 8, 2 \to 2, 0 \to 0, 6 \to 6, 6 \to 6,$
$1 \to 1, 1 \to 1, 7 \to 7, 8 \to 8, 5 \to 5, 0 \to 0, 4 \to 4, 7 \to 7, 6 \to 6, 0 \to 0, 2 \to 2,$
$5 \to 5, 3 \to 3, 1 \to 1, 5 \to 5, 6 \to 6, 7 \to 7, 5 \to 5, 4 \to 4, 1 \to 1, 9 \to 9,$
$3 \to 3, 6 \to 6, 8 \to 8, 0 \to 0, 9 \to 9, 3 \to 3, 0 \to 0, 3 \to 3, 7 \to 7, 4 \to 4,$
$4 \to 4, 3 \to 3, 8 \to 8, 0 \to 0, 4 \to 4, 1 \to 1, 3 \to 3, 7 \to 7, 6 \to 6, 4 \to 4,$
$7 \to 7, 2 \to 2, 7 \to 7, 2 \to 2, 5 \to 5, 2 \to 2, 0 \to 0, 9 \to 9, 8 \to 8, 9 \to 9,$
$8 \to 8, 1 \to 1, 6 \to 6, 4 \to 4, 8 \to 8, 5 \to 5, 8 \to 8, 0 \to 0, 6 \to 6, 7 \to 7,$
$4 \to 4, 5 \to 5, 8 \to 8, 4 \to 4, 3 \to 3, 1 \to 1, 5 \to 5, 1 \to 1, 9 \to 9, 9 \to 9,$
$9 \to 9, 2 \to 2, 4 \to 4, 7 \to 7, 3 \to 3, 1 \to 1, 9 \to 9, 2 \to 2, 9 \to 9, 6 \to 6\};$

- Use the command **Classify** to teach the program to associate the handwriting to a certain digit based on the examples. This function uses statistical criteria to assign a weight to each potential match.

In[ ]:= **digits = Classify[labeledData]**

Out[ ]= ClassifierFunction[⊞ ⠿ Input type: **Image** Number of classes: **10**]

- Copy from above the numbers 0 to 10 in a list like the following:

In[ ]:= **digits**$\left[\{0, 1, 2, 3, 4, 5, 6, 7, 8, 9\}\right]$

Out[ ]= {0, 1, 2, 3, 4, 5, 6, 7, 6, 4}

- Note that the classification has been done correctly except in the case of 6 that has been mistaken with the digit 4. The program assigns probabilities based on the stored data. You can check that the probability assigned to the symbol 6 is higher for 4 than for 6. Even a person may not be sure whether the handwriting represents a 6 or a 4.

In[ ]:= `digits[⫝, "TopProbabilities"]`

Out[ ]= {6 → 0.82}

- If 6 is written in a less equivocal way, the probability of a correct identification improves substantially.

In[ ]:= `digits[6, "TopProbabilities"]`

Out[ ]= {6 → 0.82}

1.6 Additional Resources and Supplementary Materials

Supplementary materials are available on the book's website:

http://www.mathematicabeyondmathematics.com/

The Documentation Center is the primary source about *Mathematica* and the Wolfram Language: **Help ▶ Documentation Center** or http://reference.wolfram.com/language/

You may also find the following links useful:

An Elementary Introduction to the Wolfram Language by Stephen Wolfram:
http://www.wolfram.com/language/elementary-introduction/ (there is also a print version).
Introductory tutorials:
http://reference.wolfram.com/language/tutorial/IntroductionOverview.html
Explanatory video guides: http://www.wolfram.com/broadcast
Open courses: https://www.wolfram.com/wolfram-u/
News and ideas: http://blog.wolfram.com , including a short description of the additional capabilities of new releases: https://blog.wolfram.com/version-releases/

For an overview of all the free resources that Wolfram Research makes available, visit:

https://wolfram.com/open-materials/

2

Programming: The Beauty and Power of the Wolfram Language

This chapter covers some of the concepts and functions that we need to program in Mathematica using WL. It also explains some useful notions for generating efficient code. In particular, we will learn how to code using pure functions, very important to create compact programs that take full advantage of the power of Mathematica. The upcoming sections also include examples of some of the most commonly used programming functions and describe the fundamental ideas behind package development, essential to build new applications that incorporate user-defined functionality. Finally, guidance will be provided on tools available for large projects or to develop programs that can be run in the cloud. This chapter may be the least fun, but it is important to have a basic understanding of the concepts discussed in its pages to be able to follow the rest of the book. If you are an advanced user you may skip it, although we recommend that you at least skim through its contents in case you find something useful.

2.1 *Mathematica*'s Programming Language: The Wolfram Language

As we have seen already in previous chapters, *Mathematica* is not only a program to perform symbolic and numeric calculations. It is a complete technical system that can be used to develop anything: from traffic control applications to sophisticated image manipulation interfaces. All of this is possible thanks to the power of the Wolfram Language, the high-level general-purpose programming language used by *Mathematica*. To start using it we need to know its fundamentals, and that's what we are going to learn in this chapter. This section will cover the basic components of the language.

2.1.1 Atoms and Expressions

Everything in the Wolfram Language is an expression, including notebooks. Expressions are written in the form: Head[..., {}] and are ultimately built from a small number of distinct types of atomic elements: Integer, Real, Rational, Complex, String, and Symbol. Do not get confused with the domains: Integers ($\mathbb{Z}$), Reals ($\mathbb{R}$), Rationals ($\mathbb{Q}$) and Complexes ($\mathbb{C}$).

- AtomQ can be used to test whether an expression is an atom:

In[ ]:= **10 / 3 ;**

- % gives the last result. %% gives the result before last. %% ... % (k times) gives the kth previous result.

In[ ]:= **AtomQ[%]**

Out[ ]= True

In[ ]:= **Head[%%]**

Out[ ]= Rational

Head is always capitalized and followed by a pair of enclosing square brackets containing elements separated by commas. Those elements sometimes go inside braces. Here is one example:

In[ ]:= **Head[a + b + c]**

Out[ ]= Plus

- Instead of x + y + z we write:

In[ ]:= **Plus[x, y, z]**

Out[ ]= $x + y + z$

- With FullForm we can see the basic form of expressions. One head, Plus in this case, and elements separated by commas inside brackets:

In[ ]:= **FullForm[x + y + z]**

Out[ ]//FullForm=
              Plus[x, y, z]

Cells are also expressions. Shown below is a copy of this cell displaying its contents. This was done by duplicating this cell, selecting the copy and clicking on: **Cell ▶ Show Expression**.

Cell[TextData[{"Cells are also expressions. Shown below is a copy of this cell displaying its contents. This was done by selecting the copied cell and clicking on: ",
StyleBox["Cell", FontWeight->"Bold"], " ▶ ", StyleBox["Show Expression", FontWeight->"Bold"], ". "}], "Text AM",
CellChangeTimes->{{3.6*^9, 3.6*^9}, {3.6*^9, 3.6396977*^9}, {3.6*^9, 3.6396978*^9}, {3.6*^9, 3.6*^9}, {3.6*^9, 3.6*^9}}]

All objects in *Mathematica* are one of the following types:

- Numbers: (i) Integer: Head used for integers, (ii) Real: Head of number with a decimal separator (floating-point), (iii) Rational: Head of a rational number (ratio between two integers such as 3/4 or 27/8), (iv) Complex: Head of a complex (sum of a real and an imaginary number: $3 + 2\,i$).
- Symbols: A sequence of letters, digits, or symbols not starting with a number (we will describe them later).
- Strings: Any expression between double quotation marks: "this is a *string*".
- SparseArrays: Matrices or lists in which most of the elements are 0:

In[ ]:= **MatrixForm[SparseArray[{i_, i_} :> RandomInteger[8], {4, 4}]]**

Out[ ]//MatrixForm=

$$\begin{pmatrix} 7 & 0 & 0 & 0 \\ 0 & 0 & 0 & 0 \\ 0 & 0 & 8 & 0 \\ 0 & 0 & 0 & 6 \end{pmatrix}$$

In[ ]:= **AtomQ[%]**

Out[ ]= True

In[ ]:= **Head[%%]**

Out[ ]= SparseArray

Graphs: A collection of objects, joined by vertices in pairs.

In[ ]:= **Graph[{1 ⟷ 2, 2 ⟷ 3, 3 ⟷ 1}]**

Out[ ]=

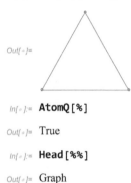

In[ ]:= **AtomQ[%]**

Out[ ]= True

In[ ]:= **Head[%%]**

Out[ ]= Graph

Mathematica internally always separates expressions first into their atomic components before computing them.

- As mentioned before, we will often use as a precaution the command **ClearAll** ["Global`*"] at the end or at the beginning of a section to avoid problems with variables previously defined:

In[ ]:= **ClearAll["Global`*"]**

2.1.2 Levels in Expressions

We've seen that *Mathematica* manipulates basic objects located at the head of each expression (**Head**).

- Remember that with **FullForm** we can see the basic form of expressions. The example below shows that a/b internally is a multiplied by b^{-1}.

In[ ]:= **FullForm[a / b]**

Out[ ]//FullForm=

Times[a, Power[b, −1]]

- Here we apply **FullForm** to a more complicated expression.

In[ ]:= **FullForm[1.5 Sin[2 Pi / 4]]**

Out[ ]//FullForm=

1.5`

- We get the result of the operation but not the structure of the initial expression as we wanted. This is because the calculation is done first and then `FullForm` is applied to the result, just a number. To avoid the execution of the calculation we use `HoldForm`.

In[ ]:= `FullForm[ HoldForm[1.5 Sin[2 Pi / 4]]]`

Out[ ]//FullForm=

    HoldForm[Times[1.5`, Sin[Times[2, Times[Pi, Power[4, −1]]]]]]

- The structure of any expression adopts a tree-like format. Here we represent the previous expression in such a format using `TreeForm`.

In[ ]:= `expr = HoldForm[1.5 Sin[2 Pi / 4]]`

Out[ ]= $1.5 \sin\left(\dfrac{2\pi}{4}\right)$

In[ ]:= `TreeForm[ expr]`

Out[ ]//TreeForm=

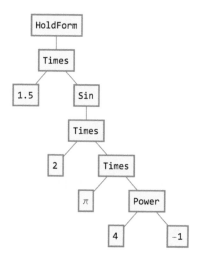

- Depth gives the total number of levels:

In[ ]:= `Depth[expr]`

Out[ ]= 7

- LeafCount gives the total number of individual atomic elements:

In[ ]:= `LeafCount[expr]`

Out[ ]= 11

- Accessing individual levels:

In[ ]:= `Level[expr, {4}]`

Out[ ]= $\left\{2, \dfrac{\pi}{4}\right\}$

If we analyze the previous tree we can see that the first expression is 1/4. Internally it is represented as 4^{-1}. Next, that expression is multiplied by Pi, the result multiplied by 2, and so on until reaching the top of the tree.

- `Trace` enables us to see the order in which operations are executed:

In[ ]:= **Trace[1.5 Sin[2 Pi / 4]]**

Out[ ]= $\left\{\left\{\left\{\left\{\frac{1}{4}, \frac{1}{4}\right\}, \frac{\pi}{4}, \frac{\pi}{4}\right\}, \frac{2\pi}{4}, \frac{2\pi}{4}, \frac{2\pi}{4}, \frac{\pi}{2}\right\}, \sin\left(\frac{\pi}{2}\right), 1\right\}, 1.5 \times 1, 1.5\right\}$

- The previous output seems to duplicate certain steps. However, with **FullForm** we can see that it's actually the basic structure that is changing. So when we see $\left\{\frac{1}{4}, \frac{1}{4}\right\}$, internally there has been a structural change: the expression has changed from (4^{-1}) to $(\frac{1}{4})$.

In[ ]:= **FullForm[%]**

Out[ ]//FullForm=

List[List[List[List[List[HoldForm[Power[4, −1]], HoldForm[Rational[1, 4]]],
        HoldForm[Times[Pi, Rational[1, 4]]], HoldForm[Times[Rational[1, 4], Pi]]],
        HoldForm[Times[2, Times[Rational[1, 4], Pi]]], HoldForm[Times[2, Rational[1, 4], Pi]],
        HoldForm[Times[Rational[1, 4], 2, Pi]], HoldForm[Times[Rational[1, 2], Pi]]],
        HoldForm[Sin[Times[Rational[1, 2], Pi]]], HoldForm[1]],
        HoldForm[Times[1.5`, 1]], HoldForm[1.5`]]

- With a simple transformation rule we can modify the previous expression so that multiplication becomes addition.

In[ ]:= **HoldForm[1.5 Sin[2 Pi / 4]] /. {Times → Plus}**

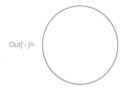

Out[ ]= $1.5 + \sin\left(2 + \left(\pi + \frac{1}{4}\right)\right)$

- The next function shows a red circumference of radius 1.

In[ ]:= **gr = Graphics[{Red, Circle[{0, 0}, 1]}]**

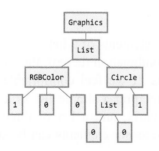

Out[ ]=

- Its internal form is as follows:

In[ ]:= **FullForm[gr]**

Out[ ]//FullForm=

Graphics[List[RGBColor[1, 0, 0], Circle[List[0, 0], 1]]]

In[ ]:= **TreeForm[gr]**

Out[ ]//TreeForm=

```
            Graphics
               |
             List
            /      \
      RGBColor     Circle
       /  \        /    \
      1    0      0   List   1
                      /  \
                     0    0
```

If you move the cursor over an individual frame you will notice that it displays the result of the operations until that frame.

- A small transformation will allow us to modify the expression so that instead of a red circumference (Circle), we get a red circle (Disk).

In[ ]:= **gr /. Circle → Disk**

Out[ ]=

Once we know the structure of an expression, we will be able to modify it. Additionally, when encountering outputs from operations that we don't understand, their internal form can give us clues about how the program works.

2.1.3 The Wolfram Language Paradigm

At the core of the Wolfram Language (or *Mathematica*) is the foundational idea that everything—data, programs, formulas, graphics, documents—can be represented as symbolic expressions. And it is this unifying concept that underlies the Wolfram Language symbolic programming paradigm.

Symbols are the ultimate atoms of symbolic data letters, digits, combinations of letters and digits not starting with a digit, and special symbols such as: \$, π, ∞, e, i, j. Upper- and lower-case letters are distinguished. Once a symbol has been created during a session, it will remain in memory unless explicitly removed. An alternative is to use functions that automatically delete a symbol once it's been used.

- If we have a list and we'd like to know the type of objects that it contains, we apply Head to each of the list elements. This can be done using Map..

In[ ]:= **Map[Head, {3, 3/2, 1.3, 3 + 4 I, I, $\sqrt{2}$, Pi, {a, b}, a, ●, 🚗}]**

Out[ ]= {Integer, Rational, Real, Complex, Complex, Power, Symbol, List, Symbol, Graphics, Image}

In[ ]:= **ClearAll["Global`*"]**

2.2 Lists Operations

A basic structure in *Mathematica* is the list. The elements of a list can be numbers, variables, functions, strings, other lists, or any combination thereof; they are written inside braces { } and are separated by commas. If we feel comfortable with the classical terminology of mathematics we can interpret a simple list as a vector, a two-dimensional list (a list of sublists) as a matrix, and a list of three or more dimensions as a tensor, keeping in mind that the tensor elements can be sublists of any dimension.

- Below we define a list consisting of three sublists.

In[ ]:= **list = {{2, 3, 6}, {3, 1, 5}, {4, 3, 7}};**

In[ ]:= **list // TreeForm**

Out[ ]//TreeForm=

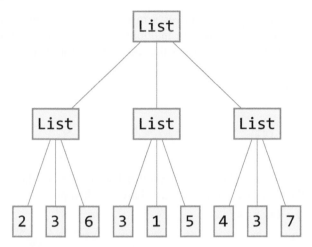

- The previous list can be shown in traditional notation as a matrix, but we should not forget that internally *Mathematica* treats it as a list.

In[ ]:= **list // TraditionalForm**

Out[ ]//TraditionalForm=

$$\begin{pmatrix} 2 & 3 & 6 \\ 3 & 1 & 5 \\ 4 & 3 & 7 \end{pmatrix}$$

If the notebook has been configured to display outputs using the traditional format (as in this case), it's not necessary to use //MatrixForm.

In[ ]:= **MatrixForm[{{1, 2}, {3, 4}}]**

Out[ ]//MatrixForm=

$$\begin{pmatrix} 1 & 2 \\ 3 & 4 \end{pmatrix}$$

- Length gives the number of elements in the list and for list with sublists Dimensions can be used to determine the length at each level:

In[ ]:= **{Length[list], Dimensions[list]}**

Out[ ]= {3, {3, 3}}

It's important to distinguish operations between lists and note that they don't behave in the same way as operations between matrices, especially with respect to multiplication.

- This is an example of list multiplication. We write * to emphasize that it's a multiplication, but if we leave an empty space instead, the result would be the same.

In[ ]:= **{{a, b}, {c, d}} * {x, y}**

Out[ ]= $\begin{pmatrix} ax & bx \\ cy & dy \end{pmatrix}$

- Here we multiply matrices, or if you prefer, a matrix by a vector. To indicate that this is a matrix multiplication we use Dot (or "**.**" which is the same). Note that "*" and "." are different operations.

In[ ]:= `{{a, b}, {c, d}}.{x, y}`

Out[ ]= $\{a\,x + b\,y,\ c\,x + d\,y\}$

- In the mathematical rule for multiplying matrices a row vector is multiplied by a column vector. *Mathematica* is less strict and allows the multiplication of two lists that do not meet the mathematical multiplication rule. It chooses the most appropriate way to perform the operation.

In[ ]:= `{x, y, z} . {a, b, c}`

Out[ ]= $a\,x + b\,y + c\,z$

In[ ]:= `{x, y, z} . {{a}, {b}, {c}}`

Out[ ]= $\{a\,x + b\,y + c\,z\}$

- In general, the sum of lists of equal dimension is another list of the same dimension whose elements are the sum of the elements of the initial lists based on their position.

In[ ]:= `{1, 2, 3} - {a, b, c} + {x, y, z}`

Out[ ]= $\{-a + x + 1,\ -b + y + 2,\ -c + z + 3\}$

- If we multiply a list by a constant, each element of the list will be multiplied by it. The same happens if we add a constant to a list.

In[ ]:= `{{1, 2}, {3, 4}} k`

Out[ ]= $\begin{pmatrix} k & 2\,k \\ 3\,k & 4\,k \end{pmatrix}$

In[ ]:= `{{1, 2}, {3, 4}} + c`

Out[ ]= $\begin{pmatrix} c+1 & c+2 \\ c+3 & c+4 \end{pmatrix}$

- It is easy to get the First or the Last element of a list:

In[ ]:= `list2 = {a, b, c};`

In[ ]:= `{First[list2], Last[list2]}`

Out[ ]= $\{a,\ c\}$

- Or

In[ ]:= `{list2[[1]], list2[[-1]]}`

Out[ ]= $\{a,\ c\}$

To extract a list element based on its position *i* in the list we use Part[*list, i*] or, in compact form, [[*i*]].

- In this example, we show two equivalent ways of extracting the second element of a list:

In[ ]:= `{Part[list2, 2], {a, b, c}[[2]]}`

Out[ ]= $\{b,\ b\}$

In the case of a list of sublists, first we identify the sublist and then the element within the sublist.

- In the following example, we extract element 3 from sublist 2:

In[ ]:= `{{a1, a2}, {b1, b2, b3}, {c1}}[[2, 3]]`

Out[ ]= b3

- In the example below, we extract elements 2 to 4. Note that the syntax `[[i;;j]]` means to extract from *i* to *j*.

In[ ]:= `list3 = {a11, 42, x² + y², Pi, ▮, "house", ⬚ };`

In[ ]:= `{list3[[2 ;; 4]], list3[[4 ;; ]], list3[[ ;; 3 ]]}`

Out[ ]= $\left\{ \{42, x^2 + y^2, \pi\}, \{\pi, ▮, \text{house}, ⬚ \}, \{a11, 42, x^2 + y^2\} \right\}$

- Here we extract elements 2 and 3 from sublists 1 and 2.

In[ ]:= `{{a1, a2, a3}, {b1, b2, b3}, {c1, c2, c3}}[[{1, 2}, {2, 3}]]`

Out[ ]= $\begin{pmatrix} a2 & a3 \\ b2 & b3 \end{pmatrix}$

- The code below, shows how to get the first column and the first row.

In[ ]:= `{{a1, a2, a3}, {b1, b2, b3}, {c1, c2, c3}}[[All, 1]]`

Out[ ]= {a1, b1, c1}

In[ ]:= `{{a1, a2, a3}, {b1, b2, b3}, {c1, c2, c3}}[[1, All]]`

Out[ ]= {a1, a2, a3}

- In this case, we use `ReplacePart` to replace a list element, specifically from sublist 2 we replace element 2 that is b2 with c2.

In[ ]:= `ReplacePart[{{a1, a2, a3}, {b1, b2, b3}}, {2, 2} → c2]`

Out[ ]= $\begin{pmatrix} a1 & a2 & a3 \\ b1 & c2 & b3 \end{pmatrix}$

- Next, we use `[[{ }]]` to rearrange the elements in a list.

In[ ]:= `list = {a, b, c, d, e, f};`

In[ ]:= `list[[{1, 4, 2, 5, 3, 6}]]`

Out[ ]= {a, d, b, e, c, f}

Lists can also be used to specify intervals for graphical plots and in other functions that require ranges as inputs.

In[ ]:= `RandomReal[{3, -3}, {2, 3}]`

Out[ ]= $\begin{pmatrix} 2.95474 & 1.10704 & 2.41065 \\ 1.09333 & 2.03312 & -1.45064 \end{pmatrix}$

The solutions (*out*) are often displayed in list format. If we need a specific part of the solution, we can use the previous command to extract it.

- To extract the second solution of this second-degree equation:

In[ ]:= `sol = Solve[1.2 x^2 + 0.27 x - 0.3 == 0, x]`

Out[ ]= {{x → −0.625}, {x → 0.4}}

- We can proceed as follows:

In[ ]:= `x /. sol[[2]]`

Out[ ]= 0.4

Other functions very useful for list manipulation are Flatten, Take, and Drop.

- With Flatten[*list*] we flatten out nested lists. With Flatten[*list,n*] we flatten the list to level *n*.

In[ ]:= **Flatten[{a, {{b, c}}, d, {{{e}}}}]**

Out[ ]= {*a, b, c, d, e*}

- With Take[*list, i*] we extract the first *i* elements of the list, and with −*i* the last *i*. With Take[*list, {i,j}*] we can extract elements *i* to *j*. In the example below, we flatten out the list before taking the first 5 elements:

In[ ]:= **Take[Flatten[{{a, b}, {c, d, e}, f, g}], 5]**

Out[ ]= {*a, b, c, d, e*}

- With Drop[*list, i*] we remove the first *i* elements of the list, and with −*i* the last *i*. Drop[*list, {i, j}*] removes elements *i* to *j*.

In[ ]:= **Drop[{a, b, c, d, e, f, g}, -3]**

Out[ ]= {*a, b, c, d*}

- There are commands particularly useful when operating on lists. It's easy to identify them since they generally contain **List** as part of their names. You can check using the help system:

> **? *List***

We omit the output due to space constraints.

There are many more functions used to manipulate lists. We recommend you consult the documentation: guide/ListManipulation.

2.3 Association and Dataset

Association and Dataset can be considered extensions to the concept of lists. With these new additions, Mathematica has a very powerful way to deal with structured data.

- Association (<|...|>) is a dictionary data structure containing key-value pairs. An association in *Mathematica* represents labeled data as key → value pairs:

In[ ]:= **<|a → 1, b → 2, c → 3, d → 4|>**

Out[ ]= <|*a* → 1, *b* → 2, *c* → 3, *d* → 4|>

- The value is directly associated with the key:

In[ ]:= **<|a → 1, b → 2, c → 3, d → 4|>[c]**

Out[ ]= 3

- This data structure can be broken down into individual lists containing just its Keys or Values. Keys must be unique.

In[ ]:= **assoc = <|a → 1, {2, 3} → 100, 4 → 200, "sphere" → |>**

Out[ ]= $\left\langle \left| a \to 1, \{2, 3\} \to 100, 4 \to 200, \text{sphere} \to \right. \right.$ $\left. \left. |\right\rangle \right.$

In[ ]:= `{Keys[assoc], Values[assoc]}`

Out[ ]:= $\begin{pmatrix} a & \{2, 3\} & 4 & \text{sphere} \\ 1 & 100 & 200 & \end{pmatrix}$

- To extract the value of an element, we can either use a single square bracket with the element's key ([key]) or Part with the position of the element in the list ([[position]]) :

In[ ]:= `{assoc["sphere"], assoc⟦4⟧}`

Out[ ]:= { , }

An alternative and more general method to extract values from association objects is to use Lookup:

In[ ]:= `Lookup[assoc, {a, "sphere"}]`

Out[ ]:= {1, }

- With Normal, the association is converted to a normal expression (in this case a list).

In[ ]:= `Normal[assoc]`

Out[ ]:= $\{a \rightarrow 1, \{2, 3\} \rightarrow 100, 4 \rightarrow 200, \text{sphere} \rightarrow$ $\}$

- AssociationMap and AssociationThread can construct associations from lists:

In[ ]:= `AssociationMap[f, {a, b, c}]`

Out[ ]:= $\langle| a \rightarrow f(a), b \rightarrow f(b), c \rightarrow f(c) |\rangle$

In[ ]:= `AssociationThread[{a, b, c}, {1, 2, 3}]`

Out[ ]:= $\langle| a \rightarrow 1, b \rightarrow 2, c \rightarrow 3 |\rangle$

In[ ]:= `Association[Thread[{"a", "b", "c"} → {1, 2, 3}]]`

Out[ ]:= $\langle| a \rightarrow 1, b \rightarrow 2, c \rightarrow 3 |\rangle$

- Common functions for list operations work directly on the keys in an association if "Key" is added to the name:

In[ ]:= `assoc = <|3 → c, 4 → d, 1 → a, 2 → b|>`

Out[ ]:= $\langle| 3 \rightarrow c, 4 \rightarrow d, 1 \rightarrow a, 2 \rightarrow b |\rangle$

In[ ]:= `KeySort[assoc]`

Out[ ]:= $\langle| 1 \rightarrow a, 2 \rightarrow b, 3 \rightarrow c, 4 \rightarrow d |\rangle$

In[ ]:= `{KeySelect[assoc, OddQ], KeyTake[assoc, {2, 3}], KeyDrop[assoc, {1, 2}] }`

Out[ ]:= $\{\langle| 3 \rightarrow c, 1 \rightarrow a |\rangle, \langle| 2 \rightarrow b, 3 \rightarrow c |\rangle, \langle| 3 \rightarrow c, 4 \rightarrow d |\rangle\}$

A dataset is formed from a list of associations:

- The example below shows a simple dataset with 2 rows and 3 columns. It consists of an association of associations. Most datasets are displayed in tabular form:

In[◦]:= `data = Dataset[<|"1" → <|"A" → 3, "B" → 4, "C" → 2|>,`
`      "2" → <|"A" → 4, "B" → 1, "C" → 3|>|>]`

Out[◦]=

	A	B	C
1	3	4	2
2	4	1	3

- You can extract parts from datasets in a similar way to the one used with lists. To extract the element from "row 2" and "column C":

In[◦]:= `data["2", "C"]`

Out[◦]= 3

- Get the elements from an entire row and from all the rows for a particular column.

In[◦]:= `{data["1"], data[All, "B"]}`

Out[◦]=
{
A	3
B	4
C	2
,
1	4
2	1
}

Wolfram documentation gives examples showing how to use a Dataset. If you want to dig deeper into Mathematica's capabilities for handling structured data, there is a vast collection of computable datasets in the Wolfram Data Repository (https://datarepository.wolframcloud.com/) that we will use later.

2.4 Matrix Operations

Matrix operations appear very often when performing algebraic calculations. The use of SparseArray speeds up computations significantly.

- Here we generate 5 sublists with 5 random integers, ranging from 0 to 10, in each one. (Note: SeedRandom is used to make the example replicable.)

In[◦]:= `SeedRandom[3142];`

In[◦]:= `mat = RandomInteger[10, {5, 5}]`

Out[◦]=
$$\begin{pmatrix} 6 & 1 & 3 & 7 & 9 \\ 3 & 3 & 9 & 4 & 9 \\ 8 & 9 & 4 & 0 & 6 \\ 10 & 7 & 10 & 0 & 0 \\ 3 & 1 & 1 & 9 & 10 \end{pmatrix}$$

- The next function displays the previous output in matrix format and evaluates its transpose, inverse, determinant, and eigenvalues. We use TabView to place the outputs under different labels visible after clicking on the corresponding tab.

```
In[ ]:= TabView[
        {"Matrix A" → MatrixForm[mat], "Transpose" → MatrixForm[Transpose[mat]],
         "Inverse" → MatrixForm[Inverse[mat]], "Determinant" → Det[mat],
         "Eigenvalues" → N[Eigenvalues[mat]]}]
```

Matrix A	Transpose	Inverse	Determinant	Eigenvalues

$$Out[ ]= \begin{pmatrix} 6 & 1 & 3 & 7 & 9 \\ 3 & 3 & 9 & 4 & 9 \\ 8 & 9 & 4 & 0 & 6 \\ 10 & 7 & 10 & 0 & 0 \\ 3 & 1 & 1 & 9 & 10 \end{pmatrix}$$

Next, we create an equation of the form **m x = b**, where **m** corresponds to the matrix **mat**. We will solve it and check that the solution is correct.

■ We generate one dimension matrix **b** randomly (a VectorQ in Mathematica notation).

```
In[ ]:= b = RandomInteger[10, {5}]
```

$$Out[ ]= \{5, 1, 6, 9, 1\}$$

■ We solve the equation using LinearSolve.

```
In[ ]:= xvec = LinearSolve[mat, b]
```

$$Out[ ]= \left\{ \frac{20191}{16684}, -\frac{3121}{8342}, -\frac{403}{8342}, -\frac{3719}{16684}, -\frac{337}{16684} \right\}$$

■ We verify that the solution is correct ("." is used for matrix multiplication).

```
In[ ]:= mat.xvec
```

$$Out[ ]= \{5, 1, 6, 9, 1\}$$

Matrix operations very frequently have matrices where only few elements have non-zero values. These kinds of matrices are called sparse arrays. In these cases, it is recommended to use SparseArray.

■ We build a tridiagonal matrix 5×5 whose main diagonal elements are 9s and its adjacent 1s.

```
In[ ]:= s = SparseArray[{{i_, i_} → 9, {i_, j_} /; Abs[i - j] == 1 → 1}, {5, 5}]
```

Out[ ]= SparseArray[⊞ ▨ Specified elements: 13
                          Dimensions: {5, 5}]

■ We show it numerically and represent it graphically using MatrixPlot.

```
In[ ]:= MatrixForm[Normal[s]]
```

Out[ ]//MatrixForm=

$$\begin{pmatrix} 9 & 1 & 0 & 0 & 0 \\ 1 & 9 & 1 & 0 & 0 \\ 0 & 1 & 9 & 1 & 0 \\ 0 & 0 & 1 & 9 & 1 \\ 0 & 0 & 0 & 1 & 9 \end{pmatrix}$$

In[ ]:= **MatrixPlot[s]**

Out[ ]=
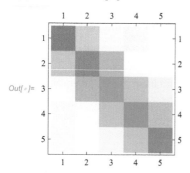

We build a linear system of 5 equations of the form **s x = b**, with **s** and **b** the matrices previously defined.

- We solve the equation and check the calculation time:

In[ ]:= **Timing[LinearSolve[s, b]]**

Out[ ]= $\left\{0., \left\{\frac{869}{1560}, -\frac{7}{520}, \frac{22}{39}, \frac{487}{520}, \frac{11}{1560}\right\}\right\}$

- We repeat the same process but for a system of 50,000 variables and 50,000 equations, of the form **s x = b**, where **s** and **b** are called **sLarge** and **bLarge**, respectively. We use ";" to avoid displaying the output since it would take too much space.

In[ ]:= **Timing[bLarge = RandomReal[1, {50 000}];]**

Out[ ]= {0., Null}

In[ ]:= **Timing[sLarge = SparseArray[**
    {{i_, i_} → 9, {i_, j_} /; Abs[i - j] == 1 → 1}, {50 000, 50 000}];]

Out[ ]= {0.59375, Null}

In[ ]:= **MatrixPlot[sLarge, ColorFunction → "Rainbow"]**

Out[ ]=
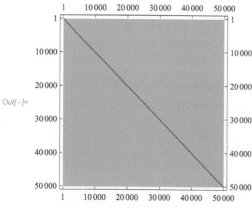

- We solve the system **sLarge x = bLarge** and see that the calculation time is extremely short.

In[ ]:= **Timing[xvec = LinearSolve[sLarge, bLarge];]**

Out[ ]= {0.03125, Null}

- We don't show the result, but we can check that the answer is correct by verifying that $m x - b = 0$. With Chop we eliminate the insignificant terms, unavoidable in numerical calculations involving decimal numbers.

In[ ]:= **Norm[sLarge.xvec - bLarge] // Chop**

Out[ ]= 0

2.5 Set, SetDelayed, and Dynamic Variables

In[ ]:= **ClearAll["Global`*"]**

A frequently asked question is how to differentiate between "=" (Set) and ":=" (SetDelayed).

In[ ]:= **x = RandomReal[]**
    y := RandomReal[]

Out[ ]= 0.921297

In[ ]:= **x - x**

Out[ ]= 0.

In[ ]:= **y - y**

Out[ ]= 0.129974

- Find out about a symbol's definitions with Information (?):

In[ ]:= **?x**

Out[ ]=

| Symbol |
| Global`x |
| Assignment |
| $x = 0.921297$ |
| Full Name Global`x |

In[ ]:= **?y**

Out[ ]=

| Symbol |
| Global`y |
| Assignment |
| $y :=$ RandomReal[] |
| Full Name Global`y |

- If we want to remove the definition assigned to **x** from the global context we can use Clear.

In[ ]:= **Clear[x]**

- Now **x** and **y** doesn't have any assignments:

In[ ]:= **x**

Out[ ]= x

What has happened? When "=" is used in **f** = **exp,** the result of evaluating **exp** is assigned to **f** immediately. From then on, whenever **f** is called, the program will always replace **f** with the value of **exp**. In the previous example, the value of **x** was stored right away, while in the case of **y**, what gets stored is the function that will be executed every time it's called. Because of that, when calculating **y−y**, RandomReal[ ] is executed twice, returning a different result each time.

- The following expression generates a list of 5 random integers between 0 and 5. We call it **data1**. These numbers have been stored and they will always be used anytime **data1** is called.

In[ ]:= **data1 = RandomInteger[{5}, 5]**

Out[ ]= {3, 1, 1, 0, 3}

- Now we type the same above expression but using ":=". We name this new list **data2**. From now on, whenever **data2** is called, the expression that generates the 5 random numbers will be executed anew. What is stored is the function RandomInteger[{5},5] not the actual numbers it generates.

In[ ]:= **data2 := RandomInteger[{5}, 5]**

- The difference can be seen by calling and executing the above defined expressions.

In[ ]:= **data1**

Out[ ]= {3, 1, 1, 0, 3}

In[ ]:= **data2**

Out[ ]= {5, 3, 3, 1, 4}

- In the following example we combine ":=" with "=". What happens?

In[ ]:= **data3a := data3b = RandomInteger[{5}, 5]**

In[ ]:= **{data3a, data3b }**

Out[ ]= $\begin{pmatrix} 0 & 1 & 4 & 0 & 1 \\ 0 & 1 & 4 & 0 & 1 \end{pmatrix}$

In[ ]:= **{data3b, data3a}**

Out[ ]= $\begin{pmatrix} 0 & 1 & 4 & 0 & 1 \\ 1 & 2 & 4 & 2 & 1 \end{pmatrix}$

In[ ]:= **{data3a, data3b}**

Out[ ]= $\begin{pmatrix} 2 & 1 & 3 & 3 & 5 \\ 2 & 1 & 3 & 3 & 5 \end{pmatrix}$

> Remember: When using ":=" (SetDelayed) in **f:=exp** the expression **exp** is not evaluated until **f** is used in a posterior calculation, and it will be executed each time **f** is called. However, if you use "=", the function **f=exp** will only be evaluated the first time it is called; after that, every time f is evaluated it will be replaced by **exp**.

The same role that " =" y ":=" play when defining functions, "→" and ":→ (or :>)" play when making replacements:

- Immediate replacement (Rule).

In[ ]:= `{x, x, x} /. x → RandomReal[]`

Out[ ]= {0.186714, 0.186714, 0.186714}

- Delayed replacement (RuleDelayed).

In[ ]:= `{x, x, x} /. x :→ RandomReal[]`

Out[ ]= {0.613595, 0.114433, 0.0048334}

In[ ]:= `Clear[x]`

We can make dynamic assignments in which the symbol (make sure that in the menu bar **Evaluation ▶ Dynamic Updating Enable** is checked) returns an entry that changes dynamically as we make new assignments to it.

- Let's create a variable.

In[ ]:= `Dynamic[dvar]`

Out[ ]= dvar

- If we now assign to **dvar** different values or expressions, you'll see that the output above keeps on changing.

In[ ]:= `dvar = `$\sqrt{25}$

Out[ ]= 5

In[ ]:= `dvar = Integrate[x^2, x]`

Out[ ]= $\dfrac{x^3}{3}$

- We can also create a box with InputField and write any *input* in it.

In[ ]:= `InputField[Dynamic[dvar]]`

Out[ ]= | dvar |

- Try to use any **f[x]** in the box above, for example **Sin[x]**, and you will see how the next graph is updated. The process will repeat itself anytime you write a new function in the box.

In[ ]:= `Dynamic[Plot[dvar, {x, 0, 5}, Filling → Axis]]`

In[ ]:= `ClearAll["Global`*"]`

2.6 Functional vs. Procedural Programming

The power of the Wolfram Language can be seen specially when we use it to create programs. Readers with experience in procedural programming using languages such as C or FORTRAN, sometimes use the same approach with *Mathematica* even though it may not be the most appropriate one. Because of this, if you are a programmer we would like to encourage you not to replicate the procedural paradigm when programming with *Mathematica*. It will be worth the effort.

- The usual method to define a function with a single argument in the Wolfram Language is:

In[ ]:= `f[x_] := x`^{2}

- Apply the function:

In[ ]:= `f[2]`

Out[ ]= 4

In[ ]:= `f@2`

Out[ ]= 4

- Function[*body*] has added new syntax since Mathematica 12.2 .

In[ ]:= `f = x ↦ x`^{2}

Out[ ]= $x \mapsto x^2$

Let's see an example for creating a function to calculate the factorial of a number using different programming styles without using the built-in factorial symbol "!":

- Using a procedural style, we could do as follows:

In[ ]:= `factorial1[n_] :=`
　　　　`(temp = 1; Do[temp = temp counter, {counter, 1, n}]; temp)`

In[ ]:= `factorial1[5]`

Out[ ]= 120

- With *Mathematica* we can create a function that virtually transcribes the definition of factorial (note the use of "=" in the first definition and of ":=" in the second one; the reason will be explained later):

In[ ]:= `factorial2[0] = 1;`
　　　　`factorial2[n_] := n factorial2[n - 1]`

In[ ]:= `factorial2[5]`

Out[ ]= 120

Notice that we combined "=" and "=:" to calculate the factorial. The use of "=" takes precedence over ":=".

It's normal to use ":="when defining functions. However, in many cases it is also possible to use " =" or ":=" without affecting the final result.

- Trace shows each step of the evaluation:

In[ ]:= `Trace[factorial2[2]]`

Out[ ]= {factorial2(2), 2 factorial2(2 − 1),
　　　　{{2 − 1, 1}, factorial2(1), 1 factorial2(1 − 1), {{1 − 1, 0}, factorial2(0), 1}, 1 × 1, 1}, 2 × 1, 2}

We can use this example to show the difference between Unset (.) and Clear:

In[ ]:= `Definition[factorial2]`

Out[ ]= factorial2(0) = 1

　　　　factorial2(n_) := *n* factorial2(*n* − 1)

- Unset removes a particular definition of factorial2:

In[ ]:= **factorial2[0] =.**

In[ ]:= **Definition[factorial2]**

Out[ ]= factorial2(n_) := *n* factorial2(*n* − 1)

- Clear removes all definitions of factorial2:

In[ ]:= **Clear[factorial2]**

In[ ]:= **Definition[factorial2]**

Out[ ]= Null

The next example consists of building, using different programming styles, a command to add the elements of a list of consecutive integer numbers and compare execution times (we use the function AbsoluteTiming to show the absolute number of seconds in real time that have elapsed, together with the result obtained). We avoid directly applying the formula for calculating the sum of an arithmetic series, which would be the easiest and most effective way.

- We start with a list consisting of the first one million integers.

In[ ]:= **list = Range$\left[10^6\right]$;**

- A procedural method (traditional) to add the elements of this list is:

In[ ]:= **AbsoluteTiming[sum = 0; Do[sum = sum + list⟦i⟧, {i, Length[list]}]; sum]**

Out[ ]= {0.87435, 500 000 500 000}

The functional method (using different instructions in each case) is much more efficient.

- We perform the same operation in three different ways and display the execution time. In the first case we use Total, a built-in function. If available, these functions are usually the fastest way to get things done in *Mathematica*. Later on in the chapter, we will explain Apply (or @@) in more detail.

In[ ]:= **AbsoluteTiming[Total[list]]**

Out[ ]= {0.000703, 500 000 500 000}

In[ ]:= **AbsoluteTiming[Apply[Plus, list]]**

Out[ ]= {0.205671, 500 000 500 000}

In this example, we build a function to calculate the sum of the square roots of *n* consecutive integers : $\{\sqrt{1} + ... + \sqrt{n}\}$, and apply it with *n* = 50.

- In the functional style, we transcribe almost literally the traditional notation in *Mathematica*. This approach is not only simpler but also more effective.

In[ ]:= **rootsum[n_] := $\sum_{i=1}^{n} \sqrt{i}$**

In[ ]:= **rootsum[50] // N**

Out[ ]= 239.036

Why do we use "//N"? To find out, see what happens after removing it from the previous function definition.

Another way of programming where *Mathematica* displays its power and flexibility is rule-based programming.

- For example: To convert $\{\{a_1, b_1, c_1\},..., \{a_i, b_i, c_i\},...\}$ into $\{a_1/(b_1 + c_1), ..., \{a_i/(b_i + c_i)\}, ...\}$ we just create the corresponding replacement rule.

In[ ]:= `{{a1, b1, c1}, {a2, b2, c2}, {a3, b3, c3}, {a4, b4, c4}} /.`
`{a_, b_, c_} → a / (b + c)`

Out[ ]= $\left\{ \dfrac{a1}{b1 + c1}, \dfrac{a2}{b2 + c2}, \dfrac{a3}{b3 + c3}, \dfrac{a4}{b4 + c4} \right\}$

This rule works fine except when the list and its sublists are both of the same size. For example, the list consists of three sublists and each of the sublists contains three elements: $\{\{a1, b1, c1\}, \{a2, b2, c2\}, \{a3, b3, c3\}\}$.

In[ ]:= `{{a1, b1, c1}, {a2, b2, c2}, {a3, b3, c3}} /. {a_, b_, c_} → a / (b + c)`

Out[ ]= $\left\{ \dfrac{a1}{a2 + a3}, \dfrac{b1}{b2 + b3}, \dfrac{c1}{c2 + c3} \right\}$

There are different ways to avoid this problem. One is to use Replace at level 1.

In[ ]:= `Replace[{{a1, b1, c1}, {a2, b2, c2}, {a3, b3, c3}},`
`{a_, b_, c_} → a / (b + c), 1]`

Out[ ]= $\left\{ \dfrac{a1}{b1 + c1}, \dfrac{a2}{b2 + c2}, \dfrac{a3}{b3 + c3} \right\}$

The effective use of pattern matching is very important for good programming. A pattern object may consist of blanks: a single Blank (_) with a single underscore character, a BlankSequence (__) with two underscores or a BlankNullSequence (___) with three underscores.

In[ ]:= `Clear[f, g]`

In[ ]:= `{1, f[], f[a], f[a, b], g[c1, c2]} /. f[_] → "replaced"`

Out[ ]= $\{1, f(), \text{replaced}, f(a, b), g(c1, c2)\}$

In[ ]:= `{1, f[], f[a], f[a, b], g[c1, c2]} /. f[__] → "replaced"`

Out[ ]= $\{1, f(), \text{replaced}, \text{replaced}, g(c1, c2)\}$

In[ ]:= `{1, f[], f[a], f[a, b], g[c1, c2]} /. f[___] → "replaced"`

Out[ ]= $\{1, \text{replaced}, \text{replaced}, \text{replaced}, g(c1, c2)\}$

In[ ]:= `ClearAll["Global`*"]`

2.7 Apply, Map, and Other Related Functions

In this section, we have selected some of the most commonly used functions for programming.

- Map[*f, list, levelspec*] or f/@*list* applies f to each of the elements of *list*. This is probably the most frequently used function in programming.

In[ ]:= `Map[g, {a, b, c}]`

Out[ ]= $\{g(a), g(b), g(c)\}$

In[ ]:= `g /@ {a, b, c}`

Out[ ]= $\{g(a), g(b), g(c)\}$

In[ ]:= `Map[Sin, func[x, y, 0, π/2, 90°]]`

Out[ ]= func(sin(*x*), sin(*y*), 0, 1, 1)

Many build-up functions are applied to different levels. Here Map is used to show how it works at different levels.

- We have the below list:

In[ ]:= `data = {{a, b}, {{1, 2, 3}, {4, 5}}};`
      `TreeForm[data]`

Out[ ]//TreeForm=

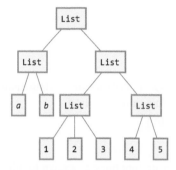

- By default, Map works at level 1:

In[ ]:= `Map[f, data]`

Out[ ]= {*f*({*a*, *b*}), *f*({{1, 2, 3}, {4, 5}})}

- Applying Map at other levels:

In[ ]:= `Map[f, data, {2}] (* applied at level 2 *)`

Out[ ]= $\begin{pmatrix} f(a) & f(b) \\ f(\{1, 2, 3\}) & f(\{4, 5\}) \end{pmatrix}$

In[ ]:= `Map[f, data, {0}] (* applied at level 0 *)`

Out[ ]= $f\left(\begin{pmatrix} a & b \\ \{1, 2, 3\} & \{4, 5\} \end{pmatrix}\right)$

- Apply[*f, list*] or f@@list applies f to list.

In[ ]:= `Apply[f, {a, b, c}]`

Out[ ]= *f*(*a*, *b*, *c*)

In[ ]:= `f @@ {a, b, c}`

Out[ ]= *f*(*a*, *b*, *c*)

In[ ]:= `Apply[f, {{a, b}, c}]`

Out[ ]= *f*({*a*, *b*}, *c*)

- We use FullForm to see how Apply works:

In[ ]:= `FullForm[f @@ {x, y, z}]`

Out[ ]//FullForm=
      f[*x*, *y*, *z*]

- Here Apply is used to multiply the elements of a list:

In[ ]:= **Times @@ {a, b, c, d}**

Out[ ]= $a\,b\,c\,d$

- Have you realized that the previous pattern can be used to compute the factorial of a number? For example, 5!:

In[ ]:= **Apply[Times, Range[5]]**

Out[ ]= 120

- Apply (@@@) at level 1:

In[ ]:= **Apply[f, {{a, b, c}, {d, e}}, {1}]**

Out[ ]= $\{f(a, b, c), f(d, e)\}$

In[ ]:= **f @@@ {{a, b, c}, {d, e}}**

Out[ ]= $\{f(a, b, c), f(d, e)\}$

An alternative way of defining functions is to use "|->" (available since Mathematica 12.2). Note that parentheses are required when entering the function definition.

In[ ]:= **({x, y} |-> x^2 + y) @@@ {{1, 2}, {3, 4}, {a, b}}**

Out[ ]= $\{3, 13, a^2 + b\}$

- Using @ in succession enables the creation of composite functions as shown in this example:

In[ ]:= **f@g@h@x**

Out[ ]= $f(g(h(x)))$

In[ ]:= **f[x_] := d x^2 ; g[y_] := c Exp[y] ; h[z_] := a Cos[b z] ;**

In[ ]:= **f@g@h@x**

Out[ ]= $c^2\, d\, e^{2\,a\cos(b\,x)}$

In[ ]:= **Clear[f, g, h]**

- Notice that **f** with Map is applied to each element of the expression while with Apply actually changes the entire expression by replacing its head.

In[ ]:= **{Map[f, {a, b, c}], Apply[f, {a, b, c}]}**

Out[ ]= $\{\{f(a), f(b), f(c)\}, f(a, b, c)\}$

- MapAll or f//@ applies f to each of the expressions.

In[ ]:= **MapAll[f, {{a, b}, {c}, {{d}}}]**

Out[ ]= $f(\{f(\{f(a), f(b)\}), f(\{f(c)\}), f(\{f(\{f(d)\})\})\})$

In[ ]:= **f //@ {{a, b}, {c}, {{d}}}**

Out[ ]= $f(\{f(\{f(a), f(b)\}), f(\{f(c)\}), f(\{f(\{f(d)\})\})\})$

Some functions, when applied to lists, operate on the individual list elements, e.g. f[{a, b, ...}] → {f[a], f[b], ...}. In this case, the function is said to be Listable.

- In the case of listable functions, Map is implicitly included in them as this example shows.

In[ ]:= **Sin[{a, b, c}]**

Out[ ]= $\{\sin(a), \sin(b), \sin(c)\}$

In[ ]:= `Sin@{a, b, c}`

Out[ ]= {sin(*a*), sin(*b*), sin(*c*)}

- Using Attributes, we can check that Sin is listable.

In[ ]:= `Attributes[Sin]`

Out[ ]= {Listable, NumericFunction, Protected}

- See also: MapAt, MapThread, Outer, Inner, Through, Convolve, and ListConvolve.

In[ ]:= `MapAt[f, {a, b, c, d}, 2]`

Out[ ]= {*a*, *f*(*b*), *c*, *d*}

In[ ]:= `MapThread[f, {{a, b, c}, {x, y, z}}]`

Out[ ]= {*f*(*a*, *x*), *f*(*b*, *y*), *f*(*c*, *z*)}

In[ ]:= `Outer[f, {a, b}, {x, y, z}]`

Out[ ]= $\begin{pmatrix} f(a, x) & f(a, y) & f(a, z) \\ f(b, x) & f(b, y) & f(b, z) \end{pmatrix}$

In[ ]:= `Inner[f, {{a, b}, {c, d}}, {x, y}, g]`

Out[ ]= {*g*(*f*(*a*, *x*), *f*(*b*, *y*)), *g*(*f*(*c*, *x*), *f*(*d*, *y*))}

In[ ]:= `Inner[Power, {a, b, c}, {x, y, z}, Times]`

Out[ ]= $a^x\, b^y\, c^z$

In[ ]:= `Through[{f, g, h}[x]]`

Out[ ]= {*f*(*x*), *g*(*x*), *h*(*x*)}

In[ ]:= `ListConvolve[{x, y}, {a, b, c}]`

Out[ ]= {*a y* + *b x*, *b y* + *c x*}

- In the next example, we apply MapIndexed together with Labeled to identify the position of each term.

In[ ]:= `MapIndexed[Labeled, D[Sin[x]^4, {x, 4}]]`

Out[ ]= $40 \sin^4(x) + 24 \cos^4(x) + -192 \cos^2(x) \sin^2(x)$
{3}   {1}   {2}

- A related function is Thread.

In[ ]:= `Thread[ {a, b, c} → {x, y, z}]`

Out[ ]= {*a* → *x*, *b* → *y*, *c* → *z*}

2.8 Iterative Functions

The most commonly used iterative functions in *Mathematica* are Nest, FixedPoint, and Fold along with their extensions NestList, NestWhile, FixedPointList, and FoldList.

People with experience in other programming languages tend to make iterations using For and Do (also available in *Mathematica*), but normally we should avoid them and use the commands in the previous paragraph. In the examples that follow and throughout the rest of the book we will do just that.

- NestList is specially useful for iterative calculations.

In[ ]:= `f[x_] := 1 + x^2`

In[ ]:= `NestList[f, x, 4]`

Out[ ]= $\left\{x, x^2 + 1, (x^2 + 1)^2 + 1, ((x^2 + 1)^2 + 1)^2 + 1, (((x^2 + 1)^2 + 1)^2 + 1)^2 + 1\right\}$

- In this example we use NestList to build an iterative formula for computing the future value of an initial investment of n at an annual interest rate i for j years and show its growth year by year. We assume an initial capital of €100 and an annual interest rate of 3% for 10 years.

In[ ]:= `interest[n_] := (1 + 0.03) n`

In[ ]:= `NestList[interest, 100, 10]`

Out[ ]= {100, 103., 106.09, 109.273, 112.551, 115.927, 119.405, 122.987, 126.677, 130.477, 134.392}

Anytime we want to do something in *Mathematica*, it's possible that the program already has a built-in function to perform the operation.

- In this example, we sum consecutively all the elements in a list.

In[ ]:= `Accumulate[{a, b, c, d, e}]`

Out[ ]= {$a, a + b, a + b + c, a + b + c + d, a + b + c + d + e$}

As mentioned previously, we will not always find the specific function for what we want to do. However, there are versatile functions useful to know that we can use to perform a specific computation. When operating with lists, FoldList is such a function.

- Here we use FoldList as an alternative to Accumulate.

In[ ]:= `Rest[FoldList[Plus, 0, {a, b, c, d, e}]]`

Out[ ]= {$a, a + b, a + b + c, a + b + c + d, a + b + c + d + e$}

In[ ]:= `ClearAll["Global`*"]`

2.9 Pure Functions

The concept of a pure function is absolutely fundamental to program properly in *Mathematica*. They are functions without explicit names. Pure functions behave like operators telling arguments what to do. Initially they may appear cryptic but their use will end up showing us how powerful they are. Therefore, it's convenient to start using them as soon as the opportunity arises.

In the Wolfram Language, you can have functions without explicit names, also known as "anonymous" functions.

- After removing all variables from memory, we are going to demonstrate several ways of defining the same pure function:

In[ ]:= `Function[u, 3 + u][x]`

Out[ ]= $x + 3$

In[ ]:= `Function[3 + #][x]`

Out[ ]= $x + 3$

In[•]:= **(3 + #) & [x]**

Out[•]= $x + 3$

In this last case, instead of Function we have used the # symbol and to finish defining the function we have added & at the end. Since the combination of # and & represent dummy variables we don't need to store them in memory, reducing the computation time. Next we are going to see several examples using this notation.

- This function squares its argument and adds 3 to it.

In[•]:= **#² + 3 & [x]**

Out[•]= $x^2 + 3$

- We define an operator, name it f, and see its behavior.

In[•]:= **f = (3 + #) &**

Out[•]= $\#1 + 3$ &

In[•]:= **{f[a], f[b]}**

Out[•]= $\{a + 3, b + 3\}$

- The same result can be obtained with Map (/@):

In[•]:= **f /@ {a, b}**

Out[•]= $\{a + 3, b + 3\}$

In[•]:= **(#^2 + 1 &) /@ func[x, y, 1, 4]**

Out[•]= $\text{func}(x^2 + 1, y^2 + 1, 2, 17)$

- Here is combined HoldForm and a pure function. (Remember that HoldForm holds an expression unevaluated):

In[•]:= $\sum_{i=1}^{10} \#1^2 \, \& [i]$

Out[•]= 385

In[•]:= $\sum_{i=1}^{10} \left(\text{HoldForm}\left[\#1^2\right] \, \& \right) [i]$

Out[•]= $1^2 + 2^2 + 3^2 + 4^2 + 5^2 + 6^2 + 7^2 + 8^2 + 9^2 + 10^2$

- Other examples of usage of pure functions:

In[•]:= **Nest[(1 + #) ^2 &, 1, 3]**

Out[•]= 676

In[•]:= **Select[{1, -1, 2, -2, 3}, # > 0 &]**

Out[•]= $\{1, 2, 3\}$

In[ ]:= **Plot3D[x^2 + y^2, {x, -1, 1}, {y, -1, 1}, ColorFunction → (Hue[#3] &)]**

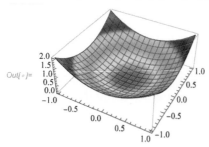

Out[ ]=

- Examples of a pure function with multiples arguments (note the use of *#1, #2, . . .*):

In[ ]:= **(#1^2 + #2^4) &[x, y]**

Out[ ]= $x^2 + y^4$

In[ ]:= **Plot3D[x^2 + y^2, {x, -1, 1}, {y, -1, 1},**
ColorFunction → (RGBColor[#1, #2, #3] &) (* (#3&) *)]

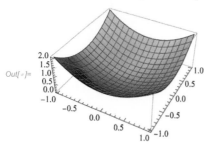

Out[ ]=

- #n, ## represent, respectively, the nth and all the arguments supplied to the pure function:

In[ ]:= **Clear[f]**

In[ ]:= **f[#] &[a, b, c]**

Out[ ]= $f(a)$

In[ ]:= **f[#2] &[a, b, c]**

Out[ ]= $f(b)$

In[ ]:= **f[##2] &[a, b, c]**

Out[ ]= $f(b, c)$

In[ ]:= **f[##] &[a, b, c]**

Out[ ]= $f(a, b, c)$

- There are many ways to define a function in *Mathematica* as shown below for the function $f(x) = x^2 + \sin(x) - 3$.

In[ ]:= **f1[x_] := x^2 + Sin[x] - 3; f1[1]**

Out[ ]= $\sin(1) - 2$

In[ ]:= **f2 = Function[{x}, x^2 + Sin[x] - 3]; f2[1]**

Out[ ]= $\sin(1) - 2$

In[ ]:= **f3 = Function[#^2 + Sin[#] - 3]; f3[1]**

Out[ ]= $\sin(1) - 2$

In[○]:= **f4 = (#^2 + Sin[#] − 3) &; f4[1]**

Out[○]= $\sin(1) - 2$

In[○]:= **(#^2 + Sin[#] − 3) &[1]**

Out[○]= $\sin(1) - 2$

- The combination of a pure function and Apply at level 1 (the short form for Apply[*f*, *expr*, {1}] is @@@) gives a powerful method to operate with a list.

In[○]:= **f = $\left(\#1 + k\, \#2^2\right)$ &;**

In[○]:= **f @@@ {{a1, b1}, {a2, b2}, {a3, b3}}**

Out[○]= $\left\{a1 + b1^2\, k,\ a2 + b2^2\, k,\ a3 + b3^2\, k\right\}$

- In this example, we want to solve the equation below step-by-step expressing the result as a function of x.

In[○]:= **y + x == 2;**

- Let's see the internal form of the previous expression.

In[○]:= **FullForm[y + x == 2]**

Out[○]//FullForm=

Equal[Plus[*x*, *y*], 2]

- We want to solve for *y* by moving *x* to the right side. That is equivalent to subtracting *x* from both sides of the equal sign (note that is necessary to use parenthesis to apply the pure function to the entire equation).

In[○]:= **# − x & /@ (y + x == 2)**

Out[○]= $y = 2 - x$

- Another way of doing it is by using the built-in function AddSides:

In[○]:= **AddSides[y + x == 2, −x]**

Out[○]= $y = 2 - x$

- Next, we generate a list of random values first.

In[○]:= **data = RandomReal[10, 20];**

- Then, we draw three lines, red, green, and blue, each one representing the median, the mean, and the geometric mean of the data, respectively. We use MapThread to define the beginning and the end of each line.

In[○]:= **lines =**

   MapThread[{#1, Line[{{0, #2}, {20, #2}}]} &, {{Red, Green, Blue},
      {Median[data], Mean[data], GeometricMean[data]}}];

- Finally, we combine the data with the generated lines.

In[ ]:= `ListPlot[data, Epilog → lines]`

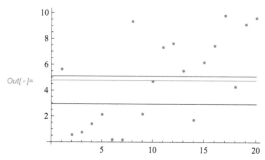

Out[ ]=

- Here we calculate $\sum_{i=1}^{n} \sqrt{i}$ for $n = 50{,}000$ using two methods: a combination of existing *Mathematica* commands and a pure function. We compare calculation times and see that in this case they are similar.

In[ ]:= `Timing[Total[ ∑ᵢ₌₁⁵⁰⁰⁰⁰ √i̅ ]] // N`

In[ ]:= $\mathtt{Timing}\left[\mathtt{Total}\left[\sum_{i=1}^{50\,000} \sqrt{i}\,\right]\right] \, / / \, \mathtt{N}$

Out[ ]= $\{0.34375, 7.45367 \times 10^{6}\}$

In[ ]:= `rootsum2[n_] := Total[ √# & /@ Range[n]] // N`

In[ ]:= `Timing[rootsum2[50 000]]`

Out[ ]= $\{0.390625, 7.45367 \times 10^{6}\}$

- In the next example, we use NestList to compute a continuous fraction.

In[ ]:= `NestList[1 + 1/ (1 + #) &, a, 4]`

Out[ ]= $\left\{a, \; \dfrac{1}{a+1} + 1, \; \dfrac{1}{\frac{1}{a+1}+2} + 1, \; \dfrac{1}{\frac{1}{\frac{1}{a+1}+2}+2} + 1, \; \dfrac{1}{\frac{1}{\frac{1}{\frac{1}{a+1}+2}+2}+2} + 1\right\}$

- The previous method, with $a = 1$, can be used to calculate $\sqrt{2}$. With Nest only the last term is shown.

In[ ]:= `Nest[1 + 1/ (1 + #) &, 1, 10] // N`

Out[ ]= `1.41421`

- FixedPointList is similar to NestList except that it will execute until the solution converges.

In[ ]:= `FixedPointList[1 + 1/ (1 + #) &, 0.1]`

Out[ ]= {0.1, 1.90909, 1.34375, 1.42667, 1.41209, 1.41458, 1.41415, 1.41422, 1.41421, 1.41421, 1.41421, 1.41421, 1.41421, 1.41421, 1.41421, 1.41421, 1.41421, 1.41421, 1.41421, 1.41421, 1.41421}

- In this example, we use FixedPoint to implement the famous Newton's method, where x0 is the initial point.
  http://mathworld.wolfram.com/NewtonsMethod.html

In[ ]:= `newton[f_, x0_] :=`
`        FixedPoint[# - f[#]/f'[#] &, N[x0], 10]`

- We apply the previous function to solve $\mathrm{Log}[x] - 5/x = 0$ (remember that Log[x] represents the natural logarithm of x).

In[ ]:= `newton[(Log[#] - 5 / #) &, 0.2]`

Out[ ]= 3.76868

In the examples below, we use different methods to simulate a random walk, also known as the drunk man's walk. Let's suppose that a drunk man tries to cross the street and each step he takes can randomly go in any direction. Many physical phenomena, such as Brownian motion, are variations of random walks (The *Mathematica* built-in function is RandomWalkProcess.)

- The function below is a very simple case of a random walk along a line. Each step has a maximum length of 1 and can go toward both ends of the line with equal probability. To simulate it, we generate random numbers between -1 and 1 (if it goes in one direction we use $+$ and in the opposite direction $-$.) What will be the distance covered after n steps?

In[ ]:= `RandomWalk[steps_] := Accumulate[RandomReal[{-1, 1}, {steps}]]`

- We create a function to graphically represent the distance covered after n steps. In this case, we assume $n = 500$.

In[ ]:= `ListPlot[RandomWalk[500], Filling → Axis]`

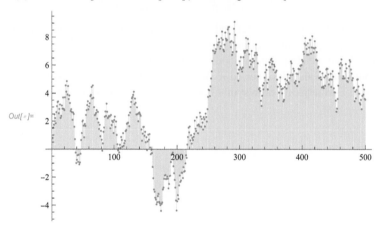

- Random walks can also be created with a combination of NestList and RandomChoice. The individual steps can be forward or backward with equal probability. We randomly generate 1s and -1s with 1 meaning a step forward and -1 indicating a step backward. Starting at an arbitrary point, we want to know the position after n steps. In this case, we have chosen ListLinePlot instead of ListPlot.

In[ ]:= `ListLinePlot[NestList[# + RandomChoice[{+1, -1}] &, 0, 500]]`

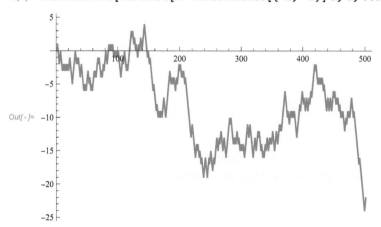

2.10 Global and Local Variables

A common problem in programming languages is how to avoid conflicts among variables. Be careful and do not confuse the mathematical concept of variable with its concept in programming, the one we use here. In this case, when we mention variables we refer to assignments or definitions made for later use. If we create a variable and later on use the same name to define a new one, conflicts may arise between them.

Don't forget that unless you exit the session, even though you create a new notebook the functions previously defined during the session will still be active.

- If you have executed all the cells until here, you'll have the following list of used names:

In[ ]:= **Names["Global`*"]**

Out[ ]= {a, a1, a11, a2, a3, a4, assoc, b, b1, b2, b3, b4, bLarge, c, c1, c2, c3, c4, counter, d,
   data, data1, data2, data3a, data3b, dvar, e, expr, f, f1, f2, f3, f4, factorial1, factorial2,
   func, g, gr, h, i, interest, j, k, lines, list, list2, list3, mat, n, newton, RandomWalk,
   rootsum, rootsum2, s, sLarge, sol, steps, sum, temp, u, x, x0, xvec, x$, y, z}

- Symbols starting with $ that you haven't defined may appear. They correspond to internal assignments done by the program. In addition, if you used Clear["Global`*"] earlier during the session, you will have deleted the previous assignments but not their names. For example: **a** will not have any value assigned to it, but RandomWalk will. You still have not cleared that assignment. Furthermore, both symbols still exist in the global context.

In[ ]:= **? a**

In[ ]:= **? RandomWalk**

- As usual, if we want to eliminate all the previously defined assignments we can use:

In[ ]:= **ClearAll["Global`*"]**

Remember that there's another way to limit the use of variables to a specific context: **Evaluation ▸ Notebook's Default Context ▸ Unique to Each Cell Group** .

- With ClearAll["Global`*"] all the assignments made will be deleted but not the variable names themselves. They can be removed as follows:

In[ ]:= **Remove["Global`*"]**

In[ ]:= **Names["Global`*"]**

Out[ ]= {}

In[ ]:= **? a**

Out[ ]= Missing[UnknownSymbol, a]

In[ ]:= **? RandomWalk**

Out[ ]= Missing[UnknownSymbol, RandomWalk]

However, a more logical way to proceed is to avoid global variables and use local variables instead, that is, variables that only have meaning inside an expression.

- Let's define the following variable in a global context.

In[ ]:= **y = 3**

Out[ ]= 3

- Let's use **y** within a local context. To create local variables we will use Module or With. For example, in the next expression, **y** appears, but in this case it's a local variable, defined using With, and its use is restricted to operations inside the defined expression.

In[ ]:= **f1[x_] := With[{y = x + 1}, 1 + y + y^2]**

- If we apply the previous function when $x = 3$, since $y = x + 1$ then $y = 4$ (local variable), that is replaced by $1 + y + y^2$.

In[ ]:= **f1[3]**

Out[ ]= 21

In[ ]:= **f2[x_] := Module[{y}, y = x + 1; 1 + y + y^2]**

In[ ]:= **f2[3]**

Out[ ]= 21

- Nevertheless, that doesn't affect the global variable y that will still be $y = 3$.

In[ ]:= **y**

Out[ ]= 3

The main difference between both functions is that with Module we explicitly indicate what variables are local while when using With, replacement rules are assigned to the local symbols, that is: **With**$[\{x = x_0, y = y_0, ...\}, expr]$. In many cases, you can use them interchangeably.

Another related function is Block, very similar to Module. The main difference is that while the latter creates temporary assignments for the specified variables during the evaluation, the former suspends any existing assignments for those variables and restores the assignments once the evaluation is done.

In[ ]:= **Module[{y}, Expand[(1 + y)^2]]**

Out[ ]= $y\$178690^2 + 2\, y\$178690 + 1$

```
In[·]:= Block[{y}, Expand[(1 + y)^2]]
```

```
Out[·]= 16
```

The behaviour of both is similar, but in some occasions With is faster than Module:

```
In[·]:= Timing[Do[Module[{x = 5}, x;], {10^5}]]
```

```
Out[·]= {0.140625, Null}
```

```
In[·]:= Timing[Do[With[{x = 5}, x;], {10^5}]]
```

```
Out[·]= {0.046875, Null}
```

You can find an interesting comparison between them using Inactive on: http://www.wolfram.com/mathematica/new-in-10/inactive-objects/optimize-code.html

- Catch returns the argument of Throw:

```
In[·]:= Catch[a = 1; b = 2; c = 3; Throw[c]; d = 4; e = 5]
```

```
Out[·]= 3
```

```
In[·]:= {a, b, c, d, e} (* the evaluation stopped after Throw[c] *)
```

```
Out[·]= {1, 2, 3, d, e}
```

- Calculate the smallest prime number greater than 200:

```
In[·]:= Catch[For[i = 1, i ≤ 100, i++, If[Prime[i] > 200, Throw[Prime[i]]]]]
```

```
Out[·]= 211
```

2.11 Conditional Expressions and Conditions

```
In[·]:= ClearAll["Global`*"]
```

A predicate is a function that can take two values: {True, False}. This type of function is called Boolean. Figure 2.1 shows all the built-in symbols in *Mathematica* used to represent Boolean functions and their corresponding Wolfram Language function.

The Wolfram Language provides various ways to set up *conditionals*, which specify that particular expressions should be evaluated only if certain conditions hold.

"x == y"	"Equal"
"x != y, x ≠ y"	"Unequal"
"x > y"	"Greater"
"x < y"	"Less"
"x ≥ y"	"GreaterEqual"
"x ≤ y"	"LessEqual"
"x && y, x ∧ y"	"And"
"x || y, x ∨ y"	"Or"
"! x, ¬ x"	"Not"
"x ⊻ y"	"Xor"
"Nand[x,y]"	"Nand"
"Nor[x,y]"	"Nor"

Figure 2.1   Boolean Functions and Logical operators in the Wolfram Language:

- Predicates are used when defining conditions. These can appear under very different situations. They can be used to logically combine multiple predicates:

In[ ]:= 2^2 == 4

Out[ ]= True

In[ ]:= 3^2 == 9 && 2 < Pi

Out[ ]= True

In[ ]:= x = 3;
IntegerQ[x] && EvenQ[x]

Out[ ]= False

- To check whether two expressions are equivalent "===" can be used. Note that 3 y 3.0 in *Mathematica* are different kind of numbers.

In[ ]:= 1 === 3 / 3

Out[ ]= True

In[ ]:= 1 === 3.0 / 3

Out[ ]= False

If and Which use predicates for tests. A predicate can return either True or False:

- The conditional statement *if-then-else* can be realized in the Wolfram Language by If:

In[ ]:= x = 5;
{If[PrimeQ[x], "Prime", "Composite"], Which[x == 1, a, x == 2, b, x == 5, c]}

Out[ ]= {Prime, c}

Which and Switch are useful when dealing with more than two branches.

- Which uses predicates, while Switch uses pattern matching:

```
In[ ]:= x = 10;
        Which[x > 0, Print["Positive"],
         x == 0, Print["Zero"], x < 0, Print["Negative"]]

        Positive

In[ ]:= x = 10;
        Switch[x, t_ /; t > 0, Print["Positive"], t_ /; t == 0, Print["Zero"],
         t_ /; t < 0, Print["Negative"], _, Print["Not a valid input"]]

        Positive
```

- Let's use f to define a function whose behavior will be different depending on whether we are using 1 or 2 arguments:

```
In[ ]:= f[x_] := 0.2 x^2
```

```
In[ ]:= f[x_, y_] := 0.2 x^2 y^3
```

```
In[ ]:= f[3]
```

```
Out[ ]= 1.8
```

```
In[ ]:= f[3, 5]
```

```
Out[ ]= 225.
```

- A frequently used format when defining conditions is "_h" to indicate that the replacement will happen for any expression whose head is h as the following examples show:

```
In[ ]:= Clear[f, g]
```

```
In[ ]:= f[x] + g[x] /. x_f → 1/x
```

$$Out[ ]= g(10) + \frac{1}{10}$$

- Here is an example of finding the square roots of only the even integers in a list:

```
In[ ]:= {4, -4, Pi, 1.41, Sqrt[2], 2/3} /. (x_ /; x > 0 && EvenQ[x]) :> Sqrt[x]
```

$$Out[ ]= \left\{2, -4, \pi, 1.41, \sqrt[4]{2}, \frac{2}{3}\right\}$$

- Notice that Sqrt[2] is interpreted as a integer because the condition is applied to the atomic expression (the integer inside of Sqrt):

```
In[ ]:= Map[Head, {4, -4, Pi, 1.41, Sqrt[2], 2/3}]
```

```
Out[ ]= {Integer, Integer, Symbol, Real, Power, Rational}
```

```
In[ ]:= Map[AtomQ, {4, -4, Pi, 1.41, Sqrt[2], 2/3}]
```

```
Out[ ]= {True, True, True, True, False, True}
```

- Here is an example of finding the square roots of the only the positive integers in a list:

```
In[ ]:= {4, -2, 3.14, 9, -25, 16} /. x_ /; x > 0 && IntegerQ[x] :> Sqrt[x]
```

```
Out[ ]= {2, -2, 3.14, 3, -25, 4}
```

In a function, the easiest way to limit a parameter **n** so that the function will not be executed unless the parameter meets the condition *cond,* is by typing **n_cond**:

- In this example **n** is required to be an integer.

```
In[ ]:= Clear[f]
```

In[ ]:= `f[x_, n_Integer] := x^n`

In[ ]:= `f[3.2, -2]`

Out[ ]= 0.0976562

In[ ]:= `f[3.2, 2.1]`

Out[ ]= $f(3.2, 2.1)$

An alternative way to establish a condition on a parameter **n** is with **n_?Test** (guide/TestingExpressions). As a matter of fact, this is best way to do it since it's the fastest one.

- In this example, **n** must be a non-negative integer ($n \geq 0$).

In[ ]:= `g[x_, (n_Integer)?NonNegative] := x^n`

In[ ]:= `g[3.2, 2]`

Out[ ]= 10.24

In[ ]:= `g[3.2, -2]`

Out[ ]= $g(3.2, -2)$

A condition can also be defined using "**/;**", the function **f:=exp/;cond** is evaluated only if the condition is met. Other equivalent forms are: **f/;cond:=exp** and **f:>exp/;cond**.

- In the next example the function takes different forms depending on the selected region (the same can be done with Piecewise, as we will see later on):

In[ ]:= `Clear[f]`

In[ ]:= `f[x_] := x^2 /; x > 0`

In[ ]:= `f[x_] := x /; -2 ≤ x ≤ 0`

In[ ]:= `f[x_] := "x must be bigger than or equal to -2" /; x < -2`

- We apply the function to different values of x.

In[ ]:= `{f[0.2], f[3], f[-3]}`

Out[ ]= {0.04, 9, x must be bigger than or equal to −2}

- MatchQ can be used to check what classes of expressions match a pattern:

In[ ]:= `MatchQ[4.18, x_ /; x > 4]`

Out[ ]= True

In the next example, we combine several conditions that must be met to apply the binomial coefficients formula $\binom{n}{r}$. There is a specific function for this calculation: Binomial, but here we want to create it ourselves.

- The function below calculates the binomial coefficients if n is an integer, r is an integer ≥ 0, and $n \geq r$ (implying that n≥0).

In[ ]:= `binomial[n_Integer, (r_Integer)?NonNegative] /; n ≥ r :=`
`    n! / (r! * (n - r)!)`

- We add a message that will be displayed if the conditions are not met.

In[ ]:= **binomial[n_, r_] :=**
　　　　"Check that n is an integer, r is a non-negative integer and n >= r"

- We test the function for several cases.

In[ ]:= **binomial[3, 2]**

Out[ ]= 3

In[ ]:= **binomial[2, 3]**

Out[ ]= Check that n is an integer, r is a non−negative integer and n >= r

- Another way to provide help and error message is the following:

In[ ]:= **binomial::usage = "This function gives the binomial coefficient (n, r)"**

Out[ ]= This function gives the binomial coefficient (n, r)

In[ ]:= **? binomial**

> Symbol
>
> *Out[ ]=*　This function gives the binomial coefficient (n, r)
>
> ⌄

In[ ]:= **binomial::badarg =**
　　　　"Check that n is an integer, r is a non-negative integer and n >= r";

In[ ]:= **binomial[n_, r_] := Message[binomial::badarg]**

In[ ]:= **binomial[-3, 2]**

　　　binomial: Check that n is an integer, r is a non−negative integer and n >= r

- In the following case, we establish the condition that *n* has to be either integer, rational, or complex to return n^2. The function will be returned unevaluated in any other case.

In[ ]:= **squareofnumber[n_Integer | n_Rational | n_Complex] := n^2**

In[ ]:= **Map[squareofnumber, {3, 3/2, 1.3, 3 + 4 I, I, √2, Pi, {a, b}, a}]**

Out[ ]= $\{9, \frac{9}{4},$ squareofnumber(1.3), $-7 + 24\,i, -1,$ squareofnumber$(\sqrt{2}),$

　　　squareofnumber(π), squareofnumber$(\{a, b\})$, squareofnumber$(a)\}$

We can also define conditions with the following functions: If, Which, Switch, and Piecewise.

- In this example, the function is defined using intervals.

In[ ]:= **Plot[**
　　　Piecewise[{{x^3 + 6, x < -2}, {x, -2 ≤ x ≤ 0}, {x^2, x > 0}}], {x, -5, 5}]

Out[ ]=

- The use of /; combined with FreeQ is particularly useful. For example, we create a function that will consider everything that is not *x* as a constant.

In[ ]:= `f[c_ x_, x_] := c f[x, x] /; FreeQ[c, x]`

In[ ]:= `{f[3 x, x], f[a x, x], f[(1 + x) x, x]}`

Out[ ]= $\{f(30, 10), a f(10, 10), f(110, 10)\}$

In[ ]:= `Clear[f, g]`

- A double underscore "__" or BlankSequence [ ] indicates a sequence of one or more expressions as shown in this example:

In[ ]:= `f[x_Symbol, p__Integer] := Apply[Plus, x^{p}]`

In[ ]:= `f[y, 1, 2, 3]`

Out[ ]= $y^3 + y^2 + y$

In[ ]:= `f[y]`

Out[ ]= $f(y)$

- A triple underscore "___" or BlankNullSequence [ ] enables the use of any number of arguments. This is especially useful when defining optional arguments as we'll see later on.

In[ ]:= `g[x_Symbol, p___Integer] := Apply[Plus, x^{p}]`

In[ ]:= `g[y, 1, 2, 3]`

Out[ ]= $y^3 + y^2 + y$

In[ ]:= `g[y]`

Out[ ]= 0

- A double underscore here is used to plot vectors and labels

In[ ]:= `Vectorplot1[pts : {{_ ?NumberQ, _ ?NumberQ}, __}] :=`
`    Graphics[{Arrow[{{0, 0}, #} & /@ pts],`
`        Inset[ToString[#], # + 0.1] & /@ pts}, Axes → True]`

In[ ]:= `Vectorplot1[{{-1, 1}, {2, 1}, {2, -1}}]`

Out[ ]=

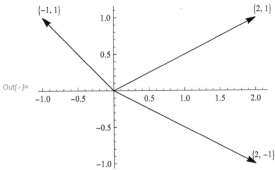

- Here it is an elegant example (from *Power Programming with Mathematica: The Kernel* by D.B. Wagner) of sorting the elements of a list using the triple underscore pattern.

In[ ]:= `sorter[{a___, b_, c_, d___} /; b > c] := sorter[{a, c, b, d}]`

In[ ]:= `sorter[x_] := x`

In[ ]:= **sorter[{5, 2, 7, 1, 4}]**

Out[ ]= {1, 2, 4, 5, 7}

- Certainly, we could have used Sort instead of sorter.

In[ ]:= **Sort[{5, 2, 7, 1, 4}]**

Out[ ]= {1, 2, 4, 5, 7}

- In this example, we use Cases to select the list elements of the form x^y.

In[ ]:= **Cases[{1, x, $\sqrt{x}$, x^y}, x^y_]**

Out[ ]= {$\sqrt{10}$, 10^y}

- The pattern "_." makes it possible to define an argument that can be omitted. By using **x^y_.** (note the dot after the underscore), **x** is also selected since the argument of **y** can be omitted.

In[ ]:= **Cases[{1, x, $\sqrt{x}$, x^y}, x^y_.]**

Out[ ]= {10, $\sqrt{10}$, 10^y}

We use DeleteCases to eliminate those elements of a list that meet certain criteria.

- We remove the elements that are decimals (we use Real, different from Reals that refers to the real numbers).

In[ ]:= **DeleteCases[{1, 1, x, 2, 3, y, 3/5, 2.7, Pi, 9, y}, _Real]**

Out[ ]= $\left\{1, 1, 10, 2, 3, y, \dfrac{3}{5}, \pi, 9, y\right\}$

- We eliminate from the list those sublists containing 0 as their last term.

In[ ]:= **DeleteCases[**
{{{1, 1}, 0}, {{2, 3}, 1}, {{y, 3/5}, .7}, {{Pi, 9}, z}}, {a_, 0}]

Out[ ]= $\begin{pmatrix} \{2, 3\} & 1 \\ \left\{y, \frac{3}{5}\right\} & 0.7 \\ \{\pi, 9\} & z \end{pmatrix}$

- Alternatives (|) allows you to specify a number of different choices so that an expression is declared a match for the pattern when it matches any one of those choices:

In[ ]:= **Cases[{a, b, 2, c, 4, 2.3, R, "Hello"}, _Integer | _Symbol]**

Out[ ]= {a, b, 2, c, 4, R}

- If you want a match where a particular kind of object is repeated a number of times in this case Repeated (..) can be used

In[ ]:= **{{}, {a, a}, {a, b}, {a, a, a}, {a}} /. {a ..} → x**

Out[ ]= {{}, 10, {a, b}, 10, 10}

- A list containing at most 5 integers:

In[ ]:= **Cases[{{1, 3, 5}, {1, 3, 5, 4, 6}, {r, s}, Range[9]},**
{Repeated[(_Integer), 5]}]

Out[ ]= {{1, 3, 5}, {1, 3, 5, 4, 6}}

- Except[c] can be used with a pattern object, which represents any expression except one that matches c.

In[ ]:= `Cases[{0, 5, "Hello", Plot}, x : Except[_String | _Symbol] :> x²]`

Out[ ]= {0, 25}

- Including a second argument in Except[*c*,*p*] allows you to represent any expression that matches *p* but not *c*. Here, all integers except 0 are returned:

In[ ]:= `Cases[{0, 5, "Hello", Plot, 7.5, 8}, Except[0, _Integer]]`

Out[ ]= {5, 8}

The functions to find patterns within strings are very similar to the ones we've seen so far. They usually have "String" at the beginning of them. The main difference is the use of "~~" to represent a sequence of string expressions.

- StringCases["*string*", *patt*] will match substrings in *string* that match *patt*. The next example selects all the substrings that start with "ab":

In[ ]:= `StringCases["aby mno abc abd bac xyz", "ab" ~~ _]`

Out[ ]= {aby, abc, abd}

- StringFreeQ["*string*", *patt*] returns True if no substrings in *string* contain *patt*. Otherwise, it returns False.

In[ ]:= `StringFreeQ["aby mno abc abd bac xyz", "ab" ~~ _]`

Out[ ]= False

- StringMatchQ["*string*", *patt*] checks whether *string* contains the pattern *patt*. The string below contains the sequence "ATG" so the function returns True:

In[ ]:= `StringMatchQ["AAGCTGCGTCGTAAATGCGAGT", ___ ~~ "ATG" ~~ ___]`

Out[ ]= True

2.12 Accuracy and Precision

The number of significant digits of a number *x* used in calculations is called precision. To know the precision used in a calculation used Precision[*x*]. Accuracy (Accuracy[*x*]) is the number of significant digits to the right of the decimal point. Precision is a relative error measure, while accuracy measures the absolute error.

- By default N uses machine precision.

In[ ]:= `ClearAll["Global`*"]`

In[ ]:= `Precision[N[π]]`

Out[ ]= MachinePrecision

In[ ]:= `Precision[N[π, 40]]`

Out[ ]= 40.

- However, if we don't use a decimal approximation, the precision of Pi is infinite.

In[ ]:= `Precision[π]`

Out[ ]= ∞

WorkingPrecision controls the precision level during internal calculations while PrecisionGoal determines the precision in the final result. Other options related to accuracy and precision are: AccuracyGoal and Tolerance.

Besides knowing how to use the options mentioned above, there's an important issue that you should know: *Mathematica* has different rules for working with accuracy and precision. An approximate quantity is identified by the presence of a decimal point. Therefore, 0.14 and 14.0 are approximate numbers, and we will refer to them as decimals. If we rationalize 0.14 we get 14/100, the ratio of two integers. When doing operations with integers or combinations of them (rationals, integer roots, etc.), *Mathematica* generates exact results.

- For example:

In[ ]:= $2 + \left(3 \times \dfrac{5}{12}\right)^7$

Out[ ]= $\dfrac{110\,893}{16\,384}$

- is different from:

In[ ]:= $2 + \left(3 \times \dfrac{5.0}{12}\right)^7$

Out[ ]= 6.76837

There are several functions to convert a decimal number into an exact one.

- For example: Round[x] returns the closest integer to x, Floor[x] the closest integer to x that is less than or equal to x, and Ceiling[x] the closest integer to x that is greater than or equal to x.

In[ ]:= `{Round[3.3], Floor[3.3], Ceiling[3.3]}`

Out[ ]= {3, 3, 4}

In the next example, we are going to see the difference between operating with integers (or expressions composed of integers, such as fractions) and operating with decimal approximations.

- We create the following function:

In[ ]:= `f[x_] := 11 x - a`

- We iterate it and check that we always get the same result:

In[ ]:= `NestList[f, a / 10, 20]`

Out[ ]= $\left\{\dfrac{a}{10}, \dfrac{a}{10}\right\}$

- However, when we replace 1/10 with its decimal approximation 0.1, we can clearly see the errors associated with machine precision.

In[ ]:= `NestList[f, 0.1 a, 20]`

Out[ ]= {0.1 a, 0.1 a, 0.1 a, 0.1 a, 0.1 a, 0.1 a, 0.1 a, 0.1 a, 0.1 a, 0.1 a, 0.100002 a, 0.100025 a, 0.100279 a, 0.103066 a, 0.133729 a, 0.471014 a, 4.18116 a, 44.9927 a, 493.92 a, 5432.12 a}

- This effect is not associated to the way *Mathematica* operates. It will happen with any program performing the same operation using machine precision. For example, try to replicate the previous calculation, replacing **a** with a positive integer (for example 3), in a program such as Microsoft® Excel, and you will see that the errors multiply.

In[ ]:= `f[x_] := 11 x - 3`

In[ ]:= **NestList[f, 0.3, 20]**

Out[ ]= {0.3, 0.3, 0.3, 0.3, 0.3, 0.3, 0.3, 0.3, 0.3, 0.3, 0.3, 0.299995, 0.299949, 0.299443,
0.293868, 0.232543, −0.442028, −7.86231, −89.4854, −987.34, −10 863.7}

- Rationalize converts a decimal number to the nearest rational with the smallest denominator.

In[ ]:= **Rationalize[0.3]**

Out[ ]= $\dfrac{3}{10}$

- When used in the previous iteration we eliminate the numerical errors since *Mathematica* uses infinite precision for calculations without decimal numbers.

In[ ]:= **NestList[f, Rationalize[0.3], 20]**

Out[ ]= $\left\{\dfrac{3}{10}, \dfrac{3}{10}\right\}$

Let's show another more complicated example, where we will see as well the effect of using decimal approximations instead of exact calculations.

Consider the following system of differential equations:

$x_1'(t) = 1 - 2.5\, x_1(t) + x_2(t)$
$x_2'(t) = 2\, x_1(t) - x_2(t)$
Initial Conditions : $x_1(0) = x_1(0) = 0$

- We solve it using DSolve. Depending on the machine precision of your computer, you may find numerical problems when visualizing the solutions. It can usually solved rationalizing the decimal coefficients, then *Mathematica* computes with integers and rationals using infinite precision.

In[ ]:= **ClearAll[x1, x2]**

In[ ]:= **{{x1[t_], x2[t_]}} = {x1[t], x2[t]} /. DSolve[**
    {x1'[t] == 1 - Rationalize[2.5] × x1[t] + x2[t],
    x2'[t] == 2 x1[t] - x2[t], x1[0] == 0, x2[0] == 0},
    {x1[t], x2[t]}, t] // ExpandAll

Out[ ]= $\left(-e^{-\frac{\sqrt{41}\,t}{4}-\frac{7t}{4}} + \dfrac{5e^{\frac{\sqrt{41}\,t}{4}-\frac{7t}{4}}}{\sqrt{41}} - e^{\frac{\sqrt{41}\,t}{4}-\frac{7t}{4}} - \dfrac{5e^{-\frac{\sqrt{41}\,t}{4}-\frac{7t}{4}}}{\sqrt{41}} + 2 \quad -2e^{-\frac{\sqrt{41}\,t}{4}-\frac{7t}{4}} + \dfrac{14e^{\frac{\sqrt{41}\,t}{4}-\frac{7t}{4}}}{\sqrt{41}} - 2e^{\frac{\sqrt{41}\,t}{4}-\frac{7t}{4}} - \dfrac{14e^{-\frac{\sqrt{41}\,t}{4}-\frac{7t}{4}}}{\sqrt{41}}\right.$

In[ ]:= **Plot[{x1[t], x2[t]}, {t, 0, 100}]**

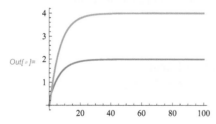

A situation where it is common to experience difficulties is when performing matrix operations (matrix inversion, resolution of systems of differential equations with decimal coefficients, etc.).

> When problems associated to operations with decimals arise, it is always a good idea to rationalize the terms containing those decimals; although you may have to consider the increase in computation time.

2.13 Choosing the Method of Computation

The power of *Mathematica* becomes particularly evident in the precision with which you can do calculations and the ability to choose the most appropriate method or algorithm given the required computation.

- NIntegrate is used for numerical integrations and like almost all other calculation-related functions, unless indicated otherwise, will choose by default a computation method. However, you can select a different method or, for a given method, choose the most appropriate parameters. In the example that follows we compare three methods after showing graphically the method of quadrature. Note the use of Defer.

In[ ]:= `f[x_] := Exp[-x]`

In[ ]:= `Show[Plot[f[x], {x, -3, 3}],`

      `DiscretePlot[f[x], {x, -3, 3}, ExtentSize → Left] ,`

      `PlotLabel → Style[Defer[∫₋₃³ Exp[-x] ⅆx], Small]]`

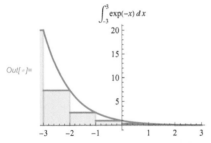

Out[ ]:=

In[ ]:= `NIntegrate[f[x], {x, -3, 3}, Method → #, PrecisionGoal → 2] & /@`
      `{Automatic, {"RiemannRule", "Type" → "Left"}, "TrapezoidalRule"}`

Out[ ]:= {20.0357, 20.2215, 20.036}

- Here we use EvaluationMonitor to display the value of x used in each iteration.

In[ ]:= `Print["x = ", x]`

      $x = x$

In[ ]:= `NIntegrate[f[x], {x, -1, 1}, Method → "TrapezoidalRule",`
      `PrecisionGoal → 2, EvaluationMonitor ⧴ Print["x = ", x]]`

x = -1.

x = -0.75

x = -0.5

x = -0.25

x = 0.

x = 0.25

x = 0.5

x = 0.75

x = 1.

Out[ ]= 2.35045

- The next function is a piecewise function, that is, it takes different values in each of the intervals.

    ClearAll["Global`*"]

In[ ]:= **fun =** $\begin{cases} \dfrac{Sin[10 x]}{\sqrt{-x}} & -\infty < x < 0 \\ \dfrac{1}{\sqrt{x}} & 0 < x < 1 \\ Sin[2000 x] x^2 & 1 < x < 2 \\ \dfrac{Cos[x]}{x} & 2 < x < \infty \end{cases}$ **;**

- To represent it graphically *Mathematica* uses different calculation methods depending on the interval.

In[ ]:= **Plot[fun, {x, -10, 10}, ImageSize → 300]**

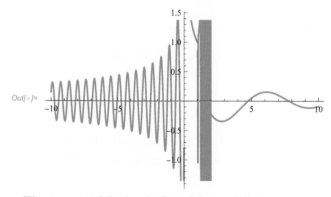

Out[ ]=

- We can control the level of precision as follows:

In[ ]:= **N[Integrate[Sin[2000 x] x^2, {x, 1, 2}], 10]**

Out[ ]= 0.001275015530

In[ ]:= **N[NIntegrate[Sin[2000 x] x^2, {x, 1, 2}], 10]**

Out[ ]= 0.00127502

In[ ]:= **NIntegrate[Sin[2000 x] x^2, {x, 1, 2}, WorkingPrecision → 10]**

Out[ ]= 0.001275015530

- Next, we present an example of a differential equation in which the solving method (tutorial/NDSolveOverview) depends on the value of δ. NDSolve chooses the best algorithm given δ. We use Manipulate, a function that we will cover in more detail in the next chapter.

```
In[ ]:= Manipulate[
       res = NDSolve[{y'[t] == y[t]² - y[t]³, y[0] == δ}, y, {t, 0, 2/δ}];
       p1 = Plot[(y /. First[res])[t], {t, 0, 2/δ}, Mesh → All],
       {{δ, 0.01}, 0.0001, 0.01}]
```

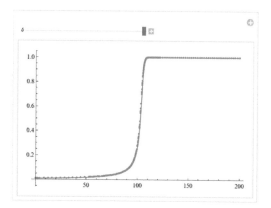

We should not worry most of the time about choosing a specific computation method. *Mathematica's* default usually chooses the most appropriate one: https://www.wolfram.com/algorithmbase/

However, we can always check if that is the case for a given problem by comparing the results of different methods.

2.14 Optimizing the Computation Time

Another interesting aspect of *Mathematica* is its use of previously calculated results. Before showing it with an example, it would be better to quit the session to remove any previous information stored in memory.

```
In[ ]:= Quit[]
```

Whenever possible, built-in functions should be used instead of recreating the function from other constructs. Built-in functions use compiled C code directly, thus are faster.

- In the next example we show the calculation time of a complicated definite integral for different integration limits.

```
In[ ]:= Timing[(f1[x1_] := Integrate[Sin[x^3], {x, 0, x1}];
       Table[f1[i], {i, 1, 10}]);]
```

```
Out[ ]= {3.48438, Null}
```

- We repeat the same exercise using "=" instead of ":=" .

```
In[ ]:= Timing[(f2[x1_] = Integrate[Sin[x^3], {x, 0, x1}];
       Table[f2[i], {i, 1, 10}]);]
```

```
Out[ ]= {1.17188, Null}
```

The calculation time is faster in this last case. It seems logical, since in the first example the integral **f1** is calculated 10 times, each time applying a different integration limit, while in the second case **f2** provides the solution to the integral as a function of the integration limit $x1$, and then replaces the value of $x1$ directly in the solution 10 times.

- We repeat the same pattern using again "=" and the calculation is even faster.

In[ ]:= **Timing[(f3[x1_] = Integrate[Sin[x^3], {x, 0, x1}];**
     Table[f3[i], {i, 1, 10}]);]

Out[ ]= {0.15625, Null}

The reason behind this behavior is that for functions such as: Integrate, Simplify and others, *Mathematica* keeps a cache of previous results that it will use later if necessary to make calculations more efficient:

http://library.wolfram.com/infocenter/Conferences/5832/

- If we now delete the cache, the calculation takes longer:

In[ ]:= **ClearSystemCache[];**

In[ ]:= **Timing[(f3[x1_] = Integrate[Sin[x^3], {x, 0, x1}];**
     Table[f3[i], {i, 1, 10}]);]

Out[ ]= {0.53125, Null}

- The code below shows a traditional definition, each time **f[5, 5, 5, 5, 5]** is computed, the consumed time is the same:

In[ ]:= **ClearAll[fOrd]**
     fOrd[a1_, a2_, a3_, a4_, a5_] := NIntegrate[1/ $\sqrt{x1 + x2 + x3 + x4 + x5}$,
     {x1, 0, a1}, {x2, 0, a2}, {x3, 0, a3}, {x4, 0, a4}, {x5, 0, a5}]

In[ ]:= **fOrd[5, 5, 5, 5, 5] // AbsoluteTiming**

Out[ ]= {0.197992, 910.019}

- Alternatively, we can combine =: and = for a faster computation. The function definition below incorporates memoization (a technique by which previously computed results are stored and reused to speed up the overall evaluation). Subsequent evaluations are much faster, since **f[5, 5, 5, 5, 5]** is stored in the definition itself (as a downvalue):

In[ ]:= **fMemo[a1_, a2_, a3_, a4_, a5_] := fMemo[a1, a2, a3, a4, a5] =**
     NIntegrate[1/ $\sqrt{x1 + x2 + x3 + x4 + x5}$, {x1, 0, a1},
     {x2, 0, a2}, {x3, 0, a3}, {x4, 0, a4}, {x5, 0, a5}]

In[ ]:= **fMemo[5, 5, 5, 5, 5] // AbsoluteTiming**

Out[ ]= {0.225405, 910.019}

In[ ]:= **fMemo[5, 5, 5, 5, 5] // AbsoluteTiming**

Out[ ]= $\{1.2 \times 10^{-6}, 910.019\}$

In many situations we can significantly reduce the computation time by using Compile. Normally it's used with functions that only need to evaluate numerical arguments. This approach enables the program to save time by avoiding checks related to symbolic operations. The function generates a compiled function that applies specific algorithms to numerical calculations.

- Example: We use Compile to calculate the following function that only accepts real values as arguments:

In[ ]:= `cf = Compile[{{x, _Real}}, x^2 - 1/(1 + x) Sin[x]]`

Out[ ]= `CompiledFunction[` ⊞ ⇄ WVM Argument count: 1 Argument types: {_Real} `]`

- CompiledFunction evaluates the expression using the compiled code:

In[ ]:= `cf[Pi]`

Out[ ]= `9.8696`

- We can now represent graphically the compiled version of the function:

In[ ]:= `Plot[cf[x], {x, -1, 3}]`

Out[ ]=

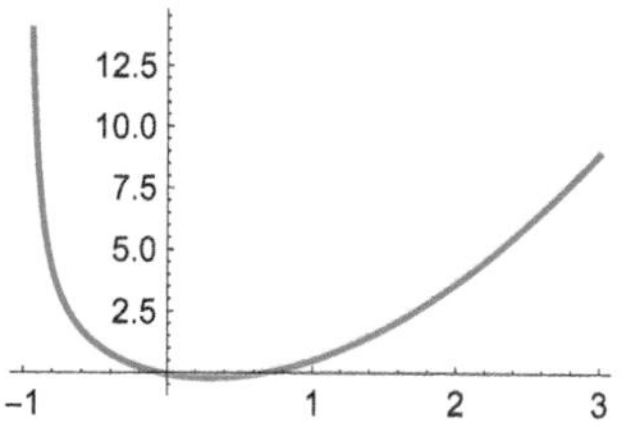

2.15 Cloud Deployment

The function CloudDeploy deploys objects to the Wolfram Cloud. Let's take a look at some examples.

- The function below creates a web page with a 100-point font message:

In[ ]:= `CloudDeploy[Style["My first cloud program", 100]]`

Out[ ]= `CloudObject[`

   `https://www.wolframcloud.com/obj/4b9408f2-6e92-4419-b55d-b959e5514c0d]`

- The first time we execute it, *Mathematica* will ask for our Wolfram ID and password as shown in Figure 2.2:

Figure 2.2   Sign-in Screen for WolframCloud.

- Clicking on the hyperlink just created will automatically take us to the web page shown in Figure 2.3:

My first cloud program

Figure 2.3　　Web page created with the CloudDeploy function.

- We can deploy any *Mathematica* function to the cloud. Figure 2.4 shows the deployment of a dynamic interface:

In[•]:= `CloudDeploy[Manipulate[Plot[Sin[x (1 + k x)], {x, 0, 6}], {k, 0, 2}]]`

Out[•]= `CloudObject[`
`https://www.wolframcloud.com/obj/baeed005-91cd-4131-86bc-bc06e4e7e75e]`

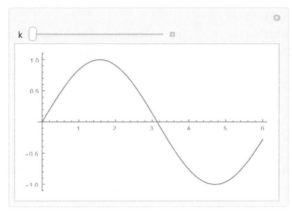

Figure 2.4　　Manipulate in the cloud.

For further information:

http://reference.wolfram.com/language/guide/CloudFunctionsAndDeployment.html

In[•]:= `ClearAll["Global`*"]`

2.16 Package Development

Mathematica's capabilities can be extended substantially with the use of packages. Packages are conceptually similar to what other programming languages call subroutines or subprograms. They basically involve the creation of functions, often complex, that can be called when needed without having to define them again.

2.16.1 The Structure of a Package

A package has the following structure:

`BeginPackage["name of the package`"]`

f`::usage = "f brief description"` **,** ... (here we would include the headings of the functions that we are going to use with a brief description, that will be seen by the user later on).

`Begin["`Private`"]`

f `[args]` `= value` **,** ... (here we would have the option to place functions that will be used by f but that will remain private. That is, the user will not be able to access them).

f `[args]` `= value` **,** ... (here we would program the package functions accessible to the user).

`End []`

```
EndPackage[]
```

2.16.2 A Simple Package

We'd like to build the function below and make sure that it's going to be available for later use in any notebook.

$$f(n, r, p) = \binom{n}{r} p^k (1-p)^{n-r}, \ n \geq 1, \ n \geq r, \ 0 \leq p \leq 1$$

- Write the following in a code cell (**Format ▶ Style ▶ Code**)

In[ ]:=
```
BeginPackage["BinomialPackage`"]

Binomialdistribution::usage =
"Binomialdistribution[n, r, p] calculates the binomial probability distribu
under the conditions that n must be an integer, n ≥ 1, n >= r and 0 <= p <:

Begin["`Private`"]
```

Out[ ]= BinomialPackage`

Out[ ]= BinomialPackage`Private`

In[ ]:=
```
Binomialdistribution[n_Integer, (r_Integer)?NonNegative,
        (p_)?NonNegative]/; n ≥r && 0≤p≤1 :=(n!/(r!*(n - r)!))*p^r*(1 - p)^(

Binomialdistribution[n_, r_, p_]:="Check that n is an integer,
r is a non-negative integer,n >= r and 0 ≤ p ≤ 1"

End[]

EndPackage[]
```

Out[ ]= BinomialPackage`Private`

- Next, from the main menu bar select **File ▶ New ▶ Package**. A new window in the package environment will appear (Figure 2.5). After that, copy the previous cell to this new notebook, turn it into an initialization cell (**Cell ▶ Cell Properties ▶ Initialization**) and, in this case, save it as a package (**File ▶ Save as ▶ Wolfram Language Package**) in the Data folder located in our working directory. Use as the name the one defined in BeginPackage: "Binomialpackage.wl" (remember to choose the extension wl or m).

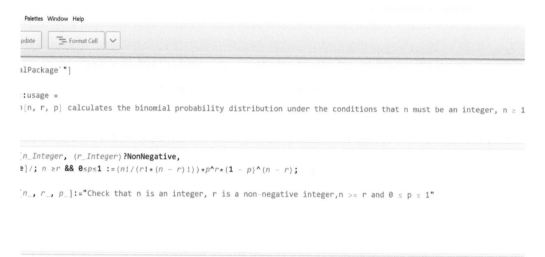

Palettes Window Help

pdate   ≣ Format Cell   ∨

ɔlPackage`"]

:usage =
ɔ[n, r, p] calculates the binomial probability distribution under the conditions that n must be an integer, n ≥ 1

n_Integer, (r_Integer) ?NonNegative,
ɔ]/; n ≥r && 0≤p≤1 := (n!/(r!*(n - r)!))*p^r*(1 - p)^(n - r);

n_, r_, p_]:="Check that n is an integer, r is a non-negative integer,n >= r and 0 ≤ p ≤ 1"

Figure 2.5   Developing packages in *Mathematica*.

- In this environment we can test, modify, and even debug the package. Once we see it is working correctly we will save it. A package is an ASCII file. It can also be modified using a text editor such as Notebook (in Windows).

- As the last step, we exit the session (to be precise: we exit the kernel) to start a new one. This can be done without closing the notebook by choosing in the menu bar **Evaluation ▶ Quit Kernel** or typing and evaluating:

In[•]:= **Quit[]**

2.16.3 Installing and Loading Packages

Best practice is to put packages (.m/.wl files and dependencies) in the directory:

In[•]:= **FileNameJoin[{$InstallationDirectory, "Applications"}]**

Out[•]= C:\Program Files\Wolfram Research\Mathematica\13.0\Applications

Then the package will be available for all users that run Mathematica installed in this directory. However, in this case we will define a specific directory.

- Let's make the folder Data our working directory.

In[•]:= **SetDirectory[FileNameJoin[{NotebookDirectory[], "Data"}]]**

Out[•]= F:\Mi unidad\Mathematica\MBN13\Data

- To use the package we need to load it first (or, for testing purposes, execute it inside a notebook). We can use either << or Needs, but in this case we choose the latter since it has the advantage of only loading the package if it hasn't been loaded previously.

In[•]:= **Needs["BinomialPackage`"]**

- One way to know what functions are available inside the package is as follows:

In[ ]:= **?BinomialPackage`***

> Symbol
>
> Binomialdistribution[n, r, p] calculates the binomial probability distribution under
>
> the conditions that n must be an integer, n ≥ 1, n >= r and 0 <= p <= 1

Out[ ]=

- At the time of loading the package, all its functions are executed, in this case just one: Binomialdistribution.

In[ ]:= **Binomialdistribution[5, 3, 0.1]**

Out[ ]= 0.0081

In[ ]:= **Binomialdistribution[5, 3, 1.4]**

Out[ ]= Check that n is an integer, r is a non−negative integer,n >= r and 0 ≤ p ≤ 1

2.17 Additional Resources

As in most cases, the best available information comes from either *Mathematica*'s own documentation or Wolfram's website:

> *Fast Introduction for Programmers,* a short introduction to the Wolfram Language for programmers, is available in the Documentation Center and also on the web: (http://www.wolfram.com/language/fast-introduction-for-programmers/)
> Wolfram's reference page for the Wolfram Language: http://www.wolfram.com/language/
> Functions and references:
> http://reference.wolfram.com/mathematica/tutorial/FunctionsAndProgramsOverview.html
> Video courses: https://www.wolfram.com/wolfram-u/catalog/language/

There are many books to learn how to program with *Mathematica* and even though some of them were written more than 20 years ago, their ideas and explanations of core concepts are still useful. Here are some of them:

> *Power Programming With Mathematica: The Kernel* by David B. Wagner (available as a free download [2022-01-05]:
> https://www.dropbox.com/s/j2dsyvptnxjd369/Wagner%20All%20Parts-RC.pdf. Even though it was written for *Mathematica* 3, its exposition of functional programming is excellent.
> *Mathematica's Programming Language* by Roman E. Maeder:
> http://library.wolfram.com/infocenter/Conferences/183/
> *The Mathematica GuideBooks* collection written in *Mathematica* 5 by Michael Trott is the "Bible" of Mathematica:
> http://www.mathematicaguidebooks.org

http://packagedata.net/ is a global repository of *Mathematica* packages that makes it easy for users to find additional functionality in a wide variety of scientific and technical fields.

3

Interactive Applications, Image Processing, and More

This chapter explores the use of dynamic functions for creating interactive simulations (demonstrations) using the **Manipulate** and **DynamicModule** functions. It also shows how to manipulate and process 2D and 3D images to detect edges, remove backgrounds or identify specific image elements. The creation of graphs and networks is also covered by giving an overview of the **Graph** function. The last section puts some of the previously discussed Mathematica's capabilities into action by explaining in detail the calculation of the period of a pendulum from a series of images. All the functionality discussed in the chapter is illustrated with relevant examples.

3.1 The Manipulate Function

One of the most interesting functions in the Wolfram Language is Manipulate (https://reference.wolfram.com/language/tutorial/IntroductionToManipulate.html). We can easily build dynamic applications with it. This function has numerous options, and in this section we show some of them. Normally we will use more than one option at a time; if you prefer you can try them separately and then combine them. Remember that if you don't know the syntax of a function you can always use the free-form input first and later modify the *Mathematica* output to suit your needs.

- We type in an *input* cell: ▤ **manipulate sin x from 0 to 10**. An input and an output are generated.

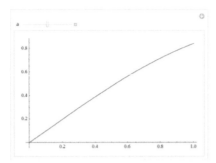

- In the input cell above we click on the floating label "Replace cell with this input" that appears right below our command. This way we obtain the function below that we can use as a basic Manipulate template. Notice that **a** is a parameter that can take values between 0 and 10 with a default value of 1. Try moving the slide bar with the cursor.

In[ ]:= `Manipulate[Plot[Sin[a x], {x, 0, 10}], {{a, 1}, 0, 10}]`

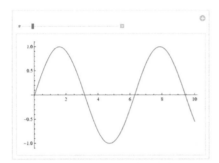

In some of the examples that follow we have added the options ContentSize and ImageSize to reduce the output size.

- The next example shows some of the many options of Manipulate. You can include a drop-down button, but it will be automatically added when a parameter contains five or more choices as shown below. In this case, we also have the possibility to choose several filling alternatives for displaying a graph. We use PlotLabel as well to show the parameter value when labeling the graph. By adding the option Appearance → *"Labeled"*, we can display to the right of the sliding bars the actual values of the parameters. If you'd like them to be shown by default, you should add the option Appearance → "Open". We can also label the parameters using the syntax: *{{a, initial a value, "a label"}, minimum a value, maximum a value}*. Another interesting option is ControlType. For example, we can use Checkbox to include check boxes. The syntax is: *{a, {a1,a2}}*, where *{a1,a2}* refers to the alternatives available, *a1* is the default option. When there are more than two choices, you can use RadioButtonBar.

In[ ]:=
```
Manipulate[
    Plot[Exp[-a x] Sin[b x], {x, 0, 10},
      PlotLabel → Style[StringForm["Exp[-a x] Sin[b x] "], 10],
      PlotStyle → {Directive[color, AbsoluteThickness[thickness]]},
      Axes → axes], {{a, 0.1, "Amplitude a"}, 0, 0.5, Appearance → "Open"},
    {{b, 3, "Amplitude b"}, 1, 10, Appearance → "Labeled"},
    {axes, {True, False}},
    {{color, Orange}, {Orange, Red, Yellow, Green, Blue}},
    {thickness, {1, 3, 5}, ControlType → RadioButtonBar}]
```

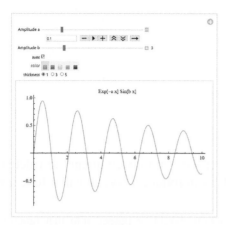

- The function "ManipulatePlot" is available in the repository:
  https://resources.wolframcloud.com/FunctionRepository/resources/ManipulatePlot/
  We can use it to generate a plot in which both, the plot ranges and function parameters can be manipulated dynamically.

In[ ]:= **ResourceFunction["ManipulatePlot"]**

Out[ ]= [▪] ManipulatePlot ✛

In[ ]:= [▪] **ManipulatePlot** **[{Sin[a*x + b], Cos[a*x + b]},**
{x, -2*Pi, 2*Pi}, {{a, 1}, 0, 2}, {{b, 1}, 0, 2}]

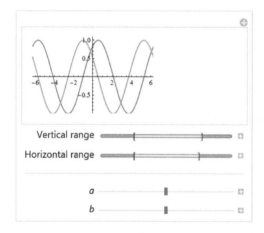

If we use Locator with Manipulate we will be able to move objects in the output with the mouse. Additionally, all the information associated to those objects will be dynamically updated. The syntax of Locator is: {{*parameter*, {*x*, *y*}}, *Locator*}.

- In this example, we can move the three vertices of a polygon and see how the triangle area is filled automatically.

In[ ]:= **Manipulate[**
Graphics[{Pink, Polygon[pts]}, PlotRange → 1.1, ImageSize → Tiny],
{{pts, {{0, 0}, {1, 0}, {0, 1}}}, Locator}]

We are going to extend the functionality of the previous command to calculate and display the area of the triangle, but first let's introduce a way to call a function from within Manipulate.

Any local variables that are used by a function should be passed into that function as an argument. Because of that, the first example doesn't work:

```
In[ ]:= ClearAll[f];
        f[x_] := Sin[a x];
        Manipulate[Plot[f[x], {x, 0, 10}], {a, 1, 5}]
```

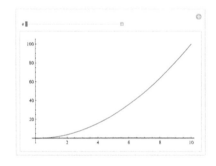

```
In[ ]:= ClearAll[f];
        f[a_, x_] := Sin[a x];
        Manipulate[Plot[f[a, x], {x, 0, 10}], {a, 1, 5}]
```

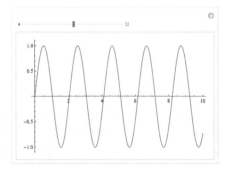

- Use the option SaveDefinitions or we can use Initialization with the function that we would like to evaluate first when Manipulate is executed. The SaveDefinitions option determines whether external definitions required to display the content area are automatically included in the output:

```
In[ ]:= ClearAll[f];
        f[x_] := x^2;
        Manipulate[f[x], {x, 0, 100}, SaveDefinitions → True]
```

In[ ]:= **Manipulate[h[y], {y, 0, 100, 1}, Initialization :→ (h[x_] := x^2)]**

- Going back to the example of the triangle, to explicitly include the value of its area we have to define a function that calculates it from the Cartesian coordinates of its three vertices: v_1, v_2, and v_3.

In[ ]:= **areaTriang[{v1_, v2_, v3_}] :=**
    Abs[Det[Join[Transpose[{v1, v2, v3}], {{1, 1, 1}}]]] / 2

In[ ]:= **areaTriang[{{3, 2}, {4, 1}, {7, -3}}]**

Out[ ]= $\dfrac{1}{2}$

- Now we include **areaTriang** in the previously typed command. Move the vertices with the mouse to see how the area is dynamically updated.

In[ ]:= **Manipulate[Graphics[{Pink, Polygon[pts]}, PlotRange → 1,**
    PlotLabel → StringForm["Triangle Area = `1`", areaTriang[pts]],
    ImageSize → Tiny],
    {{pts, {{0, 0}, {1, 0}, {0, 1/2}}}, Locator},
    Initialization :→ (areaTriang[{v1_, v2_, v3_}] :=
        Abs[Det[Join[Transpose[{v1, v2, v3}], {{1, 1, 1}}]]] / 2)]

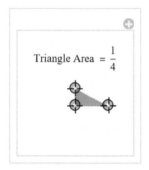

- In this example, we use BSplineFunction to generate a B-Spline curve associated with all the displayed points. If we use LocatorAutoCreate → True, we will be able to add or remove points using ⌥+click (⌘+click in OS X).

In[ ]:= **Manipulate[**
    Show[Graphics[{{Point[p], Green, Line[p]}}, ImageSize → Small],
    ParametricPlot[BSplineFunction[p][t], {t, 0, 1}]],
    {{p, pts}, Locator, LocatorAutoCreate → True},
    Initialization :→ (pts = {{1, 5}, {3, 2}, {6, 4}})]

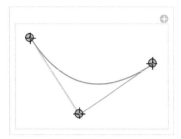

Dynamic and DynamicModule are other functions used for dynamic applications.

- Add Dynamic around bits that can be updated independently of the rest:

In[ ]:= **Manipulate[Show[ContourPlot[Cos[a x] + Cos[a y], {x, 0, 1}, {y, 0, 1}],**
    Graphics[Disk[Dynamic[p], 0.02]]], {{a, 10}, 0, 20}, {p, {0, 0}, {1, 1}}]

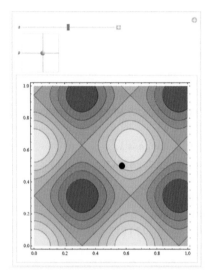

- DynamicModule exhibits many of the same scoping behaviors as Module, plus a few extra features:

In[ ]:= **DynamicModule[{x = 0.5}, {Slider[Dynamic[x]], Dynamic[x]}]**

Out[ ]= $\left\{ \underline{\hspace{5cm}} , 0.5 \right\}$

- DynamicModule stores all the definitions of its local variables by default.

In[ ]:= **DynamicModule[{x = 3, f}, f[t_] := t^2; f[x]]**

Out[ ]= 9

- The example that follows shows how to use it to display some of the different control types available in *Mathematica*.

```
In[ ]:= DynamicModule[{pua, rbba, cbba, tgba}, Grid[{
        {Style["Control Type", FontWeight → "Bold"],
         Style["Example", FontWeight → "Bold"]},
        {"Slider",
         Manipulate[x, {x, 0, 1, ControlType → Slider}, LabelStyle → 16]},
        {"2DSlider",
         Manipulate[x, {x, 0, 1, ControlType → Slider2D}, LabelStyle → 16]},
        {"PopupMenu", Row[{PopupMenu[Dynamic[pua], {a, b, c, d}],
          "      ", InputField[Dynamic[pua], FieldSize → 5]}]},
        {"RadioButtonBar", Row[{RadioButtonBar[Dynamic[rbba], Range[5]],
          "      ", InputField[Dynamic[rbba], FieldSize → 5]}]},
        {"CheckBoxBar", Row[{CheckboxBar[Dynamic[cbba], {1, 2, 3}],
          "      ", InputField[Dynamic[cbba], FieldSize → 5]}]},
        {"TogglerBar", Row[{TogglerBar[Dynamic[tgba], Range[5]],
          "      ", InputField[Dynamic[tgba], FieldSize → 7]}]},
        {"InputField", Manipulate[input,
           {input, 3.14}, ControlType → InputField, LabelStyle → 16]},
        {"ColorSlider", ColorSlider[Pink]}
       }, BaseStyle → {"TraditionalForm", FontFamily → "Helvetica"},
       Alignment → {Left}, Background → LightYellow,
       Frame → All, Spacings → {2, 0.62}]]
```

```
In[ ]:= Quit[]
```

3.2 Creating Demonstrations

The main use of Manipulate is to create eye-catching interactive visualizations with just a few lines of code. In the Wolfram Demonstrations Project site (http://demonstrations.wolfram.com/), you can find more than 12,000 of them. We will be showing here some easy examples.

3.2.1 A Random Walk

■ We use the function RandomWalkProcess to represent a random walk on a line with the probability of a positive unit step p and the probability of a negative unit step $1 - p$. By default we assume n steps = 500 and set the interval from 100 to 1,000 and $p = 0.5$. Each time we execute the command, even though the number of steps is the same, the distance covered will be different since we are assuming random walks.

In[ ]:= **Manipulate[**
    ListPlot[RandomFunction[RandomWalkProcess[p], {1, n}], Filling → Axis],
    {{n, 500, "Steps"}, 100, 1000 , 1, Appearance → "Labeled"},
    {{p, 0.5, "Probability"}, 0, 1, Appearance → "Labeled"}]

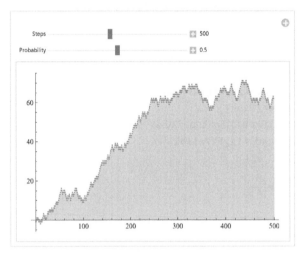

3.2.2 Zooming and the Rosetta Stone

In this example, we show how we can zoom on an image and scroll through it. In this case the image is a picture of the Rosetta Stone but any other picture would do as well.

■ We can import it directly with:

In[ ]:= **ClearAll["Global`*"] (* Remove all the variables from memory *)**

In[ ]:= **rosettaimage = Import[**
        "https://upload.wikimedia.org/wikipedia/commons/c/ca/Rosetta_Stone_BW.
          jpeg"];

In[ ]:= **rosetta = Image[rosettaimage, ImageSize → Small]**

Out[ ]=

■ Now we can copy it, as shown below, or type **rosetta** instead.

In[ ]:= **Manipulate[**

  ImageResize[ImageCrop[ **, {300, 300} / zoom, -pts],**

    {300, 300}, Resampling → ControlActive["Nearest", "Lanczos"]],
  {{pts, {0, 0}, "pts"}, {-1, -1}, {1, 1}},
  {{zoom, 1, "Zoom"}, 1, 10}]

■ We can generate a similar effect using **DynamicImage**.

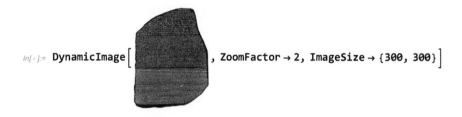

In[ ]:= `DynamicImage[` `, ZoomFactor → 2, ImageSize → {300, 300}]`

3.2.3 Derivatives in Action

In[ ]:= `ClearAll["Global`*"]`

- In this example, we use derivatives to show the tangent at the selected point for different curves. This type of visualization is useful to represent the tangent to a curve and introduce the concept of the derivative.

In[ ]:= `Manipulate[Plot[{f[x], f[x1] + f'[x1] * (x - x1)}, {x, -2 Pi, 2 Pi},`
`        PlotRange → {-2, 2}, Epilog → {PointSize[0.025], Point[{x1, f[x1]}]}],`
`    {x1, -2 Pi, 2 Pi}, { f, {Sin, Cos, Tan, Sec, Csc, Cot}}]`

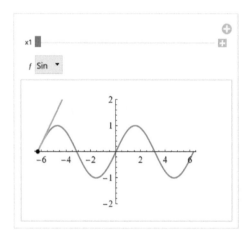

3.3 Image Processing

The image processing functionality of *Mathematica* has greatly improved in the latest versions of the program. We are going to give a brief overview of it. You can extend your knowledge by going through this tutorial: tutorial/ImageProcessing. At the end of this chapter, there are several links in case you may be interested in learning more about the topic.

In the first chapter, we saw that when you click on an image, the **Image Assistant**, a toolbar providing several options to modify the image will appear right below it. As a matter of fact, many of the operations that we will show here can be executed directly using this toolbar. However, for clarity purposes we will show the *Mathematica* commands instead of describing how to use the toolbar. Besides, if we were interested in creating a program we would have to type the required instructions.

3.3.1 Playing with The Giralda, Machu Picchu and More

- We import a JPG image (in this example, The Giralda of Seville, although with a search engine you can find many images similar to the one below or even use other images as well). The resolution of the image was preserved when downloaded. You can also cut and paste from a website.

In[ ]:= `Clear["Global`*"] (*Clear all symbols in a given context*)`

■

In[ ]:= `giralda =`  `;`

- We can check the image size in pixels. It consists of a $n \times m$ grid of triplets, each one representing a pixel.

In[ ]:= `ImageDimensions[giralda]`

Out[ ]= `{635, 635}`

- We can see the RGB value of the individual pixels:

In[ ]:= `ImageData[giralda]`

Out[ ]=
```
{{{0.607843, 0.733333, 0.886275}, {0.607843, 0.733333, 0.886275},
  {0.607843, 0.733333, 0.886275}, ⟨ 629 ⟩,
  {0.458824, 0.572549, 0.792157}, {0.458824, 0.572549, 0.792157},
  {0.458824, 0.572549, 0.792157}}, ⟨ 633 ⟩, { ⟨ 1 ⟩ }}
```

| large output | show less | show more | show all | set size limit... |

- We adjust the contrast and the brightness (if you prefer, instead of copying and pasting the image, just type "giralda").

In[ ]:= `Manipulate[ImageAdjust[`  `, {a, b}],`

`{{a, -1.5, "Contrast"}, -2, 2}, {{b, 0.5, "Brightness"}, 0, 1}]`

- The following command detects the edges of an image.

In[ ]:= **edges = EdgeDetect[giralda]**

Out[ ]=

- Here we compare both images.

GraphicsRow[{giralda, edges}]

Out[ ]=

- ImagePartition splits an image into a collection of subimages of a certain size. In this case, we apply the function to a picture of Machu Picchu:

In[ ]:= **Grid[ImagePartition[**

, 300], Frame → All]

- ImageStich enables us to stitch together a list of images by merging overlapping images into a single one. This function is one of the latest additions to *Mathematica* 13. The example below is included in the function documentation:

In[ ]:= **ImageStitch**$\left[\left\{\right.\right.$  $\left.\left.\right\}\right]$

3.3.2 Background Removal and Image Composition

- In this example, we remove the background of an image and create a new one by combining the result with an image of the Andromeda Galaxy.

In[ ]:= **foreground = RemoveBackground**$\left[\right.$ 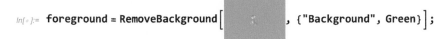 **, {"Background", Green}**$\left.\right]$**;**

In[ ]:= **ImageCompose**[ , **foreground**]

Out[ ]=

This example, available in the documentation, shows how to identify and highlight sickle cells using a combination of the Binarize, ComponentMeasurements and HighlightImage functions in *Mathematica*.

In[ ]:= **i =**  **;**

- First, binarize the image to get the foreground:

In[ ]:= **b = Binarize[i, {0, .7}]**

Out[ ]=

- Then, compute centroids and the radii of disks with the same area as those centroids only for the largest components of the image:

In[ ]:= **cells = ComponentMeasurements[b, {"Centroid", "EquivalentDiskRadius"},**
         #AdjacentBorderCount == 0 && 50 < #Area < 1200 &]

Out[ ]= {7 → {{196.016, 146.391}, 17.0101},
         8 → {{141.599, 146.897}, 16.982}, 11 → {{88.4614, 123.396}, 15.7367},
         12 → {{218.991, 111.964}, 18.412}, 13 → {{51.7714, 105.796}, 16.9163},
         14 → {{141.656, 108.643}, 16.6412}, 16 → {{178.551, 88.5622}, 16.9257},
         17 → {{91.3829, 86.464}, 16.8125}, 18 → {{223.054, 74.4934}, 15.5843},
         21 → {{115.626, 54.4298}, 14.9057}, 22 → {{78.1822, 51.4477}, 15.0121},
         23 → {{197.218, 45.1391}, 17.7697}, 25 → {{163.209, 31.0179}, 15.4818}}

- Finally, highlight the selected components:

In[ ]:= `HighlightImage[i, Circle @@@ cells〚All, 2〛]`

Out[ ]=

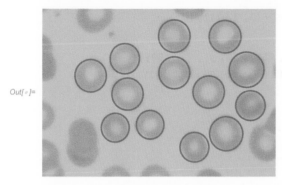

Can you identify the location of the common points in these two images? (Images available in the ImageCorrespondingPoints documentation.)

In[ ]:= `{i1, i2} = {` 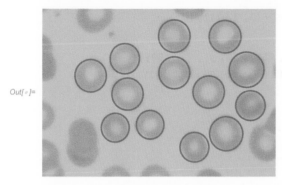 `, };`

■ The output displays the matching positions between the two images.

In[ ]:= `matches = ImageCorrespondingPoints @@ {i1, i2};`
`MapThread[`
  `Show[#1, Graphics[{Yellow, MapIndexed[Inset[#2〚1〛, #1] & , #2]}]] &,`
  `{{i1, i2}, matches}]`

Out[ ]=

■ The rotation angle between both images can also be found:

In[ ]:= `matches = ImageCorrespondingPoints[i1, i2];`
`a1 = ArcTan @@@ (# - 1 / 2 * ImageDimensions @ i1 & /@ matches〚1〛);`
`a2 = ArcTan @@@ (# - 1 / 2 * ImageDimensions @ i2 & /@ matches〚2〛);`
`RootMeanSquare[Mod[a1 - a2, 2 Pi]] / Degree`

Out[ ]= `39.6143`

3.4 Image Manipulation

3.4.1 3D Images

There are many functions (guide/3DImages) for 3D image processing, as this example shows:

In[ ]:= **ctHead = ExampleData[{"TestImage3D", "CThead"}];**

In[ ]:= **head = Image3D[ctHead, BoxRatios → {1, 1, 1}];**

In[ ]:= **Manipulate[**
       Image3D[head, ClipRange → {{x, 256}, {0, y}, {z, 100}}],
       {{x, 100}, 0, 256, 1}, {{y, 150}, 0, 256, 1}, {{z, 50}, 0, 100, 1}]

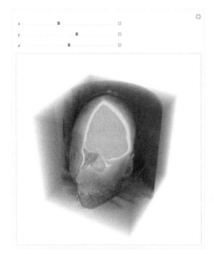

3.4.2 Eyesight Recovery

Next, we show how to improve image resolution.

- Here's an example of an image with fuzzy text (you can find the picture in the ImageDeconvolve help file):

In[ ]:= **fuzzy =** 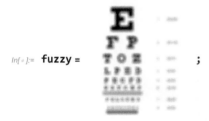 **;**

- With a deconvolution we can substantially improve the image quality.

In[ ]:= **ImageDeconvolve[fuzzy, BoxMatrix[3] / 49, Method → "TotalVariation"]**

Out[ ]=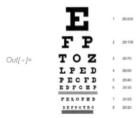

- Here is a simulation of a signal with noise.

```
In[ ]:= data = Table[Sin[i^2 + i] + RandomReal[{-.3, .3}], {i, 0, Pi, 0.001}];
        ListLinePlot[data]
```

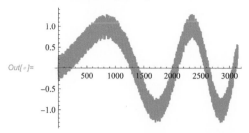

- The next function uses a filter to remove the noise component of the signal.

```
In[ ]:= Manipulate[With[{res = LowpassFilter[data, ω]},
        GraphicsColumn[{Spectrogram[res], ListLinePlot[res]}]],
        {ω, Pi, .01, Appearance → "Labeled"}]
```

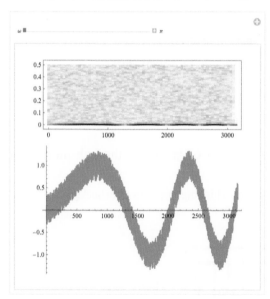

3.4.3 Edge Detection

Mathematica has several functions for detecting edges and borders that are very useful for image processing.

- In the next example, we capture the image of the author while writing this book and apply the FindFaces command to detect his face.

```
In[ ]:= picture = CurrentImage[]
```

```
In[ ]:= picture =                                    ;
```

```
In[ ]:= faces = FindFaces[picture]
```

```
Out[ ]= {Rectangle[{154.5, 38.5}, {237.5, 150.5}]}
```

In[ ]:= `Show[picture,`
         `Graphics[{EdgeForm[{Red, Thick}], Opacity[0], Rectangle @@@ faces}]]`

Out[ ]=

3.4.4 Barcode Recognition

- This example shows the use of BarcodeRecognize to identify barcodes.

In[ ]:= `boardingPass =`

`;`

In[ ]:=

In[ ]:= `BarcodeRecognize[boardingPass, "Data", "PDF417"]`

Out[ ]= `M2VANALMSICK/MARKUS    ECSEZLJ HNDDXBEK 0313 271Y018K0254`
         `136>20B1WW3271BEK 251762174509010 1    EK 116054035        CSEZLJ`
         `DXBDUSEK 0055 271Y019K0249 127251762174509011 1    EK 116054035`

3.4.5 Morphological Thinning

The function Thinning looks for the skeleton of an image through morphological analysis.

- In this example, the image is a labyrinth. Thinning replaces the background with lines showing possible paths.

In[ ]:= `labyrinth =`

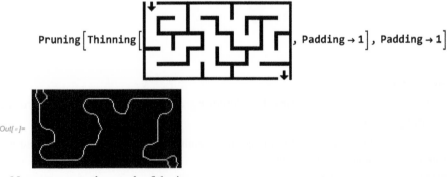

$\text{Pruning}\Big[\text{Thinning}\Big[$ [labyrinth image] $, \text{Padding} \to 1\Big], \text{Padding} \to 1\Big]$

Out[ ]=

- Now we create the graph of the image:

In[ ]:= `MorphologicalGraph[labyrinth]`

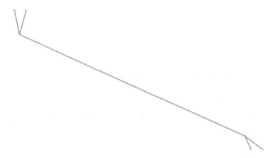

3.4.6 A Painting with a Surprise

The next example shows how to visualize an image from a certain perspective and how to select a part of it. The image was imported from a Wolfram blog:

http://blog.wolfram.com/2010/10/27/the-ambassadors

In[ ]:=   **ambassadors =**
Image[Import["https://content.wolfram.com/uploads/sites/39/2010/10/
TheAmbassadorsBlogPost01.jpg"], ImageSize → 150]

Out[ ]=

The image above corresponds to the painting *"The Ambassadors"* by Holbein. There's a skull in it. Can you see it?

- With ImagePerspectiveTransformation, we can apply a transformation to see the image from a different perspective and, combined with TransformationFunction, display the part of the image that we want to visualize.

In[ ]:= **ImagePerspectiveTransformation$\left[\text{ambassadors,}\right.$**

$$\text{TransformationFunction}\left[\begin{pmatrix} 0.5 & -0.87 & -0.05 \\ -1.34 & 2.7 & 0.43 \\ -0.62 & 0.2 & 0.525 \end{pmatrix}\right]\right]$$

Out[ ]=
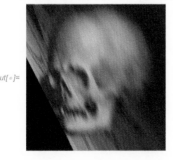

3.4.7 Video

Mathematica has added numerous capabilities for video processing.

- Generate a video of a 3D object rotating around the *z* axis:

In[ ]:= `g = ExampleData[{"Geometry3D", "SpaceShuttle"}][[1, 2]];`

In[ ]:= `VideoGenerator[Graphics3D[{EdgeForm[],`
`        GeometricTransformation[g, RotationTransform[2 π / 2 #, {0, 0, 1}]]},`
`      SphericalRegion → True, Boxed → False] &,`
`    2, RasterSize → 300, FrameRate → 10]`

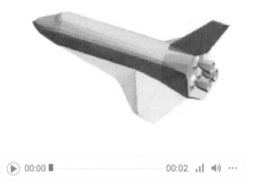

▶ 00:00 ■ ────────────────── 00:02 ,il ◀) ⋯

A basic introduction to video and audio using the Wolfram Language can be found at tutorial/VideoBasics and tutorial/AudioBasics:

https://reference.wolfram.com/language/tutorial/VideoBasics.html
https://reference.wolfram.com/language/tutorial/AudioBasics.html

3.5 Graphs and Networks

Graphs and networks is another area where the latest *Mathematica* versions incorporate new functionality: guide/GraphsAndNetworks. A graph can be represented using Graph, a function with many options. Let's see a basic example.

- We begin by defining the connections between vertices: i ⟷ j or i ⟷j.

In[ ]:= `graph1 = {"a1" ⟷ "a2", "a2" ⟷ "a3", "a2" ⟷ "a1", "a3" ⟷ "a4"};`

- Next we define the capacity of each edge and vertex. Finally we add labels to identify the vertices.

In[ ]:= **Graph[graph1, EdgeCapacity → {2, 3, 4, 5},**
   VertexCapacity → {2, 3, 4, 6}, VertexLabels → "Name"]

Out[ ]=

- We can use special symbols as labels.

In[ ]:= **g = Graph[graph1, EdgeCapacity → {2, 3, 4, 5}, VertexCapacity → {2, 3, 4, 6},**

   VertexLabels → "Name", EdgeLabels → {"a1" ↔ "a2" → ,

   "a2" ↔ "a3" → , "a2" ↔ "a1" → , "a3" ↔ "a4" → }]

Out[ ]=

- We may have a graph representation in a certain format from which we would like to extract information. We can do it as follows:

In[ ]:= **EdgeRules[g]**

Out[ ]= {a1 → a2, a2 → a3, a2 → a1, a3 → a4}

In[ ]:= **PropertyValue[g, EdgeCapacity]**

Out[ ]= {2, 3, 4, 5}

In[ ]:= **PropertyValue[g, VertexCapacity]**

Out[ ]= {2, 3, 4, 6}

In[ ]:= **PropertyValue[g, EdgeLabels]**

Out[ ]= {a2 ↔ a3 → , a3 ↔ a4 → , a2 ↔ a1 → , a1 ↔ a2 → }

- In the next example, we find the shortest way to link all the locations in Figure 3.1 with fiber optic cable:

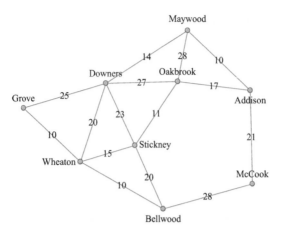

Figure 3.2   Network graph of villages in Illinois.

■ After creating the network graph with *Mathematica* (Figure 3.2, code not shown) and naming it "g", we find the minimum spanning tree containing all the villages (Figure 3.3):

In[ ]:= **fibernetwork = FindSpanningTree[g];**

In[ ]:= **HighlightGraph[g, fibernetwork, GraphHighlightStyle → "Thick"]**

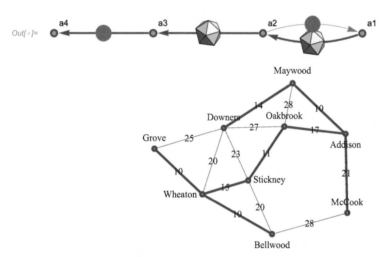

Figure 3.3   Minimum Spanning Tree.

■ In this example, we generate random graphs and search for cliques (Figure 3.4):

In[ ]:= **Grid@Table[HighlightGraph[g = RandomGraph[{8, 16}],**
          Subgraph[g, First[FindClique[g]]]], {2}, {4}]

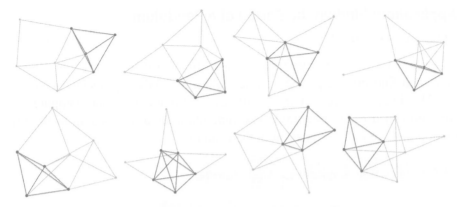

Figure 3.4   Cliques in random graphs.

■ The command below returns the edges and vertices of a dodecahedron in graph format:

```
In[•]:= {Graphics3D[
        {Opacity[.8], PolyhedronData["Dodecahedron", "Faces", "Polygon"]}],
        g = PolyhedronData["Dodecahedron", "SkeletonGraph"]}
```

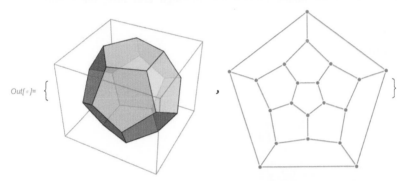

Out[•]= {   ,   }

■ Here we are trying to find the shortest path passing once through each vertex (this is the famous Hamilton's Icosian game):

```
In[•]:= h = PathGraph@First[FindHamiltonianCycle[g]];
```

```
In[•]:= {HighlightGraph[g, h, GraphHighlightStyle → "Thick"], Graphics3D[
        {Opacity[.8], PolyhedronData["Dodecahedron", "Faces", "Polygon"],
        Red, Tube[PolyhedronData["Dodecahedron", "VertexCoordinates"][[
        Append[VertexList[h], VertexList[h][[1]]]]], 0.1]}]}
```

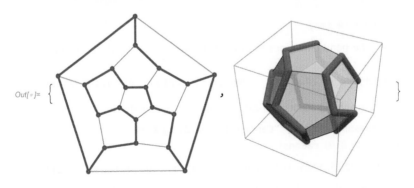

Out[•]= {   ,   }

3.6 Application: Finding the Period of a Pendulum

In[ ]:= `ClearAll["Global`*"]`

The example that follows is significantly more difficult than the previous ones. Our objective is to find out the period of a pendulum by analyzing a sequence of video images. The images were obtained with CurrentImage (if your computer has a webcam you can use this function). The function adds a time stamp to the image, information that we will use later on in the example:

`ims = {` 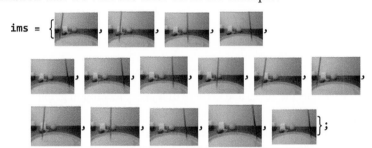 `};`

Source: Wolfram seminars for image processing, although currently, the seminar is not available.

■ Here is the video sequence:

In[ ]:= `ListAnimate[ims]`

■ We begin by highlighting the most prominent lines to identify the ones corresponding to the pendulum.

In[ ]:= `edges = EdgeDetect /@ ims;`

■ In the images above, the pendulum lines are the ones close to the vertical. We can identify them using Radon.

In[ ]:= `radons = ImageAdjust /@ (Radon[#, Automatic, {-Pi/4, Pi/4}] & /@ edges);`

■ `Binarize` detects the brightest points and extracts their x-coordinates to obtain the angle.

In[ ]:= `xs = ComponentMeasurements[#, "Centroid"][[1, 2, 1]] & /@`
            `(Binarize[#, .9] & /@ radons)`

Out[ ]= {89.5, 96., 111.5, 124., 125.5, 122.5, 112.5, 100.5, 87.5, 89., 95.5, 106., 121.5, 127.5, 124.}

■ In the next step, we obtain the relationship between the positions and the time by converting the x-coordinates to radians and combining them with the time stamps of the images.

In[ ]:= **thetas = ((xs / 215) - .5) * Pi / 4;**

In[ ]:= **times = Mod[Options[#, "MetaInformation"]⟦1, 2, 1⟧, 100] & /@ ims**

Out[ ]= {36.0593128, 36.3133382, 36.6473716, 36.9003969, 37.1534222,
37.5664635, 37.8204889, 38.1605229, 38.4995568, 38.9105979,
39.1666235, 39.4196488, 39.8376906, 40.1757244, 40.4267495}

In[ ]:= **datapoints = Transpose[{times, thetas}];**
ListPlot[datapoints]

Out[ ]=

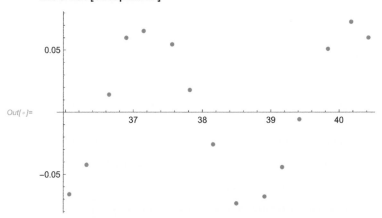

▪ We can now fit the data to a sinusoid curve using NonlinearModelFit.

In[ ]:= **nlm = NonlinearModelFit[datapoints,**
a Sin[b (t - c)] + d, {{a, .07}, {b, 3}, {c, 2}, {d, 0}}, t]

Out[ ]= FittedModel[0.0746392 sin(2.16036 (t + 10.0063)) - 0.00305312]

In[ ]:= **Show[ListPlot[datapoints],**
Plot[Normal[nlm], {t, 36, 41}], PlotRange → All]

Out[ ]=

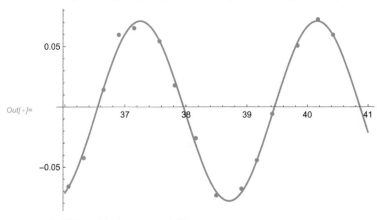

In[ ]:= **nlm["BestFitParameters"]**

Out[ ]= {a → 0.0746392, b → 2.16036, c → -10.0063, d → -0.00305312}

▪ We estimate the period, in seconds:

In[ ]:= **period = 2 Pi / (b /. %)**

Out[ ]= 2.90839

▪ Next, we use FormulaData to find the formula to calculate the length of the pendulum:

After initially typing " FormulaData ["Pend " then you will have the option to choose the full function:

In[ ]:= **FormulaData [{ "PendulumSmallOscillations", "Standard" }]**

Out[ ]= $\left\{ T = 2\pi \sqrt{\dfrac{l}{g}} , \, f = \dfrac{1}{T} , \, v_{\max} = \sqrt{2} \, \sqrt{g\, l\, (1 - \cos(\theta_0))} \right\}$

In[ ]:= **pendulumeq = FormulaData [{ "PendulumSmallOscillations", "Standard" }] [[1]]**

Out[ ]= $T = 2\pi \sqrt{\dfrac{l}{g}}$

In[ ]:= **pendulumeq // InputForm**

Out[ ]//InputForm=

```
        QuantityVariable["T", "Period"] ==
         2*Pi*Sqrt[QuantityVariable["l", "Length"]/QuantityVariable["g",
            "GravitationalAcceleration"]]
```

- Finally, we apply the formula to compute the length, in meters, of our pendulum:

In[ ]:= **Solve** [**pendulumeq** /. { **QuantityVariable** ["T", "Period"] → period,

   QuantityVariable ["l", "Length"] → l, **QuantityVariable** [

   "g", "GravitationalAcceleration"] → ⊟ g acceleration }, l]

Out[ ]= $\{\{l \rightarrow 2.1012 \text{ m/s}^2\}\}$

3.7 Additional Resources

To access the following resources, if they refer to the documentation files, write them down in a notebook, select them, and press <F1>. In the case of external links, copy them in a browser.

About Demonstrations

Manipulate: tutorial/IntroductionToManipulate
Advanced Manipulate functionality tutorial: tutorial/AdvancedManipulateFunctionality
Introduction to Dynamic tutorial: tutorial/IntroductionToDynamic
The Wolfram Demonstrations Project: http://demonstrations.wolfram.com

About Images and Videos

https://www.wolfram.com/wolfram-u/introduction-to-image-processing
Image processing tutorial: tutorial/ImageProcessing
Image Processing & Analysis guide page: guide/ImageProcessing;
Video processing: https://reference.wolfram.com/language/guide/VideoProcessing.html
A collection of video presentations and seminars is available on:
https://www.wolfram.com/wolfram-u/catalog/image-signal-processing

4

Accessing Scientific and Technical Information

This chapter introduces different ways to access real-world computable data with the help of **Entities** and computable data functions such as **ChemicalData**, **CountryData**, **FinancialData**, or **WeatherData**. Using these Mathematica's capabilities, the chapter shows how to access databases containing up-to-date information from many different fields such as: Chemistry, Climatology, Economics, Geography, History, Life Sciences and Medicine, and many more. It also provides an overview of the Wolfram data repository. The examples in the chapter cover topics as varied as how to visualize the evolution of COVID-19 in selected countries, explore molecules or generate stream plots using real-time wind data in Spain.
Later in the book, we will make extensive use of this knowledge when discussing examples.

4.1 The Wolfram Data Framework: Introducing Entities

The Wolfram Language and the Wolfram Knowledgebase through the Wolfram Data Framework™ (WDF) provide a standardized representation of real-world computable data accessible through entities. These entities are grouped into hundreds of entity types. The corresponding *Mathematica* function is EntityType, and it can return values from a wide variety of entity types such as: Chemistry, City, Country, Financial, Movie, Planet, and many more. In fact, the Wolfram Knowledgebase is continuously expanding to include new fields. When using them, *Mathematica* will connect to a server at Wolfram Research, which in turn will access a specialized database containing all the relevant data. The execution time may vary depending on the speed of your Internet connection and that of the Wolfram Research server itself.

Wolfram Knowledgebase: https://www.wolfram.com/knowledgebase/

WDF: https://www.wolfram.com/data-framework/

- Using the Inline free-form input format (Ctrl+= in a US keyboard) we can know the type of entity of the word written inside the box:

In[ ]:= **France** COUNTRY ✓

Out[ ]= France

So, France is an "Entity" of the type "Country".

If you get an error message when evaluating the previous expression it may be due to problems with your firewall or proxy configuration for accessing the Internet. Check your *Mathematica* settings in: **Edit ▸ Preferences ▸ Proxy settings**. If the issue is proxy-related, access the help system for information about how to troubleshoot it: tutorial/TroubleshootingInternetConnectivity.

- A list of entities within a given type can be found using EntityList:

In[ ]:= **EntityList["Country"] // Shallow**

Out[ ]//Shallow=

{ Afghanistan , Aland Islands , Albania , Algeria , American Samoa ,

Andorra , Angola , Anguilla , Antarctica , Antigua and Barbuda , ≪240≫}

We will use Shallow or Short to reduce the size of the output. Remove it to see the complete output.

The entity type "Country" not only includes all the individual countries of the world. It also contains classes (EntityClass ⁝⁝⁝), groups of entities with a shared characteristic. For instance, the entity class "Europe" represents all of the countries in Europe.

In[ ]:= **EntityClass["Country", "Europe"]**

Out[ ]= Europe

In[ ]:= **EntityList[%] // Shallow**

Out[ ]//Shallow=

{ Albania , Andorra , Austria , Belarus , Belgium ,

Bosnia and Herzegovina , Bulgaria , Croatia , Cyprus , Czech Republic , ≪41≫}

- Each entity has properties associated to it. Here are the ones associated to "Country":

In[ ]:= **Entity["Country"]["Properties"] // Shallow**

Out[ ]//Shallow=

{ adjusted net national income , seasonal bank borrowings from Fed, plus adjustments , regions ,

adult population , obese adults , number of aggravated assaults , rate of aggravated assault ,

aggregate home value , aggregate home value, householder 15 to 24 years ,

aggregate home value, householder 25 to 34 years , ≪741≫}

- If we want to know the *Mathematica* expression for a certain property (e.g., the adult population) we place the mouse over "adult population" in the previous output and we see that the exact syntax is:

In[ ]:= **EntityProperty["Country", "AdultPopulation"]**

Out[ ]= adult population

- The actual value of a property is given by EntityValue:

In[ ]:= `EntityValue[Entity["Country", "Japan"],`
`    EntityProperty["Country", "AdultPopulation"]]`

Out[ ]= 7.44373×10^7 people

Since the above expression is very cumbersome to type every time we want to know the value of a property, *Mathematica* offers us several approaches to simplify our queries:

- For example, combining just Entity and EntityProperty, we can obtain the same result:

In[ ]:= `Entity["Country", "Japan"][EntityProperty["Country", "AdultPopulation"]]`

Out[ ]= 7.44373×10^7 people

This expression can be further simplified:

In[ ]:= `Entity["Country", "Japan"]["AdultPopulation"]`

Out[ ]= 7.44373×10^7 people

However, it's often more convenient to just use the Inline free-form input format (Ctrl+= in a US keyboard):

In[ ]:= ▣ | **Japan** COUNTRY | [| *adult population* |] ✓

Out[ ]= 7.44373×10^7 people

- If you click on the symbol ✓ to accept the interpretation and then convert the cell to standard form, the result will be a familiar one:

In[ ]:= `Entity["Country", "Japan"][EntityProperty["Country", "AdultPopulation"]]`

Out[ ]= 7.44373×10^7 people

- WolframAlpha can help us find out the age interval of the adult population property. (You can just write == Adult population):

In[ ]:= `WolframAlpha["Adult population",`
`    IncludePods → "Input", AppearanceElements → {"Pods"}]`

Out[ ]=

Input interpretation:

| all countries, dependencies, and territories | adult population (15–65 years) |

- The following example uses EntityValue to find out the emissions of greenhouse gases (GreenhouseEmissions) in Europe:

In[ ]:= `EntityValue[` | ⊞ **Europe** COUNTRIES | `,`

| *greenhouse gas emissions* | `, "EntityAssociation"] // Short`

Out[ ]//Short=

⟨| Albania → —, Andorra → —, ≪47≫,

United Kingdom → 6.57396×10^8 metric tons of carbon dioxide equivalent /yr,

Vatican City → —|⟩

In summary, the workflow for obtaining data (e.g., the diameter of Earth) is as follows:

1. Create the entity, using Entity:

In[ ]:= `Entity["Planet", "Earth"]`

Out[ ]= Earth

2. Create the property, using EntityProperty:

In[ ]:= `EntityProperty["Planet", "Diameter"]`

Out[ ]= average diameter

3. Obtain the property value for the entity, using EntityValue:

In[ ]:= `EntityValue[Entity["Planet", "Earth"],`
`    EntityProperty["Planet", "Diameter"]]`

Out[ ]= 12 742.018 km

or using the shorter notation Entity[...][*property*]:

In[ ]:= `Entity["Planet", "Earth"]["Diameter"]`

Out[ ]= 12 742.018 km

- Summarizing :

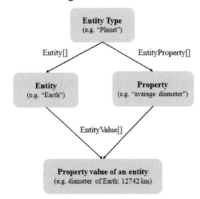

- Of course, we can also use the free-form input:

In[ ]:=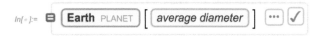

Out[ ]= 12 742.018 km

Quite often, the last approach (with Ctrl+=) is the most convenient one because we don't need to worry about the exact syntax. However, for programing purposes, the standard form is recommended to improve speed of execution and avoid ambiguity issues.

The price that we have to pay for such convenience is that *Mathematica* sometimes may not be able to "read our minds".

- Example: Let's say we want to learn more about the special powers of Mercury, the Roman god of trade, merchants, thieves, and travelers. Since mercury is an entity present in more than one entity type (planets, chemical elements, periodicals, ...), if we just write "mercury" in free form, *Mathematica*'s guess will be most probably that you are referring to a chemical element. You can use EntityTypeName to verify it:

In[ ]:= `EntityTypeName[``]`

Out[ ]= Element

- To tell the program that we refer to the mythological figure, we need to press ⟨•••⟩ and choose the appropriate option after clicking on "more":

In[•]:= ⟨ **Mercury** MYTHOLOGICAL FIGURE ⟨•••⟩ ✓ ⟩

Out[•]= Mercury

Now we can find out the actual name for that property:

In[•]:= ⟨ **Mercury** MYTHOLOGICAL FIGURE ⟩ **["Properties"]**

Out[•]= { alternate names , place of birth , body form , culture , place of death ,

   entity classes , gender , image , locations , major temples , name , native names ,

   patronages , special powers , subtype , symbols , type , literary references }

And finally learn that the ancient god was famous because he could fly and move very fast:

In[•]:= ⟨ **Mercury** MYTHOLOGICAL FIGURE ⟩ **["SpecialPowers"]**

Out[•]= {flight, swiftness}

- Sometimes, even after going through all the options, we may not see what we are looking for. The reason for this is that not all of the potential interpretations are included. To find them use:

In[•]:= **SemanticInterpretation["Mercury", AmbiguityFunction → All] // InputForm**

Out[•]//InputForm=
   AmbiguityList[{Entity["Element", "Mercury"], Entity["Planet", "Mercury"], Entity["Chemical",
   "Mercury"],
   EntityClass["MannedSpaceMission", "ProjectMercuryMannedMission"], Entity["Mineral", "Mercury"],
   Entity["Mythology", "Mercury"], Entity["Periodical", "TheMercuryAUSHobart158944"],
   Entity["Periodical", "TheMercuryZAFGreyvilleDurban156816"], Entity["Periodical",
   "TheMercuryPAPottstown14607"]}]

- Note that one of the interpretations refers to an EntityClass. A *Mathematica* function that represents a class of entities identified by a certain characteristic. In this case, the class is "ProjectMercuryMannedMission" included in the "MannedSpaceMission" entity type. "MannedSpaceMission" contains the following entity classes:

In[•]:= **EntityClassList["MannedSpaceMission"] // Shallow**

Out[•]//Shallow=
   { manned space missions , Apollo landings , Apollo crewed mission ,

   Chinese crewed mission , Earth orbit crewed mission ,

   International Space Station mission , lunar crewed mission , manned space mission ,

   Mercury Atlas crewed mission , Mercury Redstone crewed mission , ≪13≫}

- The properties of the class "ApolloLandings" are:

In[•]:= **EntityClass["MannedSpaceMission", "ApolloLandings"]["Properties"] //**
   Shallow

Out[•]//Shallow=
   { alternate names , callsign , command and service module callsign , operator countries , crew ,

   description , disciplines , distance traveled , Earth landing date , entity classes , ≪20≫}

- The entities included in the class "ApolloLandings" and the graphical representation corresponding to the "LunarLandingDate" property are:

In[ ]:= `EntityList[EntityClass["MannedSpaceMission", "ApolloLandings"]]`

Out[ ]= { Apollo 11 , Apollo 12 , Apollo 14 , Apollo 15 , Apollo 16 , Apollo 17 }

In[ ]:= `TimelinePlot[`
`    EntityList[EntityClass["MannedSpaceMission", "ApolloLandings"]] →`
`    "LunarLandingDate"]`

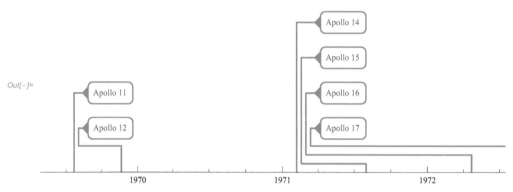

- As we can see, Apollo 11 crew:

In[ ]:= Apollo 11 `["Crew"]`

Out[ ]= { Neil Armstrong , Michael Collins , Buzz Aldrin }

- This mission was led by the astronaut Neil Armstrong born on:

In[ ]:= **Neil Armstrong** PERSON ✓ `["BirthDate"]`

Out[ ]= Tue 5 Aug 1930

Additional possibility: the Entity framework can also be used to construct and execute queries for relational databases (https://reference.wolfram.com/language/tutorial/RelationalDatabasesQuickStart.html .)

For additional information about entity types visit: https://reference.wolfram.com/language/guide/EntityTypes.html.

If you are interested in knowing the sources of Wolfram Knowledgebase (https://www.wolfram.com/knowledgebase/) there are several ways:

1. You can find them in the help for the type of entity on which you want to obtain the information. If the information is available for that entity you will find it at the end where it says "Related links". For example, if you consult the Planet entity type help "Related links", it will direct you to: https://www.wolfram.com/knowledgebase/source-information/?page=PlanetData. However, not all entity types include this information. If not, try the following:

2. Ask Wolfram Alpha (https://www.wolfram.com/knowledgebase/source-information/). For example: If you want to know where the data for the MannedSpaceMission entity type comes from, write in Wolfram Alpha: https://www.wolframalpha.com/input/?i=Manned+Space+Mission. You will notice that the sources are indicated at the end. If you then click on the icon, you will see:

Primary source:

Wolfram|Alpha Knowledgebase, 2020

External sources:

⊖ Manned space mission data

- National Aeronautics and Space Administration. "Spacecraft Query." *National Space Science Data Center.* »
- The Office of Space and Advanced Technology (OES/SAT). *U.S. Space Objects Registry.* »
- United States National Aeronautics and Space Administration. "The Apollo Missions." *Apollo.* »
- Union of Concerned Scientists. *UCS Satellite Database.* »
- The Wikimedia Foundation, Inc. *Wikipedia.* »
- Williams, D. R. "The Apollo Program (1963-1972)." *Lunar Exploration.* »

⊕ Note

3. The last option would be to contact Wolfram directly: https://www.wolfram.com/contact-us/

4.2 Computable Data Functions

The syntax described in the previous section to access entities was first introduced in version 10. Previous versions of *Mathematica* used specific functions to access computable data. They're still useful.

- Most of those functions finished with the suffix Data. You can use **?*Data** to explore some of them (output not shown). Not all functions ending in "Data" are about computable data.

In[ ]:= **?*Data**

Here are some of the most well-known:

Physics and Chemistry: ElementData, ChemicalData, IsotopeData, ParticleData ...
Earth Sciences: WeatherData, GeodesyData, GeoDistance ...
Astronomy: PlanetData, StarData ...
Life Sciences and Medicine: GenomeData, GenomeLookup, ProteinData ...
Economics and Finance: FinancialData, CountryData ...
Mathematics: FiniteGroupData, GraphData, KnotData, LatticeData, PolyhedronData ...
Linguistics: WordData, WordFrequencyData ...
Engineering: AircraftData, BridgeData, NuclearReactorData ...
People and History: PersonData, SocialMediaData, MovieData, Historical PeriodData ...

The sources used (if available) can be found in the help of the corresponding computable data type under the "Related links" section. For instance, the link below will take you to all the sources of information about WordData:

https://reference.wolfram.com/language/note/WordDataSourceInformation.html

In general, any computable data function is of the form: **Function**[*"name or Entity", "property"*], or **Function**[*"name" or "Entity", {"property", ...}*], where Entity is an entity identified by its name: Country, Element, Isotope ... and *property* is information related to such entity. Currently, we can use both syntaxes interchangeably most of the time.

Let's see with examples how these functions operate.

- To find out all the properties available for a specific function we can access the documentation or just type:

In[ ]:= `CityData["Properties"]`

Out[ ]= {AlternateNames, Coordinates, Country, Elevation, FullName, Latitude, LocationLink,
        Longitude, Name, Population, Region, RegionName, StandardName, TimeZone}

- If you type the entity, in this case "Salamanca", *Mathematica* will return all the cities in the world with that name:

In[ ]:= `CityData["Salamanca"]`

Out[ ]= { Salamanca , Salamanca , Salamanca , Salamanca , Salamanca , Salamanca }

- By hovering the mouse over the list elements, you will see the details for each of the cities. If you prefer to see them explicitly, just type the following:

In[ ]:= `salamanca =`
       `SemanticInterpretation["Salamanca",`
          `Entity["City", _], AmbiguityFunction → All]`

Out[ ]= AmbiguityList[
        { Salamanca , Salamanca , Salamanca , Salamanca , Salamanca , Salamanca }, Salamanca]

In[ ]:= `Take[First[salamanca], 6] // InputForm`

       TableForm[{Entity["City", {"Salamanca", "Salamanca", "Spain"}], Entity["City", {"Salamanca",
       "Guanajuato", "Mexico"}], Entity["City", {"Salamanca", "Coquimbo", "Chile"}],
       Entity["City",{"Salamanca", "NewYork","UnitedStates"}], Entity["City",
       {"Salamanca","NegrosOccidental", "Philippines"}], Entity["City", {"Salamanca", "Colon", "Panama"}]}]

- If we intend to refer to a particular one, we need to be more precise and tell *Mathematica* exactly what we want (if we let the program select the city for us, it may not make the right choice). For example: if we'd like to know the population of the Spanish city of Salamanca, we can use Entity with the output returned from the previous command.

In[ ]:= `CityData[Entity["City", {"Salamanca", "Salamanca", "Spain"}], "Population"]`

Out[ ]= 159 754 people

- Alternatively, we can use Interpreter (this function can be used any time there is ambiguity risk).

In[ ]:= `Interpreter["City"]["Salamanca, Spain"]`

Out[ ]=   Salamanca

- Copy the output and paste it below:

In[ ]:= `CityData[` Salamanca CITY `, "Population"]`

Out[ ]= 159 754 people

- Notice that the previous output displays the units explicitly, in this case *people*. To see the output without units, use QuantityMagnitude.

In[ ]:= `QuantityMagnitude[CityData[`
          `Entity["City", {"Salamanca", "Salamanca", "Spain"}] , "Population"]]`

Out[ ]= 159 754

- If you wish to know all the information (properties) for a given entity of a computable data function (**DF**) you can use the syntax:
  `Transpose[{DF["Properties"], DF[entities, #]& /@ DF["Properties"]}]`
- In the example below, we find all the information available in CityData for the city of Salamanca, Salamanca province, Spain: {"Salamanca","Salamanca","Spain"}.

```
In[ ]:= Transpose[{CityData["Properties"],
         CityData[Entity["City", {"Salamanca", "Salamanca", "Spain"}], #] & /@
          CityData["Properties"]}]
```

$$
Out[ ]=\begin{pmatrix}
\text{AlternateNames} & \{\text{Salamanque}\} \\
\text{Coordinates} & \{40.97,\ -5.67\} \\
\text{Country} & \boxed{\text{Spain}} \\
\text{Elevation} & 818\,\text{m} \\
\text{FullName} & \text{Salamanca, Salamanca, Spain} \\
\text{Latitude} & 40.97° \\
\text{LocationLink} & \text{http://maps.google.com/maps?q=+40.97,-5.67\&z=12\&t=h} \\
\text{Longitude} & -5.67° \\
\text{Name} & \text{Salamanca} \\
\text{Population} & 159\,754\,\text{people} \\
\text{Region} & \text{Salamanca} \\
\text{RegionName} & \text{Salamanca} \\
\text{StandardName} & \{\text{Salamanca, Salamanca, Spain}\} \\
\text{TimeZone} & 1\,\text{h}
\end{pmatrix}
$$

If we now select "http://maps.google.com/..." and press <F1> we will see the location of Salamanca in GoogleMaps.

- Moving from cities to stars, the example below shows all the different entity classes available for the star entity. We add Shallow to show only the first few lines of the result. Remove it to see the complete output.

```
In[ ]:= StarData["Classes"] // Shallow
```

Out[ ]//Shallow=

$$
\{\ \boxed{\text{stars}}\ ,\ \boxed{\text{Algol variable stars}}\ ,\ \boxed{\alpha\ 2\ \text{Canum Venaticorum variable stars}}\ ,
$$
$$
\boxed{\alpha\ \text{Centauri}}\ ,\ \boxed{\text{AM Herculis variable stars}}\ ,\ \boxed{\text{Am stars}}\ ,\ \boxed{\text{Ap stars}}\ ,
$$
$$
\boxed{\text{AR Lacertae variable stars}}\ ,\ \boxed{\text{barium stars}}\ ,\ \boxed{\text{Bayer stars}}\ ,\ \ll 178 \gg \}
$$

```
In[ ]:= StarData[ ⊞ α Centauri  STARS ]
```

$$
Out[ ]=\ \{\ \boxed{\text{Proxima Centauri}}\ ,\ \boxed{\text{Rigil Kentaurus}}\ ,\ \boxed{\text{Toliman}}\ \}
$$

- The following function displays a table with the entities, properties, and classes available for the specified collection:

```
In[ ]:= DataTable[fun_Symbol] := Module[{data},
         data = {
           {Style[fun, "Section",
             FontFamily → "Lucida Grande", FontSize → 14], SpanFromLeft},
           {"", "Length", "Examples"},
           {"Entities", Length[fun[]], Append[RandomChoice[fun[], 2], "..."]},
           {"Properties", Length[fun["Properties"]],
            Append[RandomChoice[fun["Properties"], 2], "..."]},
           {"Classes", Length[fun["Classes"]],
            Append[RandomChoice[fun["Classes"], 2], "..."]}
          };
         Grid[data, Frame → All]]
```

In[ ]:= **DataTable[CountryData]**

CountryData		
	Length	*Examples*
Entities	249	{ Barbados , Ireland , ...}
Properties	223	{ForeignRegisteredShips, HIVAIDSPopulation, ...}
Classes	337	{ australia extended , International Telecommunications Satellite Organization , ...}

Out[ ]=

Apart from asking *Mathematica* for information about many different entities, you can also query the program about formulas:

- For instance, if you need the formula for relativistic kinetic energy, you can type:

In[ ]:= **FormulaLookup["Relativistic Kinetic Energy"]**

Out[ ]= {KineticEnergyRelativistic}

- Now we can use FormulaData to obtain the actual formula:

In[ ]:= **FormulaData["KineticEnergyRelativistic"]**

Out[ ]= $\left\{ K = (c^2) \, m \, (\gamma - 1), \ \gamma = \dfrac{1}{\sqrt{(-1/c^2) \, v^2 + 1}} \right\}$

- Of course, you can always use Free-form input by typing "=" at the beginning of an input cell followed by the relevant words in plain English.

In[ ]:= **Kinetic Energy Relativistic formula** »

FormulaData["KineticEnergyRelativistic"]

Out[ ]= $\left\{ K = (c^2) \, m \, (\gamma - 1), \ \gamma = \dfrac{1}{\sqrt{(-1/c^2) \, v^2 + 1}} \right\}$

- *Mathematica* also has the capability to transcribe a given expression using SpokenString:

In[ ]:= **transcribe = SpokenString[Sqrt[x / (y + z)]]**

Out[ ]= square root of the quantity x over the quantity y plus z

- In the next example, we use DictionaryLookup to search for English (other languages are also available) words starting with "sh" and ending in "y":

In[ ]:= **DictionaryLookup[{"English", "sh" ~~ __ ~~ "y"}]**

Out[ ]= {shabbily, shabby, shadily, shadowy, shady, shaggy, shakily, shaky, shallowly, shamefacedly, shamefully, shamelessly, shandy, shanty, shapelessly, shapely, sharply, shatteringly, shay, sheeny, sheepishly, sherry, shiftily, shiftlessly, shifty, shimmery, shimmy, shinny, shiny, shirty, shiveringly, shivery, shockingly, shoddily, shoddy, shortly, shortsightedly, shorty, showery, showily, showy, shrewdly, shrilly, shrinkingly, shrubbery, shrubby, shyly}

- Using the program, we can also identify a language, transliterate expressions to different alphabets, translate between languages, and even create a button that when pressed will return a spoken representation of the argument:

In[ ]:= **text = "Привет, как дела?";**

In[ ]:= **origen = LanguageIdentify[text]**

Out[ ]= Russian

In[ ]:= `Transliterate[text, origen → "Spanish"]`

Out[ ]= Privet, kak dela?

In[ ]:= `transcribe = TextTranslation["Привет, как дела?", "Spanish"]`

Out[ ]= ¿Hola cómo estás?

In[ ]:= `Button["Press", Speak[transcribe]]`

Out[ ]= `Press`

- Next, we are going to obtain the phonetic transcription (in American English) of the first part of this sentence. We use SpokenString to split the phrase into words and the property **"PhoneticForm"**, from WordData, to get the phonetic transcription of each word. Since the output is a list containing the words' phonetic transcriptions, we apply Row[list, " "] to join the elements substituting the commas with blank spaces.

In[ ]:= `Row[WordData[#, "PhoneticForm"] & /@`
    `StringSplit["we are going to obtain the phonetic transcription"], " "]`

Out[ ]= wˈi ˈɒr gˈoʊɪŋ tˈu əbtˈeɪn ðə fənˈɛtɪk tɹˌænskrˈɪpʃən

4.3 The Wolfram Data Repository

The Wolfram Data Repository contains an expanding collection of publicly available computable datasets ready to be used for visualization, analysis and more. The datasets can be accessed directly from within *Mathematica* or downloaded using a web browser. Individual users can contribute datasets to the repository. An overview of the repository is available at:

https://datarepository.wolframcloud.com/

The most common way to access resources available in the repository is to use ResourceSearch and ResourceObject to find and retrieve a data resource and then import the actual contents using ResourceData to proceed with the required computation(s):

In[ ]:= `ResourceSearch["magna carta"]`

Name	ResourceType	ResourceObject	Description
Magna Carta	DataResource	ResourceObject["Magna Carta"]	Text of Magna Carta Libertatum (t

In[ ]:= `ResourceObject["Magna Carta"]["Properties"]`

Out[ ]= {AllVersions, Categories, ContentElementLocations, ContentElements, ContentSize, ContentTypes, ContributorInformation, DefaultContentElement, Description, Documentation, DocumentationLink, DOI, DownloadedVersion, ExampleNotebook, ExampleNotebookObject, Format, InformationElements, Keywords, LatestUpdate, Name, Originator, Properties, ReleaseDate, RepositoryLocation, ResourceLocations, ResourceType, ShortName, SourceMetadata, UUID, Version, VersionInformation, VersionsAvailable}

```
In[ ]:= ResourceObject["Magna Carta"]["SourceMetadata"]
```

Out[ ]= ⟨ | Creator → King John of England, his barons, and Stephen Langton,

Date → | Year: 1215 |, Language → English, Source → English translation of Magna Carta | ⟩

```
In[ ]:= TextCases[ResourceData["Magna Carta"], "Sentence"][[1 ;; 2]]
```

Out[ ]= {Preamble:,
John, by the grace of God, king of England, lord of Ireland, duke of Normandy and Aquitaine,
and count of Anjou, to the archbishop, bishops, abbots, earls, barons, justiciaries,
foresters, sheriffs, stewards, servants, and to all his bailiffs and liege subjects, greetings.}

We can perform many different type of analyses using the Wolfram Data Repository. The example below shows the visualization of the evolution of COVID-19 related cases and deaths in selected countries in the world.

- During the pandemic, the "Epidemic Data for Novel Coronavirus COVID-19" was updated daily with detailed information for provinces in Australia, Canada, and China and counties and states in the US. For the rest of the world, the information was available at the national level:

```
In[ ]:= ResourceUpdate["Epidemic Data for Novel Coronavirus COVID-19"]
```

Out[ ]= ResourceObject[

    Name: Epidemic Data for Novel Coronavirus COVID-19 »
    Type: DataResource

    Description: Estimated cases of novel coronavirus (COVID-19, formerly known as 2019-nCoV) inf

]

- Here is the data for all the world countries:

```
In[ ]:= ResourceObject["Epidemic Data for Novel Coronavirus COVID-19"][
        "ContentElements"]
```

Out[ ]= {AustraliaProvinces, CanadaProvinces, ChinaProvinces, Dataset,
PuertoRicoProvinces, RawData, USCounties, USStates, WorldCountries}

- Here is the data for all the world countries:

```
In[ ]:= covid19 = ResourceData[
        "Epidemic Data for Novel Coronavirus COVID-19", "WorldCountries"];
```

- The evolution of confirmed cases and deaths are shown for Spain:

```
In[ ]:= DateListLogPlot[
        Values[Normal@covid19[Select[#"Country" == Entity["Country", "Spain"] &],
            {"ConfirmedCases", "Deaths"}]],
        PlotLabel → "Evolution of Cases and Deaths in Spain",
        PlotLegends → {"Confirmed Cases", "Deaths"}]
```

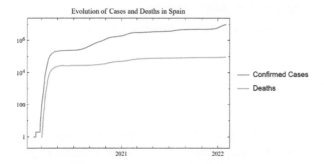

- We can also visualize the estimated deaths per 100,000 people by country (choosing the 30 countries with most deaths due to the pandemic):

```
In[•]:= DateListLogPlot[
    Callout[Tooltip[10^5 #2 / #1["Population"], CommonName[#1]], #1] & @@@
        Normal@ResourceData[
            "Epidemic Data for Novel Coronavirus COVID-19", "WorldCountries"][
        TakeLargestBy[#Deaths["LastValue"] &, 30]][
        All, {#Country, #Deaths} &], Sequence[
        PlotRange → All, GridLines → Automatic, AspectRatio → 1.5,
        ImageSize → 500, PlotLabel → "Estimated Deaths per 100,000 People (log \
    scale)"]]
```

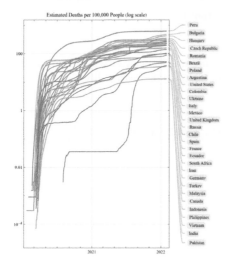

4.4 Weather Data in Real Time

```
ClearAll["Global`*"]
(*Clears values and definitions for all the previously defined symbols*)
```

With computable data we can have access to historical as well as real-time weather information: WeatherData, WeatherForecastData, AirTemperatureData, AirPressureData, WindSpeedData, WindDirectionData, WindVectorData.

The Function["*location*", "*property*",{*start, end, step*}] returns the value of the variable specified in "*property*" for a given location and period (by default it returns the most recent value). Of course you can always use the Free-form input first and then the actual function returned by *Mathematica*.

- For example, if you want to know the temperature in your location, just type:

In[•]:=

Out[•]= 17. °C

- Notice that the function used by *Mathematica* to get the temperature is AirTemperatureData. By default, it will provide you with the temperature of your location (actually your IP address location). This information is obtained from weather stations affiliated with the World Meteorological Organization (WMO). You can use the same function to find out from which weather station you got the air temperature reading.

In[•]:= **AirTemperatureData["StationName"]**

Out[•]= LESA

- Alternatively, GeoNearest can also be used to obtain the same information:

In[•]:= **GeoNearest["WeatherStation",** ▤ **Here** ⋯ ✓ **]**

Out[•]= { LESA }

- The weather station position can be found as follows:

In[•]:= **Entity["WeatherStation", "LESA"]["Position"]**

Out[•]= GeoPosition[{40.959, −5.498}]

- In this example, we get the average daily temperature in Salamanca (Spain) from January 1, 2016 to January 1, 2022 and represent it graphically. Notice the seasonality.

In[•]:= **DateListPlot[AirTemperatureData[**

Salamanca CITY ⋯ ✓ **, {{2016, 1, 1}, {2022, 1, 1}, "Day"}]]**

Out[•]=

- The graph below displays the range of daily temperatures over the past 180 days that corresponds to that station code.

```
In[ ]:= DateListPlot[(WeatherData["LESA", #1,
          {DateString[DatePlus[-180]], DateString[DatePlus[0]], "Day"}] &) /@
        {"MeanTemperature", "MaxTemperature", "MinTemperature"},
      Joined → {True, False, False}, Filling → {2 → {3}}, Mesh → All]
```

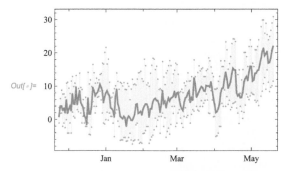

- The following function generates a series of weather-related forecasts for the week after:

```
In[ ]:= Manipulate[data = WeatherForecastData[
          Entity["City", {"Salamanca", "Salamanca", "Spain"}]];
      DateListPlot[Normal[data[g]] /. Interval[x_] :> Mean[x],
        FrameLabel → Automatic], {g, {"Temperature", "WindSpeed",
          "WindDirection", "PrecipitationAmount"}}, SaveDefinitions → True]
```

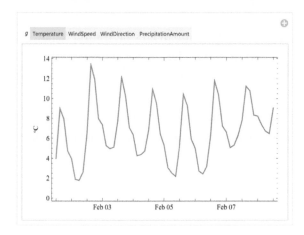

- The final example of this section generates 100 pseudo-random geo position points on the surface of Spain, obtains their most recent win vector readings from the nearest weather station, and finally generates a stream plot:

```
In[ ]:= pts = RandomGeoPosition[ Spain COUNTRY ✓ , 100]
```

```
Out[ ]= GeoPosition[ ⟨Number of points: 100
                      Lat bounds: {36.7, 43.4}
                      Lon bounds: {-8.64, 3.07}⟩ ]
```

```
In[ ]:= wind = WindVectorData[pts, Now, "DownwindGeoVector"];
```

```
In[ ]:= GeoStreamPlot[wind]
```

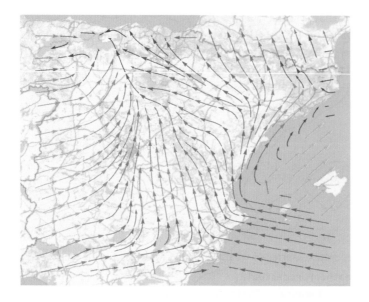

4.5 Chemical and Physical Properties of Elements and Compounds

4.5.1 Element Data

With ElementData or Entity["Element"] and ChemicalData or Entity["Chemical"], we can access a wide variety of information about chemical elements and chemical compounds. WolframAlpha's website has additional details about what exactly can be accessed:

https://www.wolframalpha.com/examples/science-and-technology/chemistry/

- Both functions use the following color scheme when representing atoms and molecules:

In[ ]:= **ClearAll["Global`*"]**

In[ ]:= **ColorData["Atoms", "Panel"]**

Out[ ]=
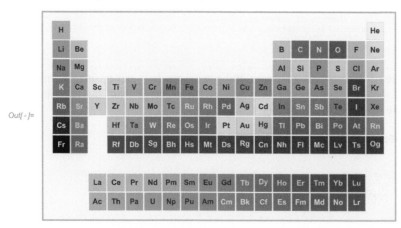

If you click on an element symbol, you will see the color code assigned by the function.

- Let's take a look at the properties available in ElementData:

In[ ]:= **ElementData["Properties"] // Shallow**

Out[ ]//Shallow=

{ adiabatic index , allotropes , allotropic multiplicities , alternate names , atomic mass ,

atomic number , atomic radius , atomic symbol , block , boiling point , ≪110≫}

ElementData [*"element"*, *"property"*] displays the requested property information for a given element.

- The next function shows the crystalline structure of an element in its solid phase:

In[ ]:=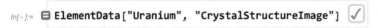

- We can use Dataset to display all the information associated with an entity and DeleteMissing to eliminate those properties without information. We can also include the option MaxItems to limit the maximum number of items displayed simultaneously.

In[ ]:= **Dataset[DeleteMissing[[carbon ELEMENT ••• ✓ ["Dataset"]], MaxItems → 10]**

allotropes	{graphite, diamond, amorphous carbon, lonsdaleite, fullerene, carbon nanotube, graphene}
allotropic multiplicities	{20, 60, 70}
alternate names	{}
atomic mass	12.011 u
atomic number	6
atomic radius	67. pm
atomic symbol	C
block	p
boiling point	4027. °C
bulk modulus	33. GPa

⊼ ∧ rows 1–10 of 98 ∨ ⊻

- The following function tells us the abundance (as a fraction of the mass) of the different elements in the human body:

In[ ]:= **humanabundance =**
Table[{QuantityMagnitude[ElementData[z, "HumanAbundance"]],
ElementData[z]}, {z, 92}];

We keep only those whose presence in the body is greater than 0.1% of the body mass. We order them by comparing the first element of each sublist using a combination of Sort and a pure function.

In[ ]:= **Sort[Cases[humanabundance, {x_, y_} /; x > 0.001], #1[[1]] > #2[[1]] &]**

Out[ ]=

$$\begin{pmatrix} 61. & \text{oxygen} \\ 23. & \text{carbon} \\ 10. & \text{hydrogen} \\ 2.6 & \text{nitrogen} \\ 1.4 & \text{calcium} \\ 1.1 & \text{phosphorus} \\ 0.20 & \text{potassium} \\ 0.20 & \text{sulfur} \\ 0.14 & \text{sodium} \\ 0.12 & \text{chlorine} \\ 0.027 & \text{magnesium} \\ 0.026 & \text{silicon} \\ 0.0060 & \text{iron} \\ 0.0037 & \text{fluorine} \\ 0.0033 & \text{zinc} \end{pmatrix}$$

4.5.2 Molecule Data

With Molecule, we can represent a molecule with atoms and bonds. We can also generate molecules from chemical entities.

In[ ]:= **Molecule[{"O", "H", "H"}, {Bond[{1, 2}, "Single"], Bond[{1, 3}, "Single"]}]**

Out[ ]= Molecule[⊞ H ∨ H Formula: H_2O Atoms: 3 Bonds: 2]

- In the example that follows, we show the caffeine molecular plot. We can move the plot with the mouse to visualize the molecule from different perspectives:

In[ ]:= **m = Molecule[caffeine CHEMICAL]**

Out[ ]= Molecule[⊞ Formula: $C_8H_{10}N_4O_2$ Atoms: 24 Bonds: 25]

In[ ]:= **MoleculePlot3D[m]**

Out[ ]=

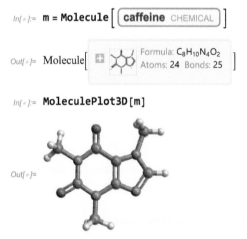

- With ChemicalReaction you can enter a reaction as a string, and then ReactionBalance is used to balance it. Finally reaction balance equation is represented in the traditional form:

In[ ]:= `ReactionBalance[ChemicalReaction["C8H10N4O2 + O2 -> CO2 + H2O + N4"]][`
    `"EquationDisplay"] // TraditionalForm`

Out[ ]//TraditionalForm=

$$2\,C_8H_{10}N_4O_2 + 19\,O_2 \longrightarrow 16\,CO_2 + 10\,H_2O + 2\,N_4$$

4.5.3 Chemical Data

We can obtain chemical data by either using ChemicalData["name","property"] or Entity["name"]["property"]. Both functions will show the information specified in "property" for a given compound.

In[ ]:= `ChemicalData["Butane", "MoleculePlot"]`

Out[ ]=

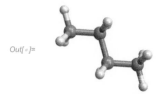

- A graph application below is shown. First the graph structure for a molecule is represented and then a 6-cycle can find:

In[ ]:= `ChemicalData[` [**caffeine** CHEMICAL ··· ✓] `, "StructureGraph"`]

Out[ ]=

In[ ]:= `FindIsomorphicSubgraph[%, CycleGraph[6]]`

Out[ ]=

- We can also compare the properties of two biomolecules:

In[ ]:=

`EntityValue[`{ [**cytosine** CHEMICAL] `,` [**glycine** CHEMICAL] }`,`

{ [*structure diagram*] `,` [*melting point*] `,` ▤ `"MolarMass"` ⌄ ✓ }`, "Dataset"`]

cytosine	Color Structure Diagram	
	Melting Point	300. °C
	MolarMass	111.10 g/mol
glycine	Color Structure Diagram	
	Melting Point	240. °C
	MolarMass	75.07 g/mol

You can now even get step-by-step answers to problems in Chemistry:

https://www.wolframalpha.com/examples/pro-features/step-by-step-solutions/step-by-step-chemistry/

4.6 Life Sciences and Medicine

The Wolfram Language provides access to an extensive collection of data related to life sciences and medicine. Computations related to genomics, food and nutrition, neuroscience, molecular biology and many other fields can be readily perform. This section covers examples about the anatomy of humans and animals, DNA and proteins.

4.6.1 Anatomic Data

Human and animal anatomical structure data are available in *Mathematica*. Their representation is based on the UMLS (Unified Medical Language System):

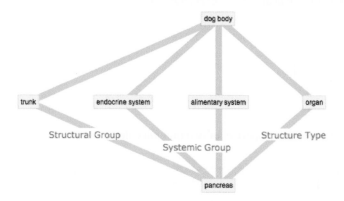

Example: Simple hierarchical network of UMLS concept

▪ Here is the list of the properties associated with the anatomical structure data:

In[ ]:= **Entity["AnatomicalStructure"]["Properties"] // Shallow**

Out[ ]//Shallow=

{ abbreviation , adjacent joint count , adjacent joints ,

adjacent structures , afferent vessel , alternate names , anatomical groups ,

anatomical regions , anatomical spaces , arterial supply , ≪127≫}

- The function below returns the parts that form the lung:

In[ ]:= **[lung ANATOMICAL STRUCTURE] [[*constitutional parts*]] // Shallow**

Out[ ]//Shallow=

{ lobe of lung , visceral pleura , anterior pulmonary nerve plexus ,

posterior pulmonary nerve plexus , root of lung ,

intrapulmonary part of tracheobronchial tree , lung parenchyma ,

intrapulmonary part of bronchial artery , intrapulmonary part of pulmonary arterial tree ,

intrapulmonary part of pulmonary venous tree organ , ≪2≫}

- We can also visualize its anatomical structure:

In[ ]:= **EntityValue[[lung ANATOMICAL STRUCTURE] ,**

 EntityProperty["AnatomicalStructure",

 "RegionalLocationHighlightedImage", {"View" → "All"}]]

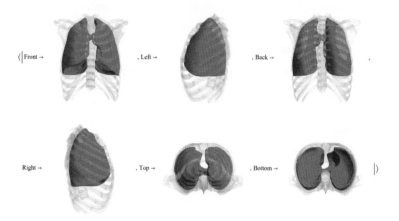

- We generate a 3-D plot of the alimentary system:

In[ ]:= **AnatomyPlot3D[[alimentary system ANATOMICAL STRUCTURE]]**

- This instruction returns all the available properties for dogs' muscles:

In[ ]:= **muscle** (dog) ANIMAL ANATOMICAL STRUCTURE **["Properties"] // Shallow**

Out[ ]//Shallow=

{ adjacent joint count , adjacent joints , anatomical groups , articulating bones ,

associated species , bilateral components , body location image ,

body side , bone articulation , bone articulation count , «36»}

- We can even answer questions such as: "What does the dog's femur bone connect to?"

In[ ]:= **femur** (dog) ANIMAL ANATOMICAL STRUCTURE [*bone articulation*]

Out[ ]= { hip bone (dog) , patella (dog) , tibia (dog) }

- Or find out the body location image of horses' organs:

In[ ]:= **organ** (domestic horse) ANIMAL ANATOMICAL STRUCTURE **["BodyLocationImage"]**

Out[ ]=

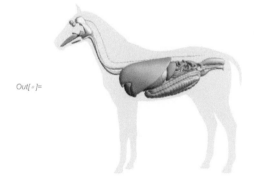

4.6.2 Genetic and Protein Data

The following examples refer to typical functions applicable to genetic and protein data.

■ We can inspect the properties of the DNA entity and visualize its structure:

In[ ]:= [**DNA** BIOMOLECULAR SEQUENCE TYPE]["Properties"] // Shallow

Out[ ]//Shallow=

{ additional bond types , allowed bonding cardinality rules ,

allowed bonding rules , alphabet , alphabet rules , alternate encoding rules ,

all associated bond types , caption , circular , complement letter rules , ≪5≫}

In[ ]:= [**DNA** BIOMOLECULAR SEQUENCE TYPE][[*alphabet rules*]]

Out[ ]= ⟨| A → adenine , C → cytosine , G → guanine , T → thymine |⟩

In[ ]:= MoleculePlot3D[[**adenine** CHEMICAL]]

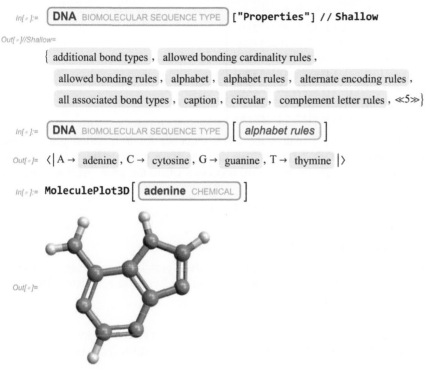

Out[ ]=

■ We can obtain even more complex representations using the new functions BioSequence and BioSequenceComplement:

In[ ]:= MoleculePlot3D[Molecule[BioSequence["DNA", "CTGA"]]]

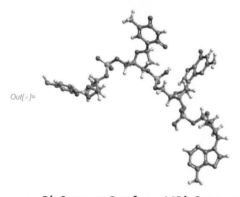

Out[ ]=

In[ ]:= BioSequenceComplement[BioSequence["DNA", "CTGA"]]

Out[ ]= BioSequence[ Type: DNA Sequence
Content: GACT (4 letters)]

■ GenomeLookup shows the chromosomes in which a specified DNA sequence appears along with its position inside of them.

ClearAll["Global`*"]

In[ ]:= **GenomeLookup["AATTTCCGTTAAAT"]**

Out[ ]=

$$\begin{pmatrix}
\{\text{Chromosome2, 1}\} & \{213\,319\,362,\ 213\,319\,375\} \\
\{\text{Chromosome3, -1}\} & \{83\,249\,225,\ 83\,249\,238\} \\
\{\text{Chromosome7, -1}\} & \{78\,235\,739,\ 78\,235\,752\} \\
\{\text{Chromosome9, -1}\} & \{5\,746\,733,\ 5\,746\,746\} \\
\{\text{Chromosome11, 1}\} & \{41\,489\,916,\ 41\,489\,929\} \\
\{\text{Chromosome11, -1}\} & \{124\,909\,525,\ 124\,909\,538\} \\
\{\text{Chromosome22, -1}\} & \{11\,132\,978,\ 11\,132\,991\} \\
\{\text{ChromosomeX, 1}\} & \{144\,568\,117,\ 144\,568\,130\}
\end{pmatrix}$$

- GenomeData["gene","property"] returns the DNA sequence for a specified gene on the reference human genome. With "property" we can specify a characteristic of the chosen gene. In this example, we show the biological processes where the gene "ACAA2" plays a role:

In[ ]:= **GenomeData["ACAA2", "BiologicalProcesses"]**

Out[ ]= {CholesterolBiosyntheticProcess,
  FattyAcidMetabolicProcess, LipidMetabolicProcess, MetabolicProcess}

- ProteinData["protein"] returns the amino acids of a given protein. In this case, we display the sequence of the protein A2M. We use Short to avoid displaying the complete output.

In[ ]:= **ProteinData["A2M"] // Short**

Out[ ]//Short=

MGKNKLLHPSLVLLLLVLLPTDASVSGKPQYMVLVPSLLHTETTEKGCVLLSYLN ...
LDKVSNQTLSLFFTVLQDVPVRDLKPAIVKVYDYYETDEFAIAEYNAPCSKDLGNA

- ProteinData["protein","property"] gives us the option to choose the property of the protein we want to know about. As an illustration, to visualize the A2M protein we can use "MoleculePlot" (this property is not yet available for all proteins at the time of writing).

In[ ]:= **ProteinData["A2M", "MoleculePlot"]**

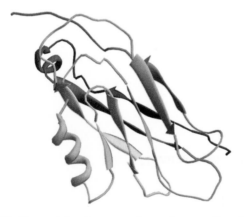

- WolframAlpha (e.g., starting a input cell ==) or the Free-form input allows us to see an overview of available results for the specified gene. (Cassel, B. J. (2018, October 16–19). *Bioinformatics in the Wolfram Language* [Conference Presentation]. Wolfram Technology Conference 2018)

> ☰ mus81 gene

In[ ]:= MUS81 structure-specific endonuclease subunit GENE

Out[ ]= MUS81 structure-specific endonuclease subunit

- We can retrieve the chromosomal context elided above:

In[ ]:= `EntityValue[%, "Context"]`

chromosome 11

Out[ ]=
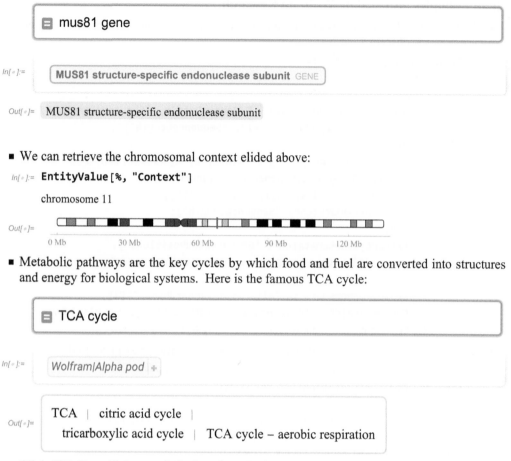
0 Mb   30 Mb   60 Mb   90 Mb   120 Mb

- Metabolic pathways are the key cycles by which food and fuel are converted into structures and energy for biological systems. Here is the famous TCA cycle:

> ☰ TCA cycle

In[ ]:= Wolfram|Alpha pod +

Out[ ]= TCA | citric acid cycle |
tricarboxylic acid cycle | TCA cycle – aerobic respiration

- Click "WolframAlpha results" choosing "PathwayTopology" and click in the upper right corner PathwayTopolog(+) "Formatted Pod"

In[ ]:= `WolframAlpha["TCA cycle",`
`    IncludePods → "PathwayTopology:MetabolicPathwayData",`
`    AppearanceElements → {"Pods"},`
`    TimeConstraint → {20, Automatic, Automatic, Automatic}]`

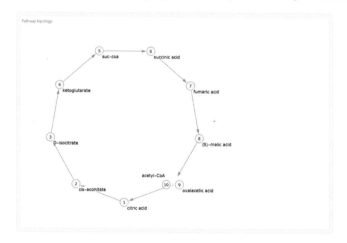

The following function is considerably more complicated. It have been adapted from examples provided by Wolfram Research.

- The function below identifies the chromosomes and the position where a DNA sequence is located inside of them :

```
In[ ]:= FindSequence[sequence_] :=
          chrpos =
            SortBy[
              Module[{chr = #, len, scale = N[GenomeData[#, "SequenceLength"] /
                      GenomeData["Chromosome1", "SequenceLength"]],
                  rls},
                rls = Flatten[{#[[1, 1, 1]] → (#[[All, 2, 1]] * scale / len) & /@
                      SplitBy[Select[GenomeLookup[sequence],
                          #[[1, 2]] == 1 &], #[[1, 1]] &], _ → {}}];
                len = GenomeData[chr, "SequenceLength"];
                {chr,
                 Extract[GenomeData[chr, "GBandScaledPositions"],
                     Position[GenomeData[chr, "GBandStainingCodes"],
                        "acen", 1, 1]][[1, 1]] * scale,
                 scale, GenomeData[#, "AlternateStandardNames"][[1]],
                 GenomeData[chr, "AlternateStandardNames"][[2]], # /. rls}] & /@
              Most@GenomeData["Chromosomes"], #[[5]] &];
```

- Here we use the previously defined function (**FindSequence[]**) to find, for a given sequence, the chromosomes and the position in which they appear.

```
In[ ]:= FindSequence["AACCTTGGGAAATT"] // Shallow
```

Out[ ]//Shallow=

{{Chromosome1, 0.471457, 1., 1, 1, {«3»}}, {Chromosome2, 0.362134, 0.982614, 2, 2, {«5»}},
 {Chromosome3, 0.354132, 0.806884, 3, 3, {«3»}},
 {Chromosome4, 0.199253, 0.773603, 4, 4, {«8»}},
 {Chromosome5, 0.166101, 0.731479, 5, 5, {«3»}},
 {Chromosome6, 0.23412, 0.691204, 6, 6, {«1»}},
 {Chromosome7, 0.227125, 0.642352, 7, 7, {«3»}},
 {Chromosome8, 0.168017, 0.591608, 8, 8, {«1»}},
 {Chromosome9, 0.180048, 0.567334, 9, 9, {}},
 {Chromosome10, 0.163645, 0.547522, 10, 10, {«1»}}, «14»}

4.7 Earth Sciences and Geographic Data

```
In[ ]:= ClearAll["Global`*"]
```

There are many functions available for the computation of data related to earth sciences and geography:

https://reference.wolfram.com/language/guide/EarthSciencesDataAndComputation.html
http://reference.wolfram.com/language/guide/GeographicData.html

4.7.1 Geodata and Geographics

- We can start getting an idea by asking *Mathematica* to list all the functions that start with the letter **Geo** (output omitted).

```
In[ ]:= ?Geo*
```

The names of the functions enable us to identify those related to geography, almost all of them except the ones starting with Geometric. Additionally, we can always access the help documentation to see their syntax. Let's take a look at some of them.

- Using FindGeoLocation[] you can find your current geolocation of your computer (you can find information about the criteria used to find your computer location in the Help) :

In[ ]:= **FindGeoLocation[]**

Out[ ]= GeoPosition[{40.97, −5.66}]

- Next we use GeoGraphics to display a map of Salamanca (a City of Spain) and its surroundings. Remember that in many cases you will find the Free-form input very useful as well (**Insert ▶ Inline Free-form Input** and type Salamanca).

In[ ]:= GeoGraphics$\left[\text{GeoDisk}\left[\boxed{\text{Salamanca \ CITY}}, \text{Quantity[50, "km"]}\right]\right]$

- An easy way of represents a dynamic, interactive, two-dimensional geographical image is with DynamicGeoGraphics.

In[ ]:= **DynamicGeoGraphics**$\left[\boxed{\text{Salamanca \ CITY \ ··· \ ✓}}, \text{GeoRange} \rightarrow \text{Quantity[10, "km"]}\right]$

- Here we use GeoNearest to see the closest locations to Salamanca:

In[ ]:= **GeoNearest["City", ▤ Salamanca , {All, 10 km}]**

Out[ ]= { Salamanca , Villamayor , Santa Marta de Tormes ,

Villares de la Reina , Carbajosa de la Sagrada , Aldeatejada , Cabrerizos ,

Carrascal de Barregas , Doninos de Salamanca , Monterrubio de Armuna ,

San Cristobal de la Cuesta , Moriscos , Arapiles , Florida de Liebana ,

Castellanos de Villiquera , Castellanos de Moriscos , Pelabravo }

- GeoElevationData returns the elevation of a geographic site above sea level in meters, informs us about the minimum and maximum elevations, and creates a relief plot:

In[ ]:= **elevation =**

GeoElevationData[GeoDisk[Salamanca CITY , Quantity[100, "km"]]]

Out[ ]= QuantityArray[▣ ▦▦ Dimensions: {361, 477}
Unit: Meters]

Data not in notebook. Store now ⇥

In[ ]:= **MinMax[elevation]**

Out[ ]= {140. m, 2426. m}

In[ ]:= **ReliefPlot[elevation, DataReversed → True,**
ImageSize → 500, ColorFunctionScaling → False,
ColorFunction → ColorData["HypsometricTints"],
PlotLegends → Placed[Automatic, Right]]

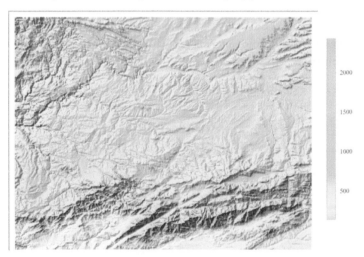

```
In[ ]:= GeoRegionValuePlot[CountryData[] → "LiteracyRate",
        ColorFunctionBinning → {"Quantile", 20},
        GeoBackground → "VectorMarketing", GeoProjection → "Mollweide"]
```

Out[ ]=
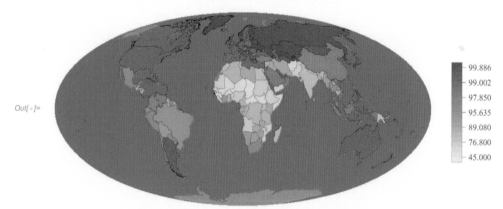

	99.886
	99.002
	97.850
	95.635
	89.080
	76.800
	45.000

We can use CloudDeploy in combination with GeoGraphics to build a web application.

- The function below generates a map of a chosen location (we will normally enter the name of a city) and its surroundings for a given radius:

```
In[ ]:= CloudDeploy[
        FormFunction[{"Location" → "Location", "Radius" → "Quantity"},
        GeoGraphics[GeoDisk[#Location, #Radius]] &, "PNG"]]
```

Out[ ]= CloudObject(https://www.wolframcloud.com/obj/047bd1e9–7599–48e5–8181–7f69dd1561a2)

- We can now copy this newly created link to our website to make it accessible from any device with web browsing capabilities:

    < a href = "https://www.wolframcloud.com/objects/..." > City Map (html) < /a >

- The application is now available to any user, although the first time we access it, we may have to enter our Wolfram ID and password. However, we don't need to have *Mathematica* installed in our computers. Once we enter the required data (Figure 4.1) we get the corresponding map (Figure 4.2).

Figure 4.1   Choosing the location and the radius.

Figure 4.2   WolframCloud generated map.

The Earth, although not flat, is not really a sphere, it is a spheroid with the polar radius being shorter than the equatorial

Terrestrial measurements are usually expressed as established by the The International Terrestrial Reference Frame (ITRF):

https://www.iers.org/IERS/EN/DataProducts/ITRF/itrf.html

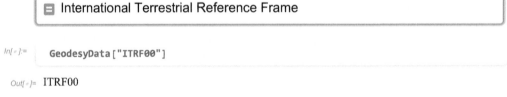

In[ ]:=   `GeodesyData["ITRF00"]`

Out[ ]= ITRF00

- Using the ITRF data the difference between the equatorial and polar radius is:

In[ ]:= `GeodesyData["ITRF00", "Semiaxes"]`

Out[ ]= $\{6.37814 \times 10^6 \text{ m}, 6.35675 \times 10^6 \text{ m}\}$

In[ ]:= `Subtract @@ %`

Out[ ]= 21 384.7 m

In this example, we show how to calculate the distance between two locations on the Earth: Where I am (Here) to Sydney

In[ ]:= `loc = GeoPosition[` **Sydney** CITY ⋯ ✓ `]`

Out[ ]= GeoPosition[{−33.87, 151.21}]

In[ ]:= `GeoDistance[Here, loc]`

Out[ ]= 17 836.4 km

■ This it is the the shortest distance (geodesic) that can be represented as follows:

```
In[ ]:= GeoGraphics[
        {PointSize[Large], Point[Here], Point[loc], Arrow@GeoPath[{Here, loc}]},
        GeoGridLines → Quantity[10, "AngularDegrees"],
        GeoProjection → "Orthographic"]
```

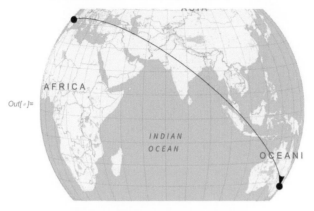

4.7.2 On Earth and Beyond

■ This example returns the location of all the earthquakes that happened in Chile between 2011-1-1 and 2021-12-31 with magnitudes between 6 and 10:

```
In[ ]:= data = EarthquakeData[Entity["Country", "Chile"],
        {6, 10}, {{2011, 1, 1}, {2020, 12, 31}}, "Position"]
```

Out[ ]= EventSeries[⊞ ▮▮▮ Time: 02 Jan 2011 to 14 Dec 2020
                        Data points: 36]

In[ ]:= **GeoListPlot[data]**

Out[ ]=

- GeoListPlot can also generate maps for locations outside the Earth. In the example below, we visualize the landing of Apollo 11, the first manned lunar landing mission:

In[ ]:= **GeoListPlot[** **Apollo 11** MANNED SPACE MISSION **] ,**

   GeoRange → All, GeoProjection → "Orthographic",

   GeoLabels → True, LabelStyle → White, ImageSize → Small]

4.7.3 Travelling Around the World

In this example, we combine several functions to show some travel routes originating in Madrid.

- We first define the cities that we want to travel to. It is better to add the country when typing the city to avoid ambiguities in case there are several cities with the same name.
  {"London, England", "Nuuk, Greenland", "Buenos Aires, Argentina", "Los Angeles, USA", "Malmo, Sweden", "Ulan Bator, Mongolia", "Sidney, Australia"}

In[ ]:= **cities =**

```
{ London CITY , Nuuk CITY ,
  Buenos Aires CITY , Los Angeles CITY ,
  Stockholm CITY , Ulaanbaatar CITY , Sydney CITY }
```

Out[ ]= { London , Nuuk , Buenos Aires , Los Angeles , Stockholm , Ulaanbaatar , Sydney }

In[ ]:= **GeoGraphPlot[Thread[Madrid CITY → cities], GeoCenter → Madrid CITY ,**
  GraphLayout → "Geodesic", VertexLabels → "Name",
  GeoBackground → "VectorDark", GeoProjection → "WinkelTripel"]

Out[ ]=
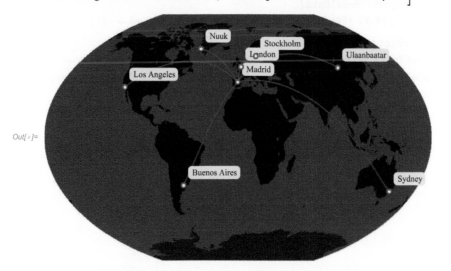

4.7.4 What Happens in the Valparaiso Geomagnetic Field?

During a visit to Universidad Técnica Federico Santa María (UFSM), a Chilean university located in Valparaíso, I observed a strange behavior in the compass of my smartphone. The magnetic field sensor indicated a value (around 20 μT) that seemed unusually low to me. Maybe it wasn't working properly?

I decided to first compare the geomagnetic information available for Valparaiso (Chile) with the one available for my place of residence (Salamanca, Spain) using GeomagneticModelData.

- Here is the resulting comparison table:

In[∘]:= {GeomagneticModelData[[**Salamanca** CITY [⋯] [✓]]],

     GeomagneticModelData[[**Valparaiso** CITY [⋯] [✓]]]} **// Dataset**

Magnitude	NorthComponent	EastComponent	DownComponent	HorizontalComponent	Declination	Inclination
45 079.8 nT	25 430.2 nT	−232.369 nT	37 221.5 nT	25 431.3 nT	−0.523527°	55.6575°
23 628.7 nT	19 500.6 nT	530.864 nT	−13 332.7 nT	19 507.8 nT	1.55938°	−34.3509°

- Then I examined the magnitude of the geomagnetic field across the entire planet. The magnitude at Valparaíso was very low indicating that there was actually no problem with the magnetic sensor of my mobile phone.

In[∘]:= **data = GeomagneticModelData[**
    {{-90, -180.}, {90, 180.}}, "Magnitude", GeoZoomLevel → -2];

In[∘]:= **GeoContourPlot[**
    Flatten@Table[GeoPosition[{(16 - i) * (45 / 8), (32 - j) * (45 / 8)}] →
        data[[33 - i, 65 - j]], {i, 1, 32}, {j, 1, 64}], ContourShading → True,
    ColorFunction → "TemperatureMap", PlotLegends → Automatic]

Out[∘]=

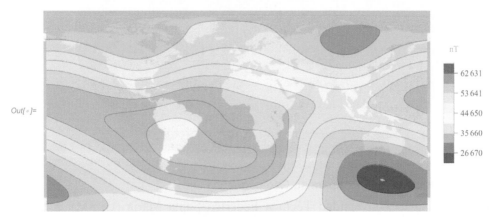

- The next series of commands generate a plot displaying the streamlines based on magnetic direction fields from the South Pole to Valparaiso:

In[∘]:= **GeoPosition[[Valparaiso CITY [⋯] [✓]]]**

Out[∘]= GeoPosition[{−33.04, −71.64}]

In[∘]:= **coords = GeoPosition[**
    Flatten[Table[{-lat, -lon}, {lat, 10, 80, 10}, {lon, 50, 90, 10}], 1]]

Out[∘]= GeoPosition[⬚ Number of points: 40
        Lat bounds: {−80, −10}
        Lon bounds: {−90, −50}]

```
gm = GeomagneticModelData[coords];
norms = QuantityMagnitude[gm[All, "Magnitude"]];
declinations = gm[All, "Declination"];
vectors = Transpose[{norms, declinations}];
```

In[ ]:= `GeoStreamPlot[GeoVector[coords → vectors],`
`  GeoBackground → "ReliefMap", StreamStyle → White]`

Out[ ]=

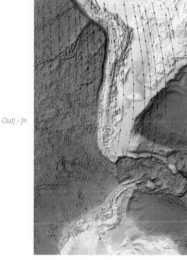

- The intensity and position of the magnetic field changes slightly over time (see GeomagneticModelData>Applications). Even the positions of the magnetic poles themselves have interchanged several times in the past:

In[ ]:= `TextSentences[WikipediaData["Geomagnetic Reversal"]][[1 ;; 5]]`

Out[ ]= {A geomagnetic reversal is a change in a planet's magnetic
field such that the positions of magnetic north and magnetic south are
interchanged (not to be confused with geographic north and geographic south).,
The Earth's field has alternated between periods of normal polarity, in which
the predominant direction of the field was the same as the present
direction, and reverse polarity, in which it was the opposite.,
These periods are called chrons., Reversal occurrences are statistically random.,
There have been 183 reversals over the last
83 million years (on average once every ~450,000 years).}

4.8 Additional Resources

To access the following resources, if they refer to the help files, write their locations in a notebook (e.g., guide/EarthSciencesDataAndComputation), select them and press <F1>. In the case of external links, copy the web addresses in a browser:

Astronomical Data: guide/AstronomicalComputationAndData
Biomolecular Sequences: guide/BiomolecularSequences
Earth Sciences: guide/EarthSciencesDataAndComputation
EngineeringData: guide/EngineeringData

Financial and Economic Data: guide/FinancialAndEconomicData
Geodesy: guide/Geodesy
GeoGraphics:tutorial/GeoGraphics
Life Sciences and Medicine: guide/LifeSciencesAndMedicineDataAndComputation
Socioeconomic Data: guide/SocioeconomicAndDemographicData
Physics and Chemistry: guide/PhysicsAndChemistryDataAndComputation
Transportation Data: guide/TransportationData
WordData: ref/WordData

5

Data Analysis and Manipulation

> *This chapter gives an overview of the software's capabilities for data analysis. The WL probably includes as many probability and statistical functions, if not more, than any other specialized computer programs for analyzing data. The downside is that, very often, users must have a deep understanding of what they are trying to accomplish to take full advantage of the program's capabilities in this area. Topics such as importing and exporting data, exploring data, curve fitting, analysis of time series, and even how to perform spatial statistics calculations are all covered. After reading this chapter, the reader will feel comfortable performing typical data analysis tasks with Mathematica.*

5.1 Importing/Exporting

A fundamental feature of *Mathematica* is its capability for importing and exporting files in different formats. The commands we need for that are `Import` and `Export`. They enable us to import or export many different types: XLS, CSV, TSV, DBF, MDB, ...

A file can be imported or exported by typing the complete source or destination path.

- In Windows, to type the file location for import or export purposes use the following syntax: "*...\\directory\\file.ext*" , as shown in the following example:

  `Import["...\\Mathbeyond\\Data\\file.xls"]`

If we want to use the working directory in which the files are located for importing or exporting purposes, normally the easiest way will be to create a subdirectory inside the directory where the notebook is.

- The following function checks the directory of the active notebook.

  In[ ]:= `NotebookDirectory[]`

  Out[ ]= F:\Mi unidad\Mathematica\MBN13\

In the directory above, create a subdirectory named **Data**.

- With the following function we set the subdirectory **Data** as our default working directory, the one that the `Import` and `Export` functions will use during this session.

In[ ]:= `SetDirectory[FileNameJoin[{NotebookDirectory[], "Data"}]]`

Out[ ]= `G:\Mi unidad\MBN2Ed\Data`

We make the previous cell an **Initialization Cell** to ensure its evaluation before any other cell in the notebook by clicking on the cell marker and choosing from the menu bar: **Cell ▶Properties ▶Initialization Cell**). This means that all the data will be imported/exported from the **Data** subdirectory. We may even consider "hiding" the initialization cell by moving it to the end of the notebook. Remember that the order of operations is not the same as it appears on screen. It is the order in which the cells are evaluated. Therefore, even if an initialization cell is located at the end, it will still be the first one to be executed.

- The total number of available formats can be seen using `$ImportFormats` and `$ExportFormats`.

In[ ]:= `{$ImportFormats // Length, $ExportFormats // Length}`

Out[ ]= `{235, 192}`

In this section, we are going to use, among others, the files available as examples in a subdirectory named **ExampleData** created during the program installation. Some of these examples are downloaded from www.wolfram.com (Internet connection required).

- We are interested in the ones related to statistics. The following function displays the name of the example, a brief description, and the dimensions of its data matrix.

In[ ]:= `Shallow[`
        `{#[[2]], ExampleData[#, "Description"], Dimensions[ExampleData[#]]} & /@`
        `ExampleData["Statistics"], {5, 3}]`

Out[ ]//Shallow=
        {{AirlinePassengerMiles, Revenue passenger miles flown by commercial airlines., {24}},
         {AirplaneGlass, Time to failure for airplane glass., {31}},
         {AnimalWeights, Brain and body weights for 28 animal species., {28, 3}},
         {AnorexiaTreatment, Anorexia data on weight change., {72, 3}},
         {AnscombeRegressionLines, Anscombe's 4 regression line data., {11, 8}}, ≪106≫}

5.1.1 Import

To import files use `Import["path/file.ext", "Elements"]`, where *file* is the name of the file you want to import. If the file is in the **Data** directory previously defined, you only need to type the file name. If it is located somewhere else you will need to enter the complete path (remember that you can use **Insert ▶ File Path...**). You can also import files directly from the Internet. To do that, specify the desired URL in `Import["http://url"]`. As an option you can import only certain file elements, the ones specified by *Elements*.

If your computer uses a non-American configuration remember that *Mathematica* uses the dot "." as a decimal separator and the comma "," to distinguish between list elements. If you have another configuration you may have problems when importing files. The program enables the configuration of the format in **Edit ▶ Preferences... ▶ Appearance** to suit the user needs. Nevertheless, it is better to modify the operating system global settings so that all programs use the same setup. In Windows that can be done by selecting in the **Control Panel** "." as decimal separator and the comma "," as list separator. In OS X, the same can be done by going to **System Preferences... ▶ Language & Region ▶ Advanced...**

- Next, we import an *xls* file containing data about the atomic elements. Note that the program can figure out the type of file from its extension. In this case, the program identifies it as a Microsoft® Excel file.

In[◦]:= `Import["ExampleData/elements.xls"]`

Out[◦]:= ({AtomicNumber, Abbreviation, Name, AtomicWeight} {1., H, Hydrogen, 1.00793} {2., He, Helium

- Notice that the previous output has the following structure {{{ ...}}}. The reason is that *Mathematica* assumes the file is an Excel spreadsheet that may contain several worksheets. Using {**"Data"**, **1**}, we tell the program to import the data from worksheet 1 (do not confuse this Data with the working subdirectory Data that we have created earlier):

In[◦]:= `examplexls = Import["ExampleData/elements.xls", {"Data", 1}]`

Out[◦]:=

AtomicNumber	Abbreviation	Name	AtomicWeight
1.	H	Hydrogen	1.00793
2.	He	Helium	4.00259
3.	Li	Lithium	6.94141
4.	Be	Beryllium	9.01218
5.	B	Boron	10.8086
6.	C	Carbon	12.0107
7.	N	Nitrogen	14.0067
8.	O	Oxygen	15.9961
9.	F	Fluorine	18.9984

- We show below how to extract the first 5 rows of the worksheet and display them in a table format.

In[◦]:= `TableForm[Take[examplexls, 5]]`

Out[◦]//TableForm=

AtomicNumber	Abbreviation	Name	AtomicWeight
1.	H	Hydrogen	1.00793
2.	He	Helium	4.00259
3.	Li	Lithium	6.94141
4.	Be	Beryllium	9.01218

- The next function imports a single column. In this case, we get the fourth column containing the atomic weights.

In[◦]:= `Import["ExampleData/elements.xls", {"Data", 1, All, 4}]`

Out[◦]:= {AtomicWeight, 1.00793, 4.00259, 6.94141, 9.01218, 10.8086, 12.0107, 14.0067, 15.9961, 18.9984}

- To remove the first sublist containing the headings (the same could have been done using **Drop**[data, 1]), we use Rest.

In[◦]:= `data = Rest[Import["ExampleData/elements.xls", {"Data", 1}]];`

- We can also visualize the atomic weights of the imported data. `Tooltip` is included so that when the cursor is over a point in the graphic, its corresponding atomic weight can be seen.

In[ ]:= `ListPlot[Tooltip[data[[All, 4]]],`
`    Filling→ Axis, PlotLabel→ "Atomic Weight"]`

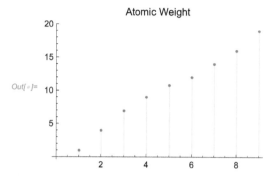

Out[ ]=

- Try to copy and paste links to any of the numerous images available in: https://earthobservatory.nasa.gov/.

  Note: The examples using files imported from the Internet may not work if their web addresses have been changed or the files have been deleted.

In[ ]:= `Import["https://eoimages.gsfc.nasa.gov/images/imagerecords/78000/78349/`
`    arctic_vir_2012147_lrg.jpg"]`

Out[ ]=

- With Import[..., "Elements"], we can find out what elements of the file are available to import (e.g., ImageSize).

In[ ]:= `Import["https://eoimages.gsfc.nasa.gov/images/imagerecords/78000/78349/`
`    arctic_vir_2012147_lrg.jpg", "Elements"]`

Out[ ]= {BitDepth, CameraTopOrientation, Channels, ColorMap, ColorProfileData, ColorSpace,
Comments, Data, DateTime, Exif, FlashUsed, GeoPosition, GPSDateTime,
Graphics, Image, ImageSize, IPTC, MakerNote, MetaInformation, RawData,
RawExif, RawIPTC, RawXMP, RedEyeCorrection, Summary, Thumbnail, XMP}

In[ ]:= `Import["https://eoimages.gsfc.nasa.gov/images/imagerecords/78000/78349/`
`    arctic_vir_2012147_lrg.jpg", "ImageSize"]`

Out[ ]= {1500, 1500}

From the URL address of a website we can limit the information selected to only what we are interested in. A good practice is to first open the URL in a browser and see what information suits our needs. After importing the web page, you will notice that it will appear in *Mathematica* as a list of lists. Analyze how the information is structured. You need to find out what lists contain the information you want and then use `[[...]]` to extract it. Alternatively, you can also use the functions: URLExecute or URLDownload.

Importing from databases is a frequent data import operation. For example: importing from an Access *mdb* file.

- We import the complete structure of the database:

In[ ]:= `Import[ "ExampleData/buildings.mdb", {"Datasets", "Buildings", "Labels"} ]`

Out[ ]= {Rank, Name, City, Country, Year, Stories, Height}

- We use `Short` or `Shallow` to display just a few data elements (this is practical if there's only one table and it's not very big).

In[ ]:= `Import[ "ExampleData/buildings.mdb" ] // Short`

Out[ ]//Short=
  (≪1≫)

- We import the records associated to several fields and display them in a table format:

In[ ]:= `Text[Style[Grid[Transpose[Import[`
`        "ExampleData/buildings.mdb", {"Datasets", "Buildings", "LabeledData",`
`        {"Rank", "Name", "City", "Height"}} ]] ], Frame → All ], Small]]`

Out[ ]=

1	Taipei 101	Taipei	508
2	Petronas Tower 1	Kuala Lumpur	452
3	Petronas Tower 2	Kuala Lumpur	452
4	Sears Tower	Chicago	442
5	Jin Mao Building	Shanghai	421
6	Two International Finance Centre	Hong Kong	415
7	CITIC Plaza	Guangzhou	391
8	Shun Hing Square	Shenzhen	384
9	Empire State Building	New York	381
10	Central Plaza	Hong Kong	374
11	Bank of China	Hong Kong	367
12	Emirates Tower One	Dubai	355
13	Tuntex Sky Tower	Kaohsiung	348
14	Aon Centre	Chicago	346
15	The Center	Hong Kong	346
16	John Hancock Center	Chicago	344
17	Shimao International Plaza	Shanghai	333
18	Minsheng Bank Building	Wuhan	331
19	Ryugyong Hotel	Pyongyang	330
20	Q1	Gold Coast	323

For big databases (ORACLE, SQLServer, etc.), with a large number of tables and thousands of records, it's better to access them with an ODBC connection as we will show later.

5.1.2 Semantic Import

The function SemanticImport detects and interprets the type of imported objects.

- Here is an example of importing CSV data as a DataSet using SemanticImport. The original source is the United Nations Sustainable Development Goals (UN SDG) Database available at https://unstats.un.org/sdgs/indicators/database. The data were originally stored in the Wolfram Cloud by Maureen Baehr and Ben Kicker.

```
In[ ]:= poverty = SemanticImport[
            "https://www.wolframcloud.com/obj/ben.kickert/sdgdata.csv"];
```

- Since the dataset is quite large, we just display its first entry:

```
In[ ]:= poverty[1]
```

Indicator	1.1.1
SeriesDescription	Proportion of population below international poverty line (%)
GeoAreaName	World
Value	9.9
Time_Detail	2015
Source	World Development Indicators database, World Bank
Units	PERCENT
[Age]	
[Location]	
[Sex]	

- We extract the element in the third position. Notice that with Dataset we use [...] instead of [[...]].

```
In[ ]:= poverty[1, 3]
```

Out[ ]= World

- If you hover the mouse over World, the following message will appear on the screen: **Entity**["Country", "World"]. Remember that in the WL the object type entities (**Entity**) have associated additional information that you can access directly, e.g., the world population (Remember that % refers to the last *output*).

```
In[ ]:= %["Population"]
```

Out[ ]= 7 794 798 729 people

- The function below shows the fields {"SeriesDescription", "Value", "Time_Detail"} for the World entity. Notice that Select[#"GeoAreaName"== "World"&] would not work since we are referring to the entity "World" and not to the word "World".

```
In[ ]:= povertyworld =
            poverty[Select[#"GeoAreaName" == Entity["Country", "World"] &],
              {"SeriesDescription", "Value", "Time_Detail"}]
```

- We now group the selected data by years ("Time_Detail") and display only the information of the related year by clicking on it:

```
In[ ]:= GroupBy[povertyworld, "Time_Detail"]
```

	SeriesDescription	Value	Time_Detail
2015	Proportion of population below international poverty line (%)	9.9	2015
	Prevalence of undernourishment (%)	10.6	2015
	Maternal mortality ratio	216.0	2015
	Number of new HIV infections per 1,000 uninfected population, by sex and age (per 1,000 uninfected population)	0.27	2015
	Tuberculosis incidence (per 100,000 population)	140.0	2015
	8 total ·		
2016	Prevalence of undernourishment (%)	10.8	2016
	Number of new HIV infections per 1,000 uninfected population, by sex and age (per 1,000 uninfected population)	0.26	2016
	Tuberculosis incidence (per 100,000 population)	136.0	2016
	Proportion of population using safely managed drinking water services, by urban/rural (%)	70.1036	2016
	Proportion of population with basic handwashing facilities on premises, by urban/rural (%)	62.2045	2016
	8 total ·		
2017	Prevalence of undernourishment (%)	10.9	2017
	Number of new HIV infections per 1,000 uninfected population, by sex and age (per 1,000 uninfected population)	0.25	2017
	Tuberculosis incidence (per 100,000 population)	134.0	2017
	Proportion of population using safely managed drinking water services, by urban/rural (%)	70.6449	2017
	Proportion of population with basic handwashing facilities on premises, by urban/rural (%)	60.1296	2017
	8 total ·		

5.1.3 Export

To export a file use the function **Export** [*"path/file.ext"*, *expr*, *format*] where *file* is the name of the file that will be created from the expression *expr*. Normally the file will be exported in the format specified by **.ext*, for example: file.xls will create an Excel file. Additionally, the format can also be specified with the third argument of the function (*format*). You can even export only parts of the data using the optional fourth argument (*elems*): **Export** [*"path/file.ext"*, *expr*, *format*, *elems*].

The variety of export formats is quite extensive: $ExportFormats. We can even export in compression formats such as *zip*.

As it is the case when importing, if we don't specify the export path, the files will be copied to our working directory (Data in this case).

- Generate a 3D graph:

```
In[ ]:= gr = BarChart3D[{RandomInteger[7, {4}], RandomInteger[{-7, -1}, {4}]},
    ChartElementFunction → "Cone"]
```

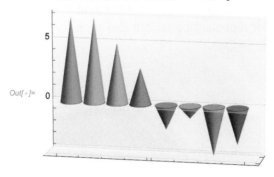

Out[ ]=

- Export the graph as a file named "cones.tif" in TIFF format.

In[ ]:= **Export["cones.tif", gr, "TIFF"];**

You can then import the image previously exported with: **Import["cones.tif"]**.

We are going to show next how to export an animation.

- First, we create an animation using Manipulate. In later chapters, we will see this interesting function in more detail.

In[ ]:= **example3DAnimation = Manipulate[**
    Plot3D[{Cos[k x y], Sin[k x y]}, {x, 0, Pi},
       {y, 0, Pi}, Filling → Automatic, PlotRange → 1], {k, 0, 2}]

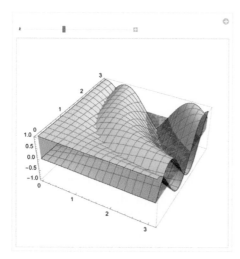

- Then, we export it to a video format, in this case MP4. (Normally this operation may last a few seconds).

In[ ]:= **Export["example3DAnimation.mp4", example3DAnimation]**

Out[ ]= example3DAnimation.mp4

- Now we can either manually locate the exported file and open it or call it directly from *Mathematica* and have the system open it using the default application for MP4 files.

In[ ]:= **SystemOpen["example3DAnimation.mp4"]**

5.1.4 Metadata

As we have previously seen, data from image files usually carry additional information about the images themselves. In this case, we download a file in the ARC Grid format, used in topography for representing terrain elevation.

- We import the following ARCGrid file.

In[ ]:= `Import[ "http://exampledata.wolfram.com/ArcGRID.zip", "ARCGrid"]`

Out[ ]=
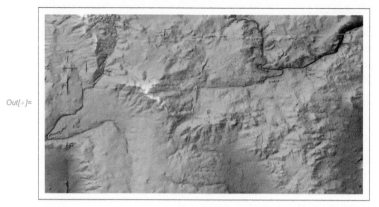

- Notice that this is a compressed file. We can decompress it to find out what files are included.

In[ ]:= `Import[ "http://exampledata.wolfram.com/ArcGRID.zip", "ZIP"]`

Out[ ]= {NED_93751926\.DS_Store, NED_93751926\info\arc.dir, NED_93751926\info\arc0000.dat,
NED_93751926\info\arc0000.nit, NED_93751926\info\arc0001.dat,
NED_93751926\info\arc0001.nit, NED_93751926\meta1.html,
NED_93751926\Metadata.xml, NED_93751926\ned_93751926\dblbnd.adf,
NED_93751926\ned_93751926\hdr.adf, NED_93751926\ned_93751926\prj.adf,
NED_93751926\ned_93751926\sta.adf, NED_93751926\ned_93751926\w001001.adf,
NED_93751926\ned_93751926\w001001x.adf, NED_93751926\NED_93751926.aux}

- Next, we represent the terrain elevations in 3D (we use MaxPlotPoints→100 to speed up the computations).

In[ ]:= `ListPlot3D[Import["http://exampledata.wolfram.com/ArcGRID.zip",`
`    {"ArcGRID", "Data"}], MaxPlotPoints → 100,`
`    ColorFunction → "SouthwestColors", Mesh → None]`

Out[ ]=
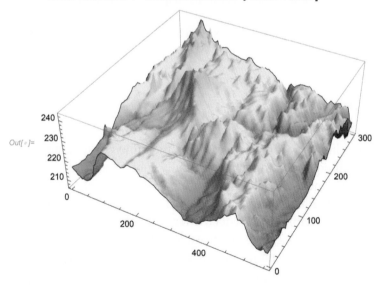

- Let's take a look at the metadata inside.

```
In[ ]:= Import["http://exampledata.wolfram.com/ArcGRID.zip",
         {"ArcGRID", "Elements"}]
```

Out[ ]= {Centering, CentralScaleFactor, CoordinateSystem, CoordinateSystemInformation, Data,
        DataFormat, Datum, ElevationRange, Graphics, GridOrigin, Image, InverseFlattening,
        LinearUnits, Projection, ProjectionName, RasterSize, ReferenceModel, ReliefImage,
        SemimajorAxis, SemiminorAxis, SpatialRange, SpatialResolution, StandardParallels}

- Now, we extract the information related to the coordinates system.

```
In[ ]:= Import["http://exampledata.wolfram.com/ArcGRID.zip",
         {"ArcGRID", "CoordinateSystemInformation"}]
```

Out[ ]= GEOGCS → {NAD83, DATUM → {North_American_Datum_1983,
               SPHEROID → {GRS 1980, 6 378 137, 298.257, AUTHORITY → {EPSG, 7019}},
               TOWGS84 → {0, 0, 0, 0, 0, 0, 0}, AUTHORITY → {EPSG, 6269}},
          PRIMEM → {Greenwich, 0, AUTHORITY → {EPSG, 8901}},
          UNIT → {degree, 0.0174533, AUTHORITY → {EPSG, 9108}},
          AXIS → {Lat, NORTH}, AXIS → {Long, EAST}, AUTHORITY → {EPSG, 4269}}

- The following function enables us to create a link to http://maps.google.com using the latitude and the longitude.

```
In[ ]:= CoordinatesToMapLink[{lat_, long_}] := Hyperlink[
         StringJoin["http://maps.google.com/maps?q=+",
          ToString@lat, "+", ToString@long, "&z=12&t=h"]
         ]
```

- We can now get the coordinates of the previous file and set up a link to Google Maps.

```
In[ ]:= coords = Import["http://exampledata.wolfram.com/ArcGRID.zip",
         {"ArcGRID", "SpatialRange"}]
```

$$Out[ ]= \begin{pmatrix} 40.0617° & 40.1447° \\ -88.3158° & -88.1619° \end{pmatrix}$$

- Since we are establishing the link with the center of the imported grid, we need to calculate the average of the latitude and longitude.

```
In[ ]:= CoordinatesToMapLink[QuantityMagnitude[Map[Mean, coords]]]
```

Out[ ]= http://maps.google.com/maps?q=+40.1032+-88.2389&z=12&t=h

5.2 Statistical Analysis

Currently, given the ease with which we can access very large amounts of data, managing and extracting insights from them can become quite challenging. Maybe we as humans are not well equipped to handle naturally more than 7 or 8 elements at once; in a famous experiment, people didn't notice when an extra person was added to a group of 8 individuals acting on a stage. However, humankind has invented statistical tools that enable us to reduce a large number of data points to a much smaller number of values (mean, median, standard deviation, and others) to make it easier to learn the most relevant information from the original data.

According to *The Oxford Dictionary of Statistical Terms*, statistics is the study of the collection, analysis, interpretation, presentation, and organization of data. Therefore, without data there can be no statistics. In statistical studies, we will frequently reduce the original data, frequently called population, to a reduced number of selected representative values, also known as samples. Their visualization will be very helpful.

5.2.1 Basic Statistical Functions

When working with big volumes of data, it's very common to perform a preliminary exploratory data analysis. In this section, we are going to see how to do it in *Mathematica*:

Basic Statistics:
https://reference.wolfram.com/language/tutorial/NumericalOperationsOnData.html

- We use `CityData` to obtain the distribution of the US population by city. We include `DeleteMissing` to eliminate those cases where the information is not complete. QuantityMagnitude is used to show only the population figure:

```
In[ ]:= usacities = QuantityMagnitude[DeleteMissing[
          CityData[#, "Population"] & /@ CityData[{All, "USA"}]]] // N;
```

The previous function is equivalent to $\mathbf{Map}[\text{func}[\text{arg}, \#] \&, \{city_1, ..., city_n\}]$, where the list of cities is generated with: `CityData[{All, "USA"}]`. We use `Shallow` to show a small sample of the total output.

```
In[ ]:= usacities // Shallow
```

Out[ ]//Shallow=
$$\{8.55041 \times 10^6, 3.97188 \times 10^6, 2.72055 \times 10^6, 2.29622 \times 10^6, 1.56744 \times 10^6,$$
$$1.56303 \times 10^6, 1.46985 \times 10^6, 1.39493 \times 10^6, 1.30009 \times 10^6, 1.02691 \times 10^6, \ll 32\,722 \gg\}$$

- We can find out the number of cities in the list:

```
In[ ]:= Length[usacities]
```

Out[ ]= 32 732

- We build a simple frequency table using `BinCounts` $[\{\{x_1, x_2, ...\}, \{\{b_1, b_2, ...\}\}]$ to count the number of x_i in the intervals $[b_1, b_2), [b_2, b_3), ...$

```
In[ ]:= TableForm[{{"0-10²", "10²-10³", "10³-10⁴",
          "10⁴-10⁵", "10⁵-10⁶", "10⁶-10⁷"}, BinCounts[usacities,
          {{0, 100, 1000, 10 000, 100 000, 1 000 000, 10 000 000}}]}]
```

Out[ ]//TableForm=

$0-10^2$	10^2-10^3	10^3-10^4	10^4-10^5	10^5-10^6	10^6-10^7
2310	13 274	12 795	4039	304	10

- We calculate the most common descriptive statistics:

```
In[ ]:= {Mean[#], StandardDeviation[#], Skewness[#], Kurtosis[#],
          Quantile[#, .6], InterquartileRange[#]} & [usacities]
```

Out[ ]= {7773.91, 65 287.1, 81.8769, 9560.86, 1775., 3718.5}

- Generate a histogram for a list of values using a log scale.

In[ ]:= `Histogram[usacities, "Log", "LogCount", Frame → True]`

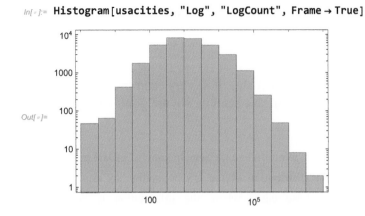

Out[ ]=

5.2.2 Graphical Representations

Graphic representations are very important in statistic analysis. We can create visualizations from custom data.

- In the following example, the word size increases based on its frequency of appearance in a text. In this case, we use the first 300 words of the *Declaration of Independence* included in ExampleData.

In[ ]:= `Text[Row[With[{data = ExampleData[`
         `{"Text", "DeclarationOfIndependence"}, "Words"][[1 ;; 200]]},`
      `MapIndexed[Style[#, 2 Count[Take[data, First[#2]], #]] &, data]], " "]]`

Out[ ]= the the the the to which the of and of to the of the which them to the to that that they are with that among these are and the of to these are among the of the of of these the of the to to and to and its powers in such to them to their and that Governments should be for and and all

- The functions below show how to create a bar chart of the frequency of characters:

In[ ]:= `letters = LetterCounts[`
      `ExampleData[{"Text", "DeclarationOfIndependence"}], IgnoreCase → True]`

Out[ ]= $\langle| e \rightarrow 928, t \rightarrow 686, o \rightarrow 578, a \rightarrow 540, n \rightarrow 537, s \rightarrow 519, i \rightarrow 495, r \rightarrow 489,$
      $h \rightarrow 393, l \rightarrow 278, d \rightarrow 266, u \rightarrow 218, c \rightarrow 204, f \rightarrow 190, m \rightarrow 178, p \rightarrow 151,$
      $g \rightarrow 150, w \rightarrow 117, b \rightarrow 105, y \rightarrow 91, v \rightarrow 76, j \rightarrow 33, k \rightarrow 21, x \rightarrow 10, q \rightarrow 6, z \rightarrow 4|\rangle$

In[ ]:= **BarChart[letters, ChartLabels → Keys[letters]]**

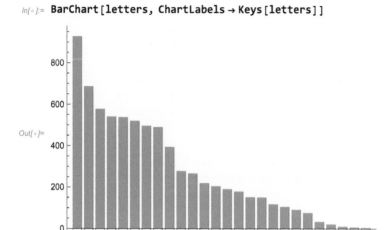

Out[ ]=

5.2.3 Global Warming

You can access a huge collection of computable datasets from the Wolfram Data Repository™:

https://datarepository.wolframcloud.com/

An introduction to the resource with examples of how it can be used is available at:

https://blog.wolfram.com/2017/04/20/launching-the-wolfram-data-repository-data-publishing-that-really-works/

- To retrieve a data collection from the data repository you can use ResourceObject["*name*"]. Among the available curated data is the evolution of the global mean temperature compared to the mean of the period 1951 — 1980. In the example below, we download the Global-mean temperatures dataset (click on ⊞ or » to obtain additional information). A detailed description, including information about all the references used, can be found at: https://data.giss.nasa.gov/gistemp/.

In[ ]:= **ResourceObject["NASA GISTEMP Global Means dTs"]**

Out[ ]= ResourceObject[

		Name: NASA GISTEMP Global Means dTs »
⊞	🏛	Type: DataResource
		Description: Global–mean monthly, seasonal, and annual temperatures (dTs) since 1880

]

- The function below imports the dataset using ResourceData and sets the number of items to be displayed at once to 10. You can use the mouse to scroll through the data.

In[ ]:= **Dataset[ResourceData["NASA GISTEMP Global Means dTs"], MaxItems → 10]**

Year	January	February	March	April	May	June	July	August	September
1880	−0.3 °C	−0.21 °C	−0.18 °C	−0.27 °C	−0.14 °C	−0.28 °C	−0.23 °C	−0.07 °C	−0.16 °C
1881	−0.09 °C	−0.14 °C	0.01 °C	−0.03 °C	−0.04 °C	−0.28 °C	−0.06 °C	−0.02 °C	−0.08 °C
1882	0.1 °C	0.09 °C	0.02 °C	−0.19 °C	−0.17 °C	−0.24 °C	−0.1 °C	0.04 °C	0. °C
1883	−0.33 °C	−0.41 °C	−0.17 °C	−0.23 °C	−0.24 °C	−0.11 °C	−0.07 °C	−0.12 °C	−0.17 °C
1884	−0.18 °C	−0.11 °C	−0.33 °C	−0.35 °C	−0.31 °C	−0.37 °C	−0.33 °C	−0.25 °C	−0.22 °C
1885	−0.64 °C	−0.28 °C	−0.22 °C	−0.43 °C	−0.4 °C	−0.49 °C	−0.27 °C	−0.26 °C	−0.18 °C
1886	−0.41 °C	−0.44 °C	−0.4 °C	−0.28 °C	−0.26 °C	−0.38 °C	−0.15 °C	−0.3 °C	−0.18 °C
1887	−0.65 °C	−0.48 °C	−0.3 °C	−0.36 °C	−0.32 °C	−0.19 °C	−0.18 °C	−0.26 °C	−0.19 °C
1888	−0.42 °C	−0.42 °C	−0.47 °C	−0.28 °C	−0.22 °C	−0.2 °C	−0.08 °C	−0.11 °C	−0.07 °C
1889	−0.19 °C	0.15 °C	0.05 °C	0.05 °C	−0.02 °C	−0.11 °C	−0.04 °C	−0.17 °C	−0.18 °C

rows 1-10 of 137   columns 1-10 of 19

- To extract the contents of specific columns, we use the following syntax: data[All, {Column Names}], in this case we choose "Year" and "JanToDec".

In[ ]:= **data = ResourceData["NASA GISTEMP Global Means dTs"][**
        All, {"Year", "JanToDec"}];

- The values are the deviations with respect to the "normal" , where "normal" always means the average over the 30-year period 1951 — 1980 for that place and time of the year. The years 1951 and 1980 correspond to rows 72 and 101, respectively. The main statistics for that period are shown below. Remember that the use of the syntax {**f1**[#], ..., **fn**[#] }&[data1] is equivalent to {**f1**[data1], ..., **fn**[data1] }

In[ ]:= **{Mean[#], Median[#], StandardDeviation[#], Max[#], Min[#] } &[**
        data[72 ;; 101, "JanToDec"]]

Out[ ]= {−0.000333333 °C, 0.01 °C, 0.109686 °C, 0.27 °C, −0.2 °C}

The mean should be zero; the small difference is due to rounding errors. The median is very close to zero, and the maximum and minimum figures indicate that during those 30 years, there was no clear trend (their absolute values are quite similar). However, when we analyze all the available data, we can see that starting in the 1970's the mean global temperature displays a clear upward trend.

In[ ]:= **DateListPlot[data, PlotTheme → "Web", AxesOrigin → {Automatic, 0}]**

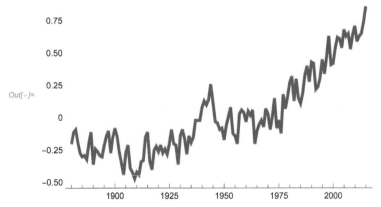

To insert the label **Min** in the graph, copy **Min** to the clipboard and double click inside the plot. Then, right-click and choose "Paste into Graphic". Finally, move the label to the appropriate location.

5.2.4 Analyzing the Weather Data of Your Location

- If you want to know your local temperature, just type in an Inline Free-format Input (or Ctrl + = in a US keyboard) : ⊟ Temperature . Then, notice that the function used to get the temperature is `AirTemperatureData`.

In[•]:= `AirTemperatureData[]`

Out[•]= 18. °C

In[•]:= `IconData["AirTemperature", %]`

Out[•]=

- By default, `AirTemperatureData` returns the most recent measurement for air temperature near your location (actually near the location associated with the IP address through which your computer connects to the Internet, which can sometimes differ greatly from your real location). You can use the same function to find out from what weather station you got the measurement:

In[•]:= `AirTemperatureData["StationName"]`

Out[•]= LESA

- We can then analyze the temperature measurements for this weather station. The function below returns all the data available for December 14, 2020 measured at 30-minutes intervals.

In[•]:= `AirTemperatureData["LESA",`
`        {DateObject[{2020, 12, 14}], DateObject[{2020, 12, 15}]}]`

Out[•]= TimeSeries[🞥 〽 Time: 14 Dec 2020 00:00:00 to 15 Dec 2020 00:00:00 / Data points: 49]

- The output is given as a time series, a format that is very useful if you want to perform additional statistical calculations. However, we can also display the actual values using `Normal`. `RandomChoice` is used in this case to show only 3 random temperature readings.

In[•]:= `RandomChoice[Normal[%], 3]`

Out[•]=

Mon 14 Dec 2020 07:30:00 GMT+2	9. °C
Mon 14 Dec 2020 18:30:00 GMT+2	9. °C
Mon 14 Dec 2020 12:00:00 GMT+2	12.6 °C

- To compute the mean temperature in a circle with a 200-km radius centered around the weather station, we first obtain LESA's coordinates and then use GeoDisk.

In[•]:= `AirTemperatureData["LESA", "Coordinates"][[1]]`

Out[•]= {40.959, −5.498}

In[•]:= `AirTemperatureData[GeoDisk[%, Quantity[200, "km"]], DateObject[], Mean]`

Out[•]= 18.4 °C

- GeoNearest can be used to find out the name of the nearest weather station.

In[ ]:= GeoNearest["WeatherStation", ▤ Here ⋯ ☑, 2]

Out[ ]= { LESA , LEVD }

In[ ]:= GeoListPlot[%, GeoRange → Spain COUNTRY ☑, GeoLabels → True]

Out[ ]=

Let's compare now the temperatures in both weather stations.

- We analyze the mean monthly temperatures for 120 months starting from January 2010. QuantityMagnitude is used to show only the amount not the unit (°C o °F). The computation may take a few minutes because the data are imported from an external source.

In[ ]:= {data1, data2} = QuantityMagnitude[
        {AirTemperatureData["LESA", DateObject[{2010, #}], Mean] & /@
         Range[120] , AirTemperatureData["LEVD",
            DateObject[{2010, #}], Mean] & /@ Range[120] }];

- We calculate the covariance and correlation between the two lists created previously.

In[ ]:= {Covariance[data1, data2], Correlation[data1, data2]}

Out[ ]= {46.2391, 0.990725}

- We visualize **data1** and **data2** using Histogram. If you position the cursor inside any of the bars, its value will be displayed.

In[ ]:= Histogram[{data1, data2}, ChartLegends → {"LESA", "LEVD"}]

Out[ ]=

▢ LESA
▢ LEVD

- We can also visualize the data using BoxWhiskerChart. When placing the cursor on any of the boxes, you will see the statistical information associated with it.

In[ ]:= `BoxWhiskerChart[{data1, data2}, ChartLabels → {"LESA", "LEVD"}]`

Out[ ]=

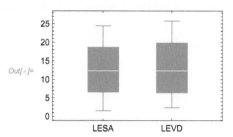

5.3 Probability Distributions

5.3.1 Data and Probability Distributions

Many processes in biology, economics, engineering, finance, and many other scientific and technical fields that involve randomness can be modeled using probability distributions. Two important concepts to master are the probability density function (pdf) and the cumulative density function (cdf).

- The Wikipedia definition:

In[ ]:= `TextSentences[WikipediaData["probability density function"]][[1 ;; 3]]`

Out[ ]= {In probability theory, a probability density function (PDF), or density of a continuous random
    variable, is a function whose value at any given sample (or point) in the sample space
    (the set of possible values taken by the random variable) can be interpreted as providing
    a relative likelihood that the value of the random variable would be close to that sample.,
    In other words, while the absolute likelihood for a continuous random variable to take
    on any particular value is 0 (since there is an infinite set of possible values to
    begin with), the value of the PDF at two different samples can be used to infer,
    in any particular draw of the random variable, how much more likely it is that
    the random variable would be close to one sample compared to the other sample.,
    In a more precise sense, the PDF is used to specify the probability of the random variable
    falling within a particular range of values, as opposed to taking on any one value.}

- The formal definition:

In[ ]:=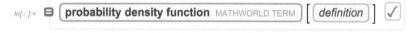

The probability density function (PDF) $P(x)$ of a continuous distribution is defined as the derivative of the (cumulative) distribution function $D(x)$,

$$D'(x) = [P(x)]_{-\infty}^{x}$$

$$= P(x) - P(-\infty)$$

Out[ ]= $$= P(x),$$

so

$$D(x) = P(X \le x)$$

$$\equiv \int_{-\infty}^{x} P(\xi)\, d\xi.$$

Different kinds of probability distribution can be used depending on the random processes involved. The number of probability distributions (guide/ParametricStatisticalDistributions) available in *Mathematica* is substantially higher than the one available in other well-known statistical programs. Additionally, *Mathematica*'s symbolic capabilities enable us to build our own distributions using variations of the available ones.

- This example shows the PDF and CDF of a Poisson distribution.

In[ ]:= **pdf = PDF[PoissonDistribution[μ], x]**

Out[ ]= $\begin{cases} \dfrac{e^{-\mu}\mu^x}{x!} & x \geq 0 \\ 0 & \text{True} \end{cases}$

In[ ]:= **cdf = CDF[PoissonDistribution[μ], x]**

Out[ ]= $\begin{cases} Q(\lfloor x \rfloor + 1, \mu) & x \geq 0 \\ 0 & \text{True} \end{cases}$

- Here several statistical parameters (mean, variance, skewness, and kurtosis) are computed.

In[ ]:= **{Mean[#], Variance[#], StandardDeviation[#], Skewness[#], Kurtosis[#]} &[**
     PoissonDistribution[μ]]

Out[ ]= $\left\{ \mu, \mu, \sqrt{\mu}, \dfrac{1}{\sqrt{\mu}}, \dfrac{1}{\mu} + 3 \right\}$

- Looking at the output above, we can see that the Poisson distribution depends on only one parameter: μ. This means that the variance is equal to the mean and therefore, the standard deviation $\sigma = \sqrt{\mu}$.

The Poisson distribution is used to model the number of events x occurring within a given time interval with the parameter μ indicating the average number of events per interval. A typical Poisson process is the radioactive disintegration.

   Example: After measuring an environmental sample of uranium over a few hours, its parameter has been determined to be $\mu = 10$ disintegrations/minute.

- The PDF and CDF of measure x = {0, 1, 2, ...} disintegrations in one minute are represented below. An option in PlotLegends is included to place the legend in the desired position.

In[ ]:= **DiscretePlot[Evaluate[{pdf, cdf} /. μ → 10], {x, 0, 15}, PlotRange → All,**
     PlotLegends → Placed[{"Probability", "Cumulative Probability"}, Below]]

Out[ ]=
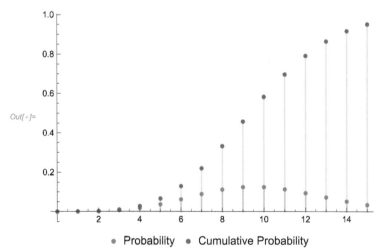

Probability[*expr, x* ≈ *dist*] (to write ≈ you can use the palettes or the shortcut: **Esc key** dist **Esc key**) calculates the probability of *expr* under the assumption that *x*

follows the probability distribution *dist*.

- The probability that there are 15 or more disintegrations in the uranium sample in a given minute is:

In[ ]:= **Probability[x ≥ 15, x ≈ PoissonDistribution[10]] // N**

Out[ ]= 0.0834585

Expectation[*expr*, $x \approx dist$] calculates the mathematical expectation of *expr* assuming that *x* follows the probability distribution *dist*.

- The moment of order 2 of a Poisson distribution is:

In[ ]:= **Expectation[x^2, x ≈ PoissonDistribution[μ]]**

Out[ ]= $\mu^2 + \mu$

- The moment of order n of a Poisson distribution is:

In[ ]:= **Expectation[x^n, x ≈ PoissonDistribution[μ],**
    Assumptions → n ∈ Integers && n > 0]

Out[ ]= $B_n(\mu)$

- BellB is-the Bell polynomial $B_{n(x)}$. (e.g., n = 4)

In[ ]:= **BellB[4, μ]**

Out[ ]= $\mu^4 + 6\mu^3 + 7\mu^2 + \mu$

5.3.2 The Normal or Gaussian Distribution

The best-known and probably the most commonly occurring probability distribution in nature is the Normal or Gaussian (although it was not Carl F. Gauss who discovered it).

- Using Inline Free-form Input (⊟ normal distribution [⊟ definition]) we can get information about it.

In[ ]:= [**normal distribution** MATHWORLD TERM] [*definition*]

Out[ ]=
A normal distribution in a variate X with mean μ and variance σ^2 is a statistic distribution with probability density function

$$P(x) = \frac{1}{\sigma\sqrt{2\pi}}\, e^{-(x-\mu)^2/(2\sigma^2)}$$

on the domain $x \in (-\infty, \infty)$. While statisticians and mathematicians uniformly use the term "normal distribution" for this distribution, physicists sometimes call it a Gaussian distribution and, because of its curved flaring shape, social scientists refer to it as the "bell curve." Feller uses the symbol $\varphi(x)$ for $P(x)$ in the above equation, but then switches to n (x) in Feller.

One of the main features of probability distributions is that if certain experimental data follow a specific distribution, the difference between the theoretical distribution and the experimental one will get smaller and smaller as the number of experimental data points gets larger and larger.

- In this example, we use Histogram to display a histogram of experimental data (in this case simulated data) following a normal probability distribution with mean 0 and standard deviation 1, commonly written as N(0,1), and compare it with its theoretical distribution. The experimental data are simulated using RandomVariate [*dist*], a *Mathematica* function that generates random numbers that follow a distribution *dist* . Strictly speaking, we should be talking about pseudorandom numbers since they are generated by algorithms. However, for most practical applications they can be considered random numbers. We place the function inside `Manipulate` using *n,* the number of data points, as the parameter. We can clearly see that the shape of the histogram gets closer to the probability density function (PDF) of the theoretical N(0,1) as the number of simulated data points increases.

```
In[ ]:= Manipulate[
        Histogram[RandomVariate[NormalDistribution[0, 1], n], Automatic, "PDF",
          PlotRange → {{-4, 4}, {0, 0.45}}, AxesOrigin → {-4, 0}, Epilog → First@
            Plot[PDF[NormalDistribution[0, 1], x], {x, -4, 4}, PlotStyle → Red]],
        {{n, 1000, "Number of data points"}, 50, 10000}]
```

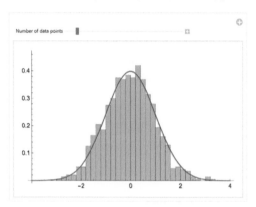

5.3.3 Automatically Finding Probability Distributions

You can use FindDistribution to find the distribution of your data.

Example: A pacemaker has a redundant system consisting of independent units so that when a single unit fails, the pacemaker does not need to be replaced.

- The data below represents the failing times of 100 pacemakers.

```
In[ ]:= pacemaker = {13.6, 9.3, 6.5, 31., 4.2, 12.5, 15.3, 19.1, 6.5, 30.5, 77.2,
          21.7, 8.1, 19.7, 8.3, 24.9, 23., 22.3, 24.8, 18.5, 24., 13.1, 2.9,
          10.4, 19.1, 25.3, 12.2, 3.9, 7.3, 14., 8.2, 10.2, 8.4, 6.4, 5.9,
          13.6, 25.1, 19., 7.1, 7.3, 3.3, 6.3, 24.1, 14.9, 16.1, 28.5, 17.1,
          8.6, 5.3, 23.2, 27.9, 17.1, 15.8, 12.4, 5.9, 26.4, 3.4, 29.5,
          28.4, 10.7, 7.5, 15.1, 11., 7.5, 9.7, 7., 59.1, 2.8, 7.4, 21.7,
          10.9, 15.9, 6.4, 16., 0.3, 11.4, 20.1, 28.1, 12.7, 27.6, 10.,
          5.9, 4.4, 2.7, 6.5, 30., 19.2, 7.9, 16.5, 18.4, 15.9, 6.6, 6.5};
```

- We can now proceed to find the underlying distribution. By default, FindDistribution will return the one that fits the data best. However, we can change that behaviour by adding a second argument to the function as we do in the example below. In this case, we ask the program to show us the best two distributions:

```
In[ ]:= pacedistribution = FindDistribution[pacemaker, 2]
```

```
Out[ ]= {GammaDistribution[1.79873, 8.73275], WeibullDistribution[1.34491, 17.1987]}
```

```
In[ ]:= {gamma, weibull} = pacedistribution;
```

- In fact, the most commonly used distributions for modeling equipment failure are the Gamma distribution (GammaDistribution) and the Weibull distribution (WeibullDistribution). We can even use the function with a third argument to generate a Dataset containing the values of all possible properties associated to them:

In[ ]:= **FindDistribution[pacemaker, 2, All]**

	PearsonChiSquare	CramerVonMises	BIC	AIC
GammaDistribution[1.79873, 8.73275]	0.159992	0.611598	−7.24975	−7.23625
WeibullDistribution[1.34491, 17.1987]	0.0437283	0.384223	−7.29446	−7.28096

- The function below displays the graphs of their probability density functions along with the histogram of the data:

In[ ]:= **Show[Histogram[pacemaker, Automatic, "PDF"],**
     Plot[{PDF[gamma, x], PDF[weibull, x]}, {x, 0, 50},
        Filling → Axis, PlotLegends → {"Gamma PDF", "Weibull PDF"}]]

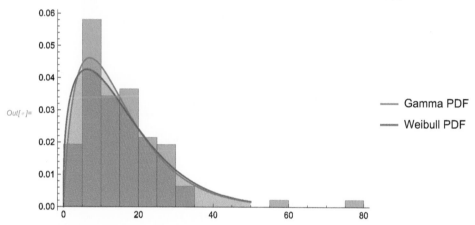

- The probability that the pacemaker will still be working after 20 years is:

In[ ]:= **Probability[x ≥ 20, x ≈ gamma] // N**

Out[ ]= 0.279733

5.3.4 Mixture Distributions

In this subsection, we're going to analyze the "OldFaithful" dataset related to geyser emissions. Let's see a detailed description of the file.

In[ ]:= **ExampleData[{"Statistics", "OldFaithful"}, "Properties"]**

Out[ ]= {ApplicationAreas, ColumnDescriptions, ColumnHeadings,
      ColumnTypes, DataElements, DataType, Description, Dimensions, EventData,
      EventSeries, LongDescription, Name, ObservationCount, Source, TimeSeries}

In[ ]:= **ExampleData[{"Statistics", "OldFaithful"}, "LongDescription"]**

Out[ ]= Data on eruptions of Old Faithful geyser, October 1980. Variables are the
      duration in seconds of the current eruption, and the time in minutes to the next
      eruption. Collected by volunteers, and supplied by the Yellowstone National Park
      Geologist. Data was not collected between approximately midnight and 6 AM.

```
In[ ]:= Transpose[{ExampleData[{"Statistics", "OldFaithful"}, "ColumnHeadings"],
        ExampleData[{"Statistics", "OldFaithful"},
         "ColumnDescriptions"]}] // TableForm
```

Out[ ]//TableForm=

Duration	Eruption time in minutes
WaitingTime	Waiting time to next eruption in minutes

- We assign names to the column data and create histograms for each column:

```
In[ ]:= OldFaithfuldata = ExampleData[{"Statistics", "OldFaithful"}];
```

```
In[ ]:= {duration, waitingtime} = Transpose[OldFaithfuldata];
```

```
In[ ]:= GraphicsRow[
        {Histogram[duration, 12, "PDF", ChartElementFunction → "FadingRectangle",
          PlotLabel → Style["Duration", 10, Blue]], Histogram[waitingtime,
          12, "PDF", ChartElementFunction → "GradientRectangle",
          PlotLabel → Style["Waiting Time", 10, Blue]]}, ImageSize → Medium]
```

Out[ ]=

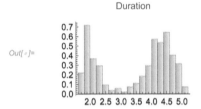 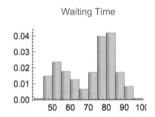

- We can measure the correlation between the two variables.

```
In[ ]:= Correlation[duration, waitingtime]
```

Out[ ]= 0.900811

- They are highly correlated. To visualize it we can use PairedHistogram after rescaling the data.

```
In[ ]:= sclData1 =
        With[{dat1 = MinMax[#] &[duration], dat2 = MinMax[#] &[waitingtime]},
         Rescale[#, dat1, dat2] & /@ duration];
```

```
In[ ]:= PairedHistogram[sclData1, waitingtime, 12, "PDF",
        ChartElementFunction → "GlassRectangle", ChartLabels →
         {Style["Scaled Duration", "Text"], Style["Waiting Time", "Text"]}]
```

Out[ ]=

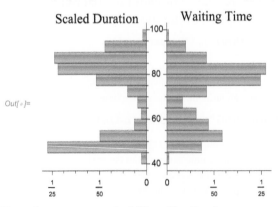

If we have two probability distributions, we can obtain a mixture distribution with the following function:

`MixtureDistribution[{p1, p2}, {N[m1, s1] , N[m2, s2]}]`

where p1, p2 represent the proportion of N[m1, s1] and N[m2, s2], respectively.

- Let's build a mixture distribution using the "OldFaithful" dataset again.

In[ ]:= `mixDist1 = EstimatedDistribution[waitingtime, MixtureDistribution[{p1, p2},`
    `{NormalDistribution[m1, s1], NormalDistribution[m2, s2]}]]`

Out[ ]= MixtureDistribution[{0.639116, 0.360884},
    {NormalDistribution[80.091, 5.86777], NormalDistribution[54.6148, 5.87117]}]

- Now let's plot together both, the original data and the mixture distribution, approximating it:

In[ ]:= `Show[{Histogram[waitingtime, 25, "PDF"],`
    `Plot[PDF[mixDist1, x], {x, 41, 100}, PlotStyle → Red]}]`

Out[ ]=

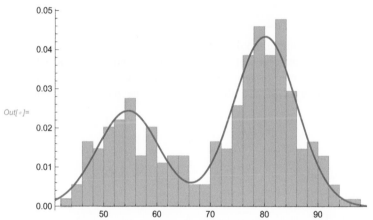

The new distribution can be used for statistical computations such as:

- The probability that the time between two emissions will be 80 minutes or more:

In[ ]:= `Probability[x ≥ 80, x ≈ mixDist1]`

Out[ ]= 0.323516

5.3.5 Empirical and Adjusted Distributions

In this example, based on a case study from a Wolfram virtual seminar, WeatherData is used to investigate the feasibility of using wind to generate electricity in Salamanca (Spain).

- Let's find the name of the weather station near Salamanca first:

In[ ]:= `wsSalamanca = WeatherData[{"Salamanca", "Spain"}, "StationName"]`

Out[ ]= LESA

- We can now get the average daily wind speed, in km/h, for the area beginning in 2000 and eliminate incomplete records. We use DeleteMissing[list, 1, 2] to remove any sublists containing missing data within the first 2 levels. We also use "Path" to extract Time-value pairs:

In[ ]:= `wsData =`
`    WeatherData[wsSalamanca, "MeanWindSpeed", {{2000, 1, 1}, Date[], "Day"}]`

Out[ ]= TimeSeries[⊞ ∿ Time: 01 Jan 2000 to 19 May 2022
                        Data points: 8121]

        Data not in notebook. Store now →

In[ ]:= `wData = DeleteMissing[wsData["Path"], 1, 2];`

- Let's display the first 5 elements of our list:

In[ ]:= `Take[wData, 5]`

Out[ ]= $\begin{pmatrix} 3\,155\,673\,600 & 2.78\,\text{km/h} \\ 3\,155\,760\,000 & 2.78\,\text{km/h} \\ 3\,155\,846\,400 & 2.04\,\text{km/h} \\ 3\,155\,932\,800 & 2.04\,\text{km/h} \\ 3\,156\,019\,200 & 2.22\,\text{km/h} \end{pmatrix}$

- The entries are ordered pairs containing the date and the corresponding average wind speed. The figures assigned to the dates correspond to the number of seconds elapsed since January 1, 1900. The reason behind this way of representing time is that *Mathematica* often uses AbsoluteTime. By default, it gives us the absolute time at the moment the function is executed.

In[ ]:= `AbsoluteTime[]`

Out[ ]= $3.8622060040812979 \times 10^9$

- We can show it in a date format:

In[ ]:= `DateList[AbsoluteTime[]]`

Out[ ]= {2022, 5, 22, 11, 0, 4.0962878}

- We need to extract the wind speeds from the data. We use the Part function to get a list of the daily average speeds, convert them to new units (meters/second), and visualize the result with histograms.

In[ ]:= `wndData = 1000 wData[[All, 2]] / 60^2;`
`    GraphicsRow[{histplotpdf, histplotcdf} = {Histogram[wndData, 45,`
`        "PDF", PlotLabel → Style["Daily average speed", "Title", 12]],`
`      Histogram[wndData, 45, "CDF", PlotLabel → Style[`
`        "Accumulated daily average speed", "Title", 12]]}, ImageSize → Large]`

Out[ ]=

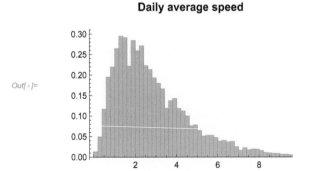
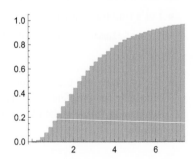

- We can now fit the previous data to an existing distribution. However, we can also create a new one based on the data. This last operation can be done automatically with the function SmoothKernelDistribution.

In[•]:= $\mathcal{D}_e$ **= SmoothKernelDistribution[QuantityMagnitude@wndData]**

Out[•]= DataDistribution[⊞ 〽 Type: SmoothKernel
Data points: 8072]

- While the ability to estimate a distribution is a valuable tool in statistical analysis, one must be careful when actually using it. It is possible for estimated distributions to have very small negative values in the tails. Estimated distributions of this sort behave in a similar manner to interpolating functions and can show unexpected behavior.

In[•]:= **Show[{**
  histplotpdf,
  estplot = Plot[PDF[$\mathcal{D}_e$, x], {x, 0, 13}, PlotStyle → {Thick, Red}]
  }]

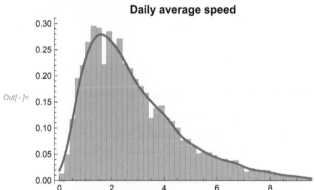

- Assuming that the generator we are using for our analysis requires at least a wind speed of 3.5 m/s to generate electricity, we can use the Probability function to estimate the proportion of time the generator will be contributing to the grid.

In[•]:= **Probability[x > 3.5, x ≈ $\mathcal{D}_e$]**

Out[•]= 0.298209

- An alternative method would be to estimate the distribution parameters and construct the distribution. The graphical representation of the data suggests that the wind speed could be approximated using a LogNormalDistribution. Zeros are eliminated because they are not compatible with such distribution.

In[•]:= **params = FindDistributionParameters[**
  DeleteCases[QuantityMagnitude[wndData], 0], LogNormalDistribution[μ, σ]]

Out[•]= {μ → 0.836653, σ → 0.7039}

In[•]:= $\mathcal{D}_p$ **= LogNormalDistribution[μ, σ] /. params**

Out[•]= LogNormalDistribution[0.836653, 0.7039]

- Visualize our distribution estimate:

```
In[ ]:= GraphicsRow[{Show[{
          histplotpdf,
          estplot = Plot[PDF[𝒟ₚ, x], {x, 0, 14}, PlotStyle → {Thick, Red}]
        }], Show[{
          histplotcdf,
          estplot = Plot[CDF[𝒟ₚ, x], {x, 0, 14}, PlotStyle → {Thick, Red}]}]
     }]
```

Daily average speed **Accumulated daily average speed**

Out[ ]=
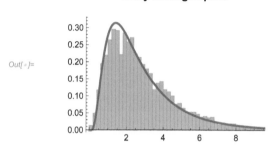   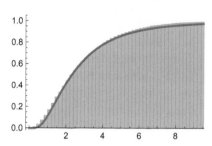

- We perform a test to assess the quality of the fit:

```
In[ ]:= tstData =
          DistributionFitTest[Drop[Sort[wndData], 1], 𝒟ₚ, "HypothesisTestData"];
```

```
In[ ]:= Column[{Style[tstData["AutomaticTest"], "Text"],
          Style[tstData["TestConclusion"], "Text"]}, Frame → True]
```

Out[ ]=
> CramerVonMises
> The null hypothesis that the data is distributed according to the
>   QuantityDistribution[LogNormalDistribution[0.836653, 0.7039], km/h]
> is rejected at the 5 percent level based on the Cramér-von Mises test.

- Get the test statistic and *p*-value:

```
In[ ]:= tstData["TestDataTable"]
```

Out[ ]=

	Statistic	P-Value
Cramér-von Mises	2.48879	1.03184×10^{-6}

- The test tells us that the fit is not good, but when there is a large number of data points, the best possible test is the graphical representation of the data alongside the density function of the estimated distribution. In that case, the fit is quite reasonable. We can recalculate the probability of the wind speed being at least 3.5 m/s and see that the result is very close to the one previously obtained.

```
In[ ]:= Probability[x > 3.5, x ≈ 𝒟ₚ]
```

Out[ ]= 0.27721

5.3.6 Automatic Probability Calculations

In the next couple of examples, we use the following functions related to probability calculations: NProbability, Expectation, and OrderDistribution:

A person is standing on a road counting the number of cars that pass before seeing a black one. Assuming that 10% of the cars are black, find the expected number of cars that will pass before a black one appears.

■ We use the geometric distribution (GeometricDistribution[p]) to model the number of trials until success, given a probability of success of p. The expected number of cars will be the mean of the GeometricDistribution with $p = 0.1$. It can be computed in two different ways:

In[ ]:= `{Mean[GeometricDistribution[0.1]],`
`    Expectation[x, x ≈ GeometricDistribution[0.1]]}`

Out[ ]= {9., 9.}

Find the probability of counting 8 or fewer cars before seeing a black one.

In[ ]:= `NProbability[x ≤ 9, x ≈ GeometricDistribution[0.1]]`

Out[ ]= 0.651322

Three friends are standing on a road in different places counting the number of cars until a black one appears. The winner will be the first one to spot a black car. Assuming that 10% of the cars are black, find the expected number of cars that will pass before one of the friends sees a black car.

■ We use OrderDistribution[{*dist*, *n*}, *k*], where k represents the kth-order statistics distribution for *n* observations from the distribution *dist*.

In[ ]:= `Expectation[x, x ≈ OrderDistribution[{GeometricDistribution[0.1], 3}, 3]]`

Out[ ]= 16.9006

The output indicates the average total number of cars that the friends will count before a black one appears. Since it's an average, the number doesn't have to be an integer.

■ What is the probability that the winner will count less than 10 cars and at least 4?

In[ ]:= `Probability[Conditioned[x < 10, x ≥ 4],`
`    x ≈ OrderDistribution[{GeometricDistribution[0.1], 3}, 3]]`

Out[ ]= 0.245621

There are many more statistical functions available. You may be interested in exploring some of the latest ones, such as: CopulaDistribution, MarginalDistribution, and TransformedDistribution.

5.3.7 Application: An Election Tie

The example in this section refers to a real scenario that happened in Spain during an extraordinary party congress discussing a highly controversial proposal. The party decided to put the proposal to a secret ballot and the participants had to either accept it or reject it. The result was a tie: 1,515 people voted Yes and 1,515 voted No. The outcome was so unexpected that the media speculated whether the result had been rigged, and the blogosphere fueled a debate that included the use of statistical software about how likely a tie was in a voting process involving two options and 3,030 participants. In our analysis, we make the assumption that all the participants vote for either option so there are no null or blank votes even though those kinds of votes would not affect the result.

Next, we answer the following question: What is the probability that a Yes/No vote with 3,030 people results in a tie?

These types of processes with only two possible outcomes are known as Bernoulli processes and can be modeled using a binomial distribution: $X \sim \text{Binomial}(n, p)$

with *n* in this case being the number of voters and *p* the probability of a Yes vote. The answer will depend on our previous knowledge regarding the value of *p*.

- If we don't know anything about p in advance, then the probability can take any value between 0 and 1. This is equivalent to the assumption that p follows a uniform distribution between 0 and 1. For any value of n, the solution is:

```
In[ ]:= sol1[n_] := Probability[x == n/2,
    x ≈ ParameterMixtureDistribution[
        BinomialDistribution[n, p], p ≈ UniformDistribution[{0, 1}]]]
```

- With *n* = 3030, the probability of a tie is:

```
In[ ]:= sol1[3030]
```

$$Out[\ ]= \frac{1}{3031}$$

A more direct way of calculating p is to enumerate all the different possibilities: {{3030 Yes, 0 No}, {3029 Yes, 1 No},...., {1 Yes, 3029 No},{ 0 Yes, 3030 No}}. There are a total of 3,031 possible cases, and since we know nothing about p, we assume that all of them have the same probability of happening. Therefore, p = 1/3,031 for any result and a tie is one of the potential outcomes.

- In the actual case, there was some previous knowledge. There had been earlier votes with outcomes very close to a tie. Let's assume that $\hat{p}$ = 0.49 with a standard deviation of 0.02. Then we can calculate the probability assuming that *p* follows a Normal probability distribution: X ~ Normal(0.49, 0.02):

```
In[ ]:= NIntegrate[PDF[NormalDistribution[0.49, 0.02], p] *
    PDF[BinomialDistribution[3030, p], 1515],
    {p, 0, 1}, WorkingPrecision → 10]
```

```
Out[ ]= 0.005402504702
```

- If we'd like to perform a Bayesian analysis, it's convenient to assume that *p* follows a Beta distribution with parameters α and β, whose actual values depend on our *prior* knowledge of *p*.

```
In[ ]:= sol2[n_, α_, β_] := Probability[x == n/2,
    x ≈ ParameterMixtureDistribution[
        BinomialDistribution[n, p], p ≈ BetaDistribution[α, β]]]
```

- Without previous knowledge about *p*, then α = 1 and β = 1 and the result is the same as the one previously obtained.

```
In[ ]:= sol2[3030, 1, 1]
```

$$Out[\ ]= \frac{1}{3031}$$

- For the secret ballot scenario, we have some prior knowledge since the results of the previous 3 or 4 votes had been very close to a tie. We can therefore assume that *p* is approximately 1/2. This is equivalent to α and β taking very large values. An extreme case:

```
In[ ]:= sol2[3030, 100 000, 100 000] // N
```

```
Out[ ]= 0.0143853
```

- If we knew for sure that *p* was exactly 1/2, this problem would be equivalent to computing the probability that after tossing a fair coin 3,030 times, we would get the same number of heads and tails.

In[ ]:= **PDF[BinomialDistribution[3030, 1/2], 1515] // N**

Out[ ]= 0.0144938

In summary, the probability that a tie could happen was not that small (keeping in mind that earlier votes were close affairs). Highly unlikely events happen all the time. An extreme example occurred shortly after the vote, also in Spain. The same person won the biggest prize in the national lottery for two consecutive weeks. In this game, the probability of winning the highest prize ("El Gordo") is 1/100,000. This means that a priori, the probability that a person purchasing just one ticket each time will win twice in a row is $1/100,000 \times 1/100,000 = 10^{-10}$ quite a small likelihood. Actually, the odds of dying from lightning are higher. It's not unusual that with so much going on around us at any given moment, the small probability events are the ones drawing our attention.

5.4 Exploratory Data Analysis

5.4.1 Customized Graphical Representations

This section will present several types of graphs for displaying multivariate data. For additional information see: guide/StatisticalVisualization.

> To build statistical graphs load the palette: **Palettes ▶ Chart Elements Schemes**.

Many empirical data are represented by more than one random variable; we will refer to them as multivariate data.

- The function below plots the PDF *Student's t* distribution for two variables.

In[ ]:= **Plot3D[PDF[MultivariateTDistribution[{{1, 1/2}, {1/2, 1}}, 20], {x, y}],**
    {x, -3, 3}, {y, -3, 3}]

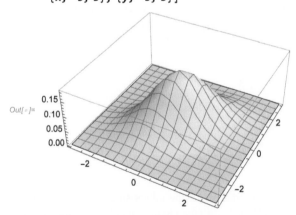

Some statistical studies may contain simulated instead of empirical data. In those cases, random number generators are used. *Mathematica* has the functions RandomReal and RandomInteger to generate random numbers. Since these functions use algorithms, strictly speaking they don't provide truly random data. However, in most situations we can assume that they do.

- The next function generates random numbers from a multivariate Poisson distribution. If you remove the ";" you will notice that the result is a list of sublists, that can be interpreted as a matrix with three columns:

In[ ]:= **multidata =**

    RandomVariate[MultivariatePoissonDistribution[1, {2, 3, 4}], 125];

- We represent each sublist with a point. We include a red point to show the mean.

In[ ]:= **Show[{ListPointPlot3D[multidata],**

    Graphics3D[{Red, PointSize[Large], Point[Mean[multidata]]}]}]

Out[ ]=
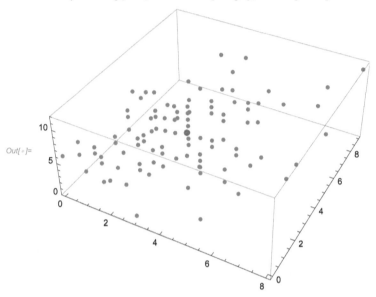

- The next graph shows a histogram of the first two dimensions of the random data previously generated (**multidata**) along with a histogram of 5,000 pairs of normally distributed random numbers. (Remember that to extract list elements we use: Part or [[...]]).

In[ ]:= **GraphicsRow[{Histogram3D[multidata⟦All, 1 ;; 2⟧, 5], Histogram3D[**

    RandomReal[NormalDistribution[], {5000, 2}]]}, ImageSize → Medium]

Out[ ]=
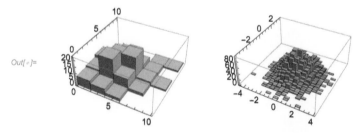

- An alternative way of representing the same data with the edges smoothed out is by using SmoothHistogram3D and SmoothDensityHistogram (the change in color is associated to the actual data value).

In[ ]:= `Row[{SmoothHistogram3D[multidata[[All, 1 ;; 2]], ImageSize → Medium],`
        `SmoothDensityHistogram[multidata[[All, 1 ;; 2]], ImageSize → Medium]}]`

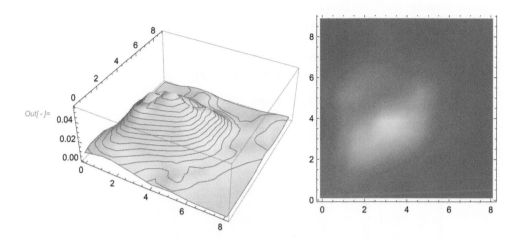

Out[ ]=

5.4.2 Life Expectancy

▪ Here we obtain the life expectancy in Spain, US, France, and Mexico using CountryData. The result is displayed in which each country is represented by its flag. Use Outer to get several properties for a set of entities.

In[ ]:= `Outer[f, {a, b}, {x, y, z}]`

Out[ ]= $\begin{pmatrix} f(a, x) & f(a, y) & f(a, z) \\ f(b, x) & f(b, y) & f(b, z) \end{pmatrix}$

In[ ]:= `data = Outer[CountryData, {"Spain", "UnitedStates", "France", "Mexico"},`
        `{"LifeExpectancy", "FemaleLifeExpectancy", "MaleLifeExpectancy"}];`

In[ ]:= `flags = {`
        `ImageResize[CountryData["Spain", "Flag"], 50],`
        `ImageResize[CountryData["UnitedStates", "Flag"], 50],`
        `ImageResize[CountryData["France", "Flag"], 50],`
        `ImageResize[CountryData["Mexico", "Flag"], 50]};`

```
In[ ]:= BarChart3D[data,
    ChartElementFunction → "Cylinder", ChartStyle → "Pastel",
    ChartLegends → {"Life Expectancy(\!\(\*Cell[\"♀ + ♂\"]\))",
       "Female (♀)", "Male (♂) "},
    ChartLabels → {Placed[flags, {.5, 0, 1.15}],
       Placed[{"\!\(\*Cell[\"♀+♂\"]\)", "♀", "♂"}, Center]},
    BarSpacing → {.1, .6}, PlotLabel → Style["Life expectancy",
       16, FontFamily → "Times"], ImageSize → Large]
```

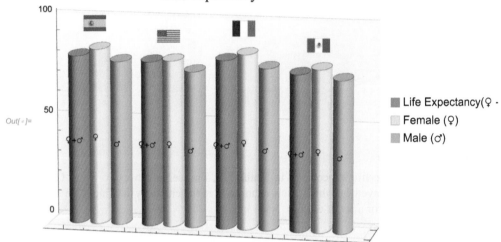

- With CountryData we get not only the Life Expectancy but also the Gross Domestic Product (GDP) per Capita and Population for all the countries in South America:

```
In[ ]:= data1 = Table[Tooltip[{{ (CountryData[c, #1] &) /@
          {"LifeExpectancy", "GDPPerCapita", "Population"}},
          Column[{CountryData[c, "Name"], CountryData[c, "Population"]}]]],
          {c, CountryData["SouthAmerica"]}];
```

- The countries that don't have available information are deleted:

```
In[ ]:= data2 = DeleteMissing[data1, 1, 3]
```

```
Out[ ]= {( 78.07 yr  $8441.92 per person per year   45 195 777 people ),
         ( 70.7 yr   $3143.05 per person per year   11 673 029 people ),
         ( 74.98 yr  $6796.84 per person per year   212 559 409 people ),
         ( 79.57 yr  $13 231.7 per person per year  19 116 209 people ),
         ( 76.91 yr  $5332.77 per person per year   50 882 884 people ),
         ( 77.76 yr  $5600.39 per person per year   17 643 060 people ),
         ( 77.9 yr   $35 400. per person per year   3483 people ),
         ( 77.121 yr $8300. per person per year     298 682 people ),
         ( 71.59 yr  $6955.94 per person per year   786 559 people ),
         ( 78.13 yr  $4949.75 per person per year   7 132 530 people ),
         ( 74.96 yr  $6126.87 per person per year   32 971 846 people ),
         ( 73.57 yr  $6491.14 per person per year   586 634 people ),
         ( 78.19 yr  $15 438.4 per person per year  3 473 727 people ),
         ( 72.22 yr  $16 055.6 per person per year  28 435 943 people )}
```

- The result is displayed as a bubble chart. The diameters of the bubbles are proportional to the population size. You can see the country name and its corresponding population by putting the mouse over each bubble:

In[ ]:= `BubbleChart[data2, FrameLabel → {"Life expectancy", "GDP/capita"}]`

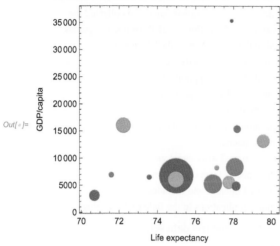

Out[ ]=

5.4.3 Fisher's Iris Data

Ronald Aylmer Fisher (1890 –1962) was one of the leading scientists of the 20th century making major contributions to fields such as statistics, evolutionary biology, and genetics. The next example is a multivariate analysis case study that analyses four characteristics of three iris species. The data come from one of Fisher's classic articles, and they are available in the Wolfram Data Repository.

In[ ]:= `fisheriris = ResourceObject["Sample Data: Fisher's Irises"]`

Out[ ]= ResourceObject []

- In 1936 he introduced the Iris flower data set as an example of discriminant analysis. Please read carefully the following sentences describing the experiment:

In[ ]:= `TextSentences[WikipediaData["Iris_flower_data_set"]][[ ;; 6]]`

Out[ ]= {The Iris flower data set or Fisher's Iris data set is a multivariate data set introduced by the British
statistician and biologist Ronald Fisher in his 1936 paper The use of multiple
measurements in taxonomic problems as an example of linear discriminant analysis.,
It is sometimes called Anderson's Iris data set because Edgar Anderson collected the
data to quantify the morphologic variation of Iris flowers of three related species.,
Two of the three species were collected in the Gaspé Peninsula "all from the
same pasture, and picked on the same day and measured at
the same time by the same person with the same apparatus".,
The data set consists of 50 samples from each of three species of Iris
(Iris setosa, Iris virginica and Iris versicolor).,
Four features were measured from each sample: the length and the
width of the sepals and petals, in centimeters.,
Based on the combination of these four features, Fisher developed a linear
discriminant model to distinguish the species from each other.,
Fisher's paper was published in the Annals of Eugenics and includes
discussion of the contained techniques' applications to the field of Phrenology.}

- We use ResourceData to find out the source of the Iris data set:

In[ ]:= `ResourceData[fisheriris, "Source"]`

Out[ ]= Fisher,R.A. "The use of multiple measurements in taxonomic problems" Annual Eugenics, 7, Part II,
179–188 (1936); also in "Contributions to Mathematical Statistics" (John Wiley, NY, 1950).

This paper is freely available online in the following site:
https://onlinelibrary.wiley.com/doi/epdf/10.1111/j.1469-1809.1936.tb02137.x

- Here we show the three species of Iris. (At the time of this writing, the images were located in positions 3 to 5. However, as Wikipedia is subject to frequent changes, if the function below doesn't return the correct images modify the numbers accordingly.)

In[ ]:= `WikipediaData["Iris_flower_data_set", "ImageList"][[3 ;; 5]]`

In this case, we are going to do a simple exploratory data analysis (EDA). Later on in the book, we will use the same data to show how to perform some machine learning tasks.

- Here we can see the elements in the file:

In[ ]:= `fisheriris["ContentElements"]`

Out[ ]= {ColumnDescriptions, ColumnTypes, Content, DataType,
Dimensions, ObservationCount, RawData, Source, TrainingData, TestData}

- The function below retrieves the description of the columns:

In[ ]:= `ResourceData[fisheriris, "ColumnDescriptions"]`

Out[ ]= {Sepal length in cm., Sepal width in cm., Petal length in cm., Petal width in cm., Species of iris}

- Now, we display the data of the 50 samples from each of three species of Iris (Iris setosa, Iris virginica, and Iris versicolor):

In[ ]:= `FisherIrisData = ResourceData["Sample Data: Fisher's Irises"]`

Species	SepalLength	SepalWidth	PetalLength	PetalWidth
setosa	5.1 cm	3.5 cm	1.4 cm	0.2 cm
setosa	4.9 cm	3. cm	1.4 cm	0.2 cm
setosa	4.7 cm	3.2 cm	1.3 cm	0.2 cm
setosa	4.6 cm	3.1 cm	1.5 cm	0.2 cm
setosa	5. cm	3.6 cm	1.4 cm	0.2 cm

To select the data only for the setosa species we proceed as follows:

In[ ]:= `setosadata = FisherIrisData[Select[#"Species" == "setosa" &]]`

Species	SepalLength	SepalWidth	PetalLength	PetalWidth
setosa	5.1 cm	3.5 cm	1.4 cm	0.2 cm
setosa	4.9 cm	3. cm	1.4 cm	0.2 cm
setosa	4.7 cm	3.2 cm	1.3 cm	0.2 cm
setosa	4.6 cm	3.1 cm	1.5 cm	0.2 cm
setosa	5. cm	3.6 cm	1.4 cm	0.2 cm
setosa	5.4 cm	3.9 cm	1.7 cm	0.4 cm
setosa	4.6 cm	3.4 cm	1.4 cm	0.3 cm

- The following table contains information about some of the basic statistics used in EDA applied to the setosa dataset.

In[ ]:= `sepalpetal = {"SepalLength", "SepalWidth", "PetalLength", "PetalWidth"};`

In[ ]:= `TableForm[`
`    Map[{setosadata[Max, #], setosadata[Min, #], setosadata[Mean, #],`
`        setosadata[Median, #], setosadata[InterquartileRange, #],`
`        setosadata[StandardDeviation, #] } &, sepalpetal],`
`    TableHeadings → {sepalpetal, {"Max", "Min", "Mean", "Median",`
`        "Interquartile Range", "StandardDeviation"}}]`

	Max	Min	Mean	Median	Interquartile Range	StandardDeviation
SepalLength	5.8 cm	4.3 cm	5.006 cm	5. cm	0.4 cm	0.35249 cm
SepalWidth	4.4 cm	2.3 cm	3.428 cm	3.4 cm	0.5 cm	0.379064 cm
PetalLength	1.9 cm	1. cm	1.462 cm	1.5 cm	0.2 cm	0.173664 cm
PetalWidth	0.6 cm	0.1 cm	0.246 cm	0.2 cm	0.1 cm	0.105386 cm

- The same statistical information for all the species based on their sepal lengths:

In[ ]:= `FisherIrisData[GroupBy["Species"],`
`    {Max[#] , Min[#], Mean[#], Median[#], InterquartileRange[#],`
`        StandardDeviation[#]} &, "SepalLength"]`

setosa	5.8 cm
	4.3 cm
	5.006 cm
	5. cm
	0.4 cm
	0.35249 cm
versicolor	7. cm
	4.9 cm
	5.936 cm
	5.9 cm
	0.7 cm
	0.516171 cm
virginica	7.9 cm
	4.9 cm
	6.588 cm
	6.5 cm
	0.7 cm
	0.63588 cm

▪ Here we compare the sepal length histograms of the three species:

```
In[·]:= FisherIrisData[GroupBy["Species"],
          Histogram[#, {3, 8, 0.5}] &, "SepalLength"]
```

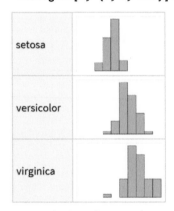

▪ Next, we calculate the covariances and correlations to explore the relationships between the variables.

```
In[·]:= sepalpetaldata = Transpose[Normal[FisherIrisData[All, #] & /@ sepalpetal]];
```

```
In[·]:= Through[{Covariance, Correlation}[sepalpetaldata]]
```

$$Out[\cdot]= \{\{\{0.685694 \text{ cm}^2, -0.042434 \text{ cm}^2, 1.27432 \text{ cm}^2, 0.516271 \text{ cm}^2\},$$
$$\{-0.042434 \text{ cm}^2, 0.189979 \text{ cm}^2, -0.329656 \text{ cm}^2, -0.121639 \text{ cm}^2\},$$
$$\{1.27432 \text{ cm}^2, -0.329656 \text{ cm}^2, 3.11628 \text{ cm}^2, 1.29561 \text{ cm}^2\},$$
$$\{0.516271 \text{ cm}^2, -0.121639 \text{ cm}^2, 1.29561 \text{ cm}^2, 0.581006 \text{ cm}^2\}\},$$
$$\{\{1., -0.11757, 0.871754, 0.817941\}, \{-0.11757, 1., -0.42844, -0.366126\},$$
$$\{0.871754, -0.42844, 1., 0.962865\}, \{0.817941, -0.366126, 0.962865, 1.\}\}\}$$

The results indicate that sepal length is strongly correlated with both, petal length and petal width. They also show that petal length is strongly associated to petal width. However, we can see that the linear association between sepal length and sepal width is quite weak.

5.5 Bootstrapping and Confidence Estimates

The following data are the total daily precipitation amounts in cm during 2020 from the weather station LESA:

```
In[ ]:= precipitationData = DeleteMissing[
          Transpose[Normal[WeatherData["LESA", "TotalPrecipitation",
              {{2020, 1, 1}, {2020, 12, 31}, "Day"}]]][[2]]];
```

- We can assume that these data represent the entire population. The average precipitation, in cm/day, during that year in that meteorological station is:

```
In[ ]:= Mean[precipitationData]
```

```
Out[ ]= 0.0947443
```

- Let's supposed we don't know all the data for that year and instead we only have precipitation information for just 30 random days.
  We can use SeedRandom to get repeatable random values:

```
In[ ]:= sample1 = (SeedRandom[2514]; RandomChoice[precipitationData, 30])
```

```
Out[ ]= {0.2, 0., 0., 0., 0.02, 0., 0., 0., 0., 0., 0., 0., 0.05, 0., 0.,
         0.28, 0., 0.02, 0., 0.43, 0.02, 0., 0.28, 0., 0., 0., 0., 1.45, 0., 0.05}
```

- We want to estimate the daily mean precipitation during that year along with its standard deviation based on the sample:

$$In[ ]:= \{ms1, ss1\} = \left\{Mean[sample1], \frac{StandardDeviation[sample1]}{\sqrt{Length[sample1]}}\right\}$$

```
Out[ ]= {0.0933333, 0.0506176}
```

- A 68% confidence interval for the population mean is:

```
In[ ]:= {ms1 - ss1, ms1 + ss1}
```

```
Out[ ]= {0.0427157, 0.143951}
```

- We can also get directly a confidence interval around the mean as follows:

```
In[ ]:= MeanAround[sample1]
```

```
Out[ ]= 0.09 ± 0.05
```

- This confidence interval indicates that 68.27% of the time the true population mean will be within its bounds. We can make such a statement because according to the central limit theorem, the mean statistic of the sample follows a normal distribution regardless of the underlying population distribution.

```
In[ ]:= Probability[μ - σ < x < μ + σ, x ≈ NormalDistribution[μ, σ]] // N
```

```
Out[ ]= 0.682689
```

- If instead of a single weather station we have precipitation information from several stations located in the area, even though we may not have the corresponding dates for such data, we can still estimate the mean daily precipitation for the year. We use again SeedRandom to make the example reproducible:

```
In[ ]:= samples = { (SeedRandom[1234]; RandomChoice[precipitationData, 34]),
          (SeedRandom[7341]; RandomChoice[precipitationData, 27]),
          (SeedRandom[3211]; RandomChoice[precipitationData, 27]),
          (SeedRandom[9247]; RandomChoice[precipitationData, 25]),
          (SeedRandom[3782]; RandomChoice[precipitationData, 34])};
```

- We group the data to simulate a bigger single sample and find out that the mean confidence interval gets smaller:

In[ ]:= **MeanAround[Flatten[samples]]**

Out[ ]= 0.077 ± 0.026

- We can obtain a very similar result if we calculate the mean of the sample means:

In[ ]:= **MeanAround[Mean[#] & /@ samples]**

Out[ ]= 0.077 ± 0.027

In a real-world situation, we may only have a single sample from a population. In this case, it's very likely that we will still be able to improve our confidence interval estimate for the population mean by using bootstrapping. This statistical technique is similar to the one used in the example above where we had several samples but with a fundamental difference: the successive samples are generated from the original one and not from the actual population. Using the original sample, we generate first a new one using sampling with replacement so that some of the data points may not be sampled and others may be sampled more than once. We calculate its mean and use it to approximate the mean of the population. Since this value would not be very precise, we repeat the previous two steps many times, in the process obtaining a new mean estimate each time. We can then construct a distribution with all those values, a bootstrap distribution, representing a proxy for the actual distribution of the population mean. Finally, using the bootstrap distribution we can calculate its mean (a point estimate) and confidence intervals around it the same way as before.

- Now we put our newly acquired knowledge about bootstrapping into practice. The function below creates a collection of 100 datasets from the data using sampling with replacement. In theory, we could have created our own bootstrapping function, but since that would have been time-consuming, we decided to see if someone had already created it by searching the Wolfram Function Repository. In this case, using the keyword "bootstrap", we got a link to the "BootstrapStatistics" function page:

  https://resources.wolframcloud.com/FunctionRepository/resources/BootstrapStatistics/

  That page also contains information about the syntax of the function, and a link to the source notebook where we can see the actual *Mathematica* code. Remember that to use the functions available in the Wolfram Function Repository you need an Internet connection since those functions are not part of *Mathematica*.

In[ ]:= **MeanAround[**
    Flatten[ResourceFunction["BootstrapStatistics"][sample1, 100, Mean]]]

Out[ ]= 0.099 ± 0.004

  The accuracy of this technique depends on how representative of the population the original sample is. The less representative it is, the less accurate the computed approximation will be. For example, if our original sample had contained an unusually high number of rainy days, we would have obtained an inaccurate result.

- Let's see another example. This one is related to the calculation of the probability of catastrophic flooding by analyzing a river flow. This type of analyses often use data related to the maximum flows registered over a certain period of time. As in previous examples, we are going to use real data. In this case, the maximum instantaneous annual flows in the Tormes river in Salamanca (Spain) from 1965 — 2018 (the data is not available for all the years):

```
In[*]:= flowTormes = { 68.2, 604.8, 139.4, 95.7, 327., 357.7, 134.3, 99.6, 300.2,
       208.7, 48.6, 43.1, 128.5, 553.9, 45.4, 119.9, 63.5, 243.8,
       274.8, 117., 134.4, 179.1, 29.6, 168.9, 23.3, 13.7, 510.2,
       48.5, 420.6, 330.9, 576.9, 22.3, 73., 750.2, 33.6, 471.5, 138.2,
       53.4, 67.6, 250.9, 216.7, 242.5, 188.8, 66.2, 43.1, 322.4};
```

- Using bootstrapping, we can get a larger set to give us a better idea of how much this measure varies. We also calculate its 95% confidence interval:

```
In[*]:= flowTormes1 =
       ResourceFunction["BootstrapStatistics"][flowTormes , 10000, Mean];
       Histogram[flowTormes1]
```

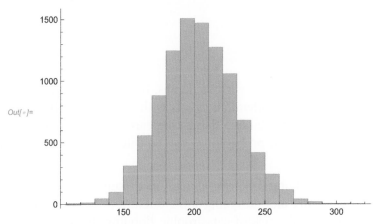

```
In[*]:= Quantile[flowTormes1, {0.025, 0.975}]
```

Out[*]= {153.887, 256.615}

- The function to compute the maximum and minimum flows during that period is shown below:

```
In[*]:= Through[{ Min, Max}[flowTormes1]]
```

Out[*]= {112.543, 312.302}

- Next, we approximate the data using a smooth kernel distribution. In statistics, kernel density estimation (KDE) is a non-parametric way to estimate the probability density function of a random variable. KDE is a fundamental data-smoothing problem where inferences about the population are made, based on a finite data sample.

```
In[*]:= 𝒟 = SmoothKernelDistribution[flowTormes ];
```

```
In[*]:= bSamp = ResourceFunction["BootstrapStatistics"][flowTormes, 250];
```

- We now smooth over each bootstrapped sample and obtain the confidence estimates:

```
In[*]:= 𝒟ₐ = SmoothKernelDistribution[#] & /@ bSamp;
```

- Next, we calculate the confidence band from the bootstrapped examples (we choose a range close to the minimum and maximum flows of the river):

```
In[*]:= rng = Range[10., 1000];
       pdf = Table[PDF[i, rng], {i, 𝒟ₐ}];
```

```
In[*]:= High = Table[Quantile[i, .975], {i, Transpose[pdf]}];
       Low = Table[Quantile[i, .025], {i, Transpose[pdf]}];
```

- Finally, we visualize the estimate of the PDF with 95% confidence bands:

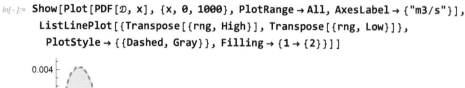

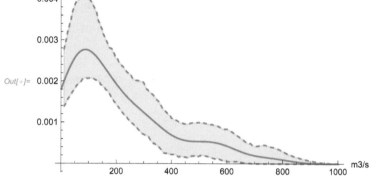

When designing big engineering projects such as damns, bridges, or nuclear power stations, is necessary to take into consideration events with very low probability of happening but high negative impact. Massive floods, big earthquakes, and tsunamis are instances of such events, and their probabilities can be calculated using extreme value theory.

Going back to our example, studying Tormes' annual maximum flows since 1965 would very useful when planning and designing infrastructure (e.g., road location) along the river path to minimize the risk of catastrophic flooding. In cases like this one, the most frequently used distributions are generalized extreme value distributions such as the Gumbel distribution. [Gumbel, E. J. (1958). *Statistics of Extremes*. Irvington, New York: Columbia University Press.]

Before showing how to analyze the Tormes' flow using the Gumbel distribution, there are a couple of concepts from extreme value theory that we need to explain first: exceedance and return period.

The concept of exceedance refers to the number of times in N future observations that we will observe a more extreme observation m than a certain threshold ξ (determined from n historical observations). That number can be modelled as a continuous random variate following a distribution $w(n, m, N)$.

Regarding the second concept, one of the most important objectives of frequency analysis is to calculate the recurrence interval of events, a.k.a the return period. If the variable (x), equal to or greater than an event of magnitude x_T, occurs once in T years, then the probability of occurrence $P(X \geq x)$ in a given year of the variable is:

$$P(x \geq x_T) = \frac{1}{T}$$

$$T = \frac{1}{1 - P(x \leq x_T)}$$

Based on the previous two equations, we can now defined the return period as the average length of time between events of the same magnitude or greater. This can also be interpreted as the number of observations such that, on average, there is one observation equalling or exceeding x. For example, we may be interested in estimating the maximum annual rainfall that can be expected to occur every 100 years.

Now that we have defined the concepts of exceedance and return period, we can see how it can be applied in our example, example that has actually been possible thanks to Jim Baldwin's answer in https://community.wolfram.com/groups/-/m/t/2187431

The calculation of the mean. the standard deviation and other properties of a Gumbel distribution: Gumbel, E. J. (1954). *Statistical Theory of Extreme Values and Some Practical Applications. A Series of Lectures*. Bureau of Standards, Applied Mathematics Division.

To calculate the maximum expected flow $x(T)$ for a return period T we use the following equation:

$$x(T) = \hat{x} - \frac{\text{LogLog}\left(\frac{T}{T-1}\right) - y_n}{\sigma_n} \sigma_X \tag{5.2}$$

where: $\hat{x}$ = mean and σ_X = standard deviation of the variate X, in our case the river flow series.

- That is:

```
In[ ]:= {xmean, sd, size} =
        {Mean[#], StandardDeviation[#], Length[#]} & [flowTormes]

Out[ ]= {203.274, 181.765, 46}
```

- The values for y_n and σ_n are estimated as follows:

```
In[ ]:= Y[n_] := Mean[Table[-Log[-Log[m/ (n + 1)]], {m, 1, n}]]

In[ ]:= S[n_] := StandardDeviation[
        Table[-Log[-Log[m/ (n + 1)]], {m, 1, n}]] Sqrt[(n - 1)/n]
```

- The final step is the application of (5.1) to calculate the expected maximum flows for different return periods:

```
In[ ]:= TableForm[Table[{T, xmean - (Log[Log[T/T-1]] - Y[size]) sd/S[size]},

        {T, 100, 1000, 100}], TableHeadings →

        {None, {"Return Period\n(years)", "Maximun Flow\n(m3/s)"}}] // N
```

```
Out[ ]//TableForm=
```

Return Period (years)	Maximun Flow (m3/s)
100.	1014.15
200.	1123.75
300.	1187.76
400.	1233.15
500.	1268.34
600.	1297.09
700.	1321.4
800.	1342.45
900.	1361.02
1000.	1377.62

5.6 Curve Fitting

When performing experimental studies, it's common to fit the results to a model. *Mathematica* has several functions for doing that: FindFit, FindFormula, Fit, FittedModel, GeneralizedLinearModelFit, LinearModelFit, NonlinearModelFit, and TimeSeriesModelFit. For more details, please see:

https://reference.wolfram.com/language/guide/StatisticalModelAnalysis.html.

In[ ]:= `ClearAll["Global`*"] (*Clear all symbols as usual *)`

In this section, we are going to see some of those functions through basic examples. We use two files containing the data we are going to fit: noisydata.xls and pulsar1257.dat. These two files must be in a subdirectory, named "Data", inside the directory where the notebook is located.

- Remember that in the first section of this chapter, we discussed how to create such a folder and make it the working directory. In case you haven't done so yet, just execute the following function after creating the "Data" subdirectory in your computer:

In[ ]:= `SetDirectory[FileNameJoin[{NotebookDirectory[], "Data"}]]`

Out[ ]= F:\Mi unidad\Mathematica\MBN13\Data

5.6.1 A Linear Regression Model

We are going to fit the data in noisydata.xls to a third-degree polynomial using LinearModelFit, the *Mathematica* function most commonly used when fitting linear regression models and nonlinear ones such as polynomials. This function will create a linear model of the form $\hat{y} = \beta_0 + \beta_1 f_1 + \beta_2 f_2 + \ldots$ (even though the model is linear, the f_i don't need to be linear functions; they can be, for example, polynomials.)

The assumptions required for this type of fitting to be valid are that the original data y_i must be independent and normally distributed, with mean $\hat{y}_i$ and a constant standard deviation s.

- We import the data from the *xls* file.

In[ ]:= `data = Import["noisydata.xls", {"Data", 1}];`
`dataplot = ListPlot[data]`

- We fit a third-degree polynomial.

In[ ]:= `fit = LinearModelFit[data, {1, x, x², x³}, x];`

▪ This is the ANOVA table:

In[ ]:= **fit["ANOVATable"]**

	DF	SS	MS	F-Statistic	P-Value
x	1	165 320.	165 320.	1626.9	3.47061×10^{-32}
x^2	1	58 481.7	58 481.7	575.515	3.78263×10^{-24}
x^3	1	114 956.	114 956.	1131.27	2.41932×10^{-29}
Error	37	3759.8	101.616		
Total	40	342 517.			

Out[ ]=

▪ Now we graphically show both the data and the fitted function.

In[ ]:= **Show[dataplot,**
    Plot[fit[x], {x, -5, 15}, PlotStyle → Red]]

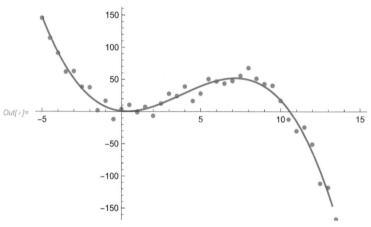

Out[ ]=

5.6.2 A Nonlinear Regression Model

In this example, we fit the experimental light measurement data collected over a three-year period from pulsar PSR1257+12 (originally obtained by Alex Wolszczan and available in the file pulsar1257.dat) to the following function:

$f(t) = \epsilon + \beta \cos(t\,\theta) + \delta \cos(t\,\phi) + \alpha \sin(t\,\theta) + \gamma \sin(t\,\phi).$

For nonlinear fit we use NonlinearModelFit.

▪ We import the file:

In[ ]:= **data = Import["pulsar1257.dat"];**

▪ Represent it graphically:

In[ ]:= **dataplot = ListPlot$\left[\text{data, AspectRatio} \to \frac{1}{4}, \text{PlotStyle} \to \text{Red}\right]$**

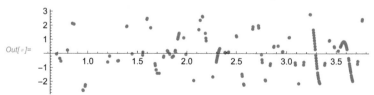

Out[ ]=

▪ We fit the function $f(t)$ previously mentioned:

In[ ]:= **pulsarfit =**
    NonlinearModelFit[data, α Sin[θ t] + β Cos[θ t] + γ Sin[φ t] + δ Cos[φ t] + ϵ,
    {{α, 1}, {β, 1}, {γ, 1}, {δ, 0}, {θ, 23.31}, {φ, 34.64}, {ϵ, 0}}, t]

Out[ ]:= FittedModel[1.33261 cos(23.3869 t) + 0.209533 cos(34.5111 t) + ≪19≫ sin(≪18≫ t) − 1.29803 sin(34.5111 t) + 0.0769581]

- We put the data and the function on the same graph to check the quality of the fit:

In[ ]:= **Show[Plot[pulsarfit[t], {t, 0.68, 3.76}],**
    dataplot]

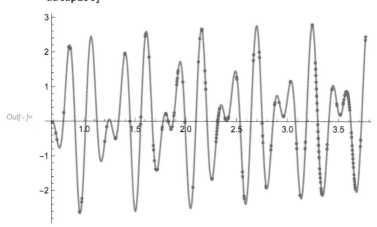

- Once the model has been created, we can perform all kinds of statistical analyses. However, it would be useful to list all the properties available for the model:

In[ ]:= **pulsarfit["Properties"] // Shallow**

Out[ ]//Shallow=
{AdjustedRSquared, AIC, AICc, ANOVATable,
    ANOVATableDegreesOfFreedom, ANOVATableEntries, ANOVATableMeanSquares,
    ANOVATableSumsOfSquares, BestFit, BestFitParameters, ≪40≫}

- For example, here we display the residuals:

In[ ]:= **ListPlot[pulsarfit["FitResiduals"], GridLines → {None, Automatic}]**

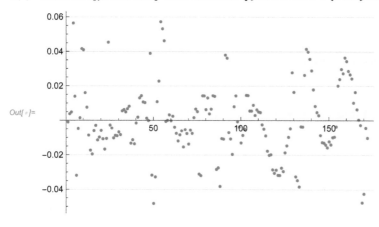

5.6.3 Analysis of the Evolution of Two Cell Populations

In this subsection, we are going to analyze the evolution over time of two cell populations {time (in hours), density (thousands of cells per square millimeter) of cells type 1, and type 2}

```
ClearAll["Global`*"]

cells = {{"time", "Population 1", "Population 2"},
    {0.5`, 0.7`, 0.1`}, {1.`, 1.3`, 0.2`}, {1.5`, 1.9`, 0.3`},
    {2.`, 2.3`, 0.4`}, {2.5`, 2.7`, 0.5`}, {3.`, 3.`, 0.6`},
    {3.5`, 3.3`, 0.7`}, {4.`, 3.6`, 0.7`}, {4.5`, 3.8`, 0.8`},
    {5.`, 4.`, 0.8`}, {5.5`, 4.2`, 0.9`}, {6.`, 4.3`, 1.`},
    {6.5`, 4.4`, 1.`}, {7.`, 4.5`, 1.1`}, {7.5`, 4.6`, 1.1`},
    {8.`, 4.6`, 1.2`}, {8.5`, 4.7`, 1.2`}, {9.`, 4.8`, 1.3`}};
```

- We remove the headings and reorganize the data to get two separate lists representing the evolution of the two cell populations.

```
In[ ]:= data = Drop[cells, 1];
```

```
In[ ]:= {ti, p1, p2} = Transpose[data];
```

```
In[ ]:= {population1, population2} = {Transpose[{ti, p1}], Transpose[{ti, p2}]}
```

$$Out[ ]= \begin{pmatrix} \{0.5, 0.7\} & \{1., 1.3\} & \{1.5, 1.9\} & \{2., 2.3\} & \{2.5, 2.7\} & \{3., 3.\} & \{3.5, 3.3\} & \{4., 3.6\} & \{4.5, 3.8\} & \{5., 4.\} \\ \{0.5, 0.1\} & \{1., 0.2\} & \{1.5, 0.3\} & \{2., 0.4\} & \{2.5, 0.5\} & \{3., 0.6\} & \{3.5, 0.7\} & \{4., 0.7\} & \{4.5, 0.8\} & \{5., 0.8\} \end{pmatrix}$$

- We plot the evolution of both populations.

```
In[ ]:= ListPlot[{population1, population2}, Joined → True,
        PlotLabel → Style["Cell Populations Evolution", Bold]]
```

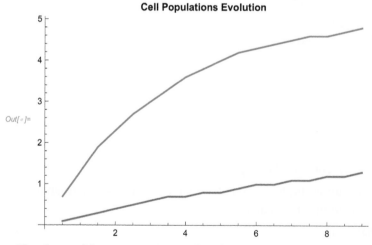

$Out[ ]=$

- The shape of the curves suggests that they could be approximated by functions of the form: $a + b \, \mathrm{Exp}[-k \, t]$. We use the function FindFit to find the best-fit parameters for the fitting functions.

```
In[ ]:= fitpopulation1 = FindFit[population1, a1 + b1 Exp[-k1*t], {a1, b1, k1}, t]
```

$Out[ ]= \{a1 \to 5.11361, b1 \to -5.14107, k1 \to 0.303326\}$

```
In[ ]:= fitpopulation2 = FindFit[population2, a2 + b2 Exp[-k2*t], {a2, b2, k2}, t]
```

```
Out[ ]= {a2 → 1.9681, b2 → -1.9703, k2 → 0.114289}
```

The next example, showing how to customize a graph, includes options that we haven't covered yet (we will explain them later in the book). As usual, if any of the instructions are not clear, consult *Mathematica* help by selecting the function and pressing **F1**. You can use this code as a template for similar cases. Alternatively, you might find it easier to create the labels and plot legends with the drawing tools (⌃CTRL+D).

- We display the fitted curves along with the original data after having replaced the symbolic parameters in the curves with the values obtained from FindFit. We also customize the size of the points representing the experimental data.

```
In[ ]:= fitsol1 = a1 + b1 Exp[-k1*t] /. fitpopulation1;
     fitsol2 = a2 + b2 Exp[-k2*t] /. fitpopulation2;
     Plot[{fitsol1, fitsol2}, {t, 0, 10}, AxesLabel → {"t(hours)",
       "Concentration"}, PlotLegends → Placed[{"Population 1", "Population 2"},
       Center], Epilog → {{Hue[0.3], PointSize[0.02], Map[Point,
         population1 ] }, {Hue[0], PointSize[0.02], Map[Point, population2]}},
      PlotLabel → Style[Framed["Evolution of 2 cell populations"],
        12, Blue, Background → Lighter[Cyan]]]
```

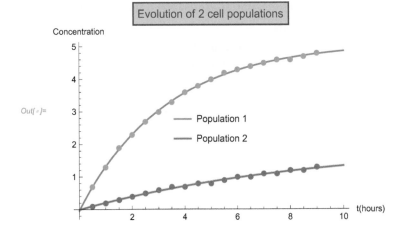

5.6.4 Global Primary Energy Consumption by Source

In the next example, we are going to use what we have learned so far to analyze the evolution of energy consumption globally.

```
In[ ]:= ClearAll["Global`*"]
```

- The following function will take you to a website that contains the file with historical production data for different types of energy by country and region.

```
In[ ]:= SystemOpen[
       "http://www.bp.com/en/global/corporate/energy-economics/statistical-
         review-of-world-energy.html"]
```

Download it by clicking on "Statistical Review of World Energy - all data, 1965-2021" and save it without changing its name in the subdirectory **Data**. The file that we are going to use corresponds to the 2016 (the last one available when this was written) report containing data until the end 2021 (Report 2022).

We are going to conduct several studies using this *XLSX* file. In principle, we could modify directly the data using Microsoft® Excel and create a new workbook with only the required information for our analyses. This method would probably be the most suitable one, but in this case, for educational purposes, we will perform all the required transformations in *Mathematica*. This will help us to have a better understanding of the file manipulation capabilities of the program, although it will require that we pay careful attention to the *Mathematica* syntax.

The file contains many sheets. Before importing it, open the spreadsheet in Excel or any other similar program and examine its contents. Note that each worksheet has a label describing its contents. This is normally very useful.

> Before importing an Excel file containing several worksheets, it's better to rename each sheet using a descriptive label replacing the ones used by default: Sheet1, Sheet2, ...

With the help of the following functions, we are going to import part of the file:

- We can see the labels for the worksheets by using the argument "Sheets".

```
In[*]:= Import[
         "https://www.bp.com/content/dam/bp/business-sites/en/global/corporate/
            xlsx/energy-economics/statistical-review/bp-stats-review-2022-all-
            data.xlsx", {"Sheets"} ] // Short
```

```
Out[*]//Short=
         {Contents, Primary Energy Consumption,
          Primary Energy - Cons by fuel, ≪78≫, Definitions, Methodology}
```

We analyze the global consumption of primary energy (including all types: electricity, transportation, etc.) in 2021, from different sources (oil, gas, hydropower, nuclear, etc.)

- From the sheet labels we conclude that the information we want is in the sheet "Primary Energy - Cons by fuel".

```
In[*]:= primaryenergyDatabyFuel = Import[
            "https://www.bp.com/content/dam/bp/business-sites/en/global/corporate/
               xlsx/energy-economics/statistical-review/bp-stats-review-2022-all
               -data.xlsx",
            {"Sheets", "Primary Energy - Cons by fuel"}];
```

- We use TableView to see the imported data. You can either increase the number of visible rows with ImageSize or explore the data like in a spreadsheet, using the slide bars on the right and at the bottom to move around the sheet and clicking on specific cells.

```
In[*]:= TableView[primaryenergyDatabyFuel]
```

   The output is not shown

- Notice that the information we want is in row 3 (energy types) and row 91 (the total global consumption, in Exajoules).

```
In[*]:= {energytype1, totalcons } =
         { primaryenergyDatabyFuel[[3]], primaryenergyDatabyFuel[[91]] }
```

Out[ ]= {{Exajoules, Oil, Natural Gas, Coal, Nuclear energy,
  Hydro electric, Renew- ables, Total, Oil, Natural Gas,
  Coal, Nuclear energy, Hydro electric, Renew- ables, Total},
 {Total World, 174.171, 138.441, 151.07, 24.4415, 41.0898, 34.7991,
  564.012, 184.214, 145.349, 160.104, 25.3128, 40.2596, 39.9126, 595.151}}

In[ ]:= **{energytype, totalcons2021} = {energytype1⟦9 ;; 14⟧, totalcons⟦9 ;; 14⟧}**

Out[ ]= {{Oil, Natural Gas, Coal, Nuclear energy, Hydro electric, Renew- ables},
 {184.214, 145.349, 160.104, 25.3128, 40.2596, 39.9126}}

- We now visualize the data using a pie chart showing the percentage share of the total for each energy source. An interesting observation is the fact that fossil fuels represent the vast majority of the world primary energy consumption with renewables (excluding hydroelectric) playing a very small role (wind, sun, biomass, etc.).

In[ ]:= **PieChart[totalcons2021 / Total[totalcons2021],**
  ChartStyle → "DarkRainbow", ChartLabels → energytype, LabelingFunction →
   (Placed[Row[{NumberForm[100 #, {3, 1}], "%"}], "RadialCallout"] &),
  PlotLabel → Style["Sources of Energy (%)", Bold]]

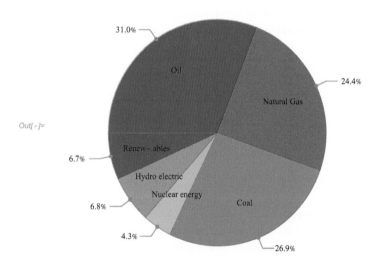

Out[ ]=

Sources of Energy (%)

To highlight an individual source of energy just click on its wedge.

> If you'd like to customize a graph, use the information available in the help system. For example, go to the documentation page for PieChart, and look for an example that suits your needs.

Now we are going to analyze the evolution of the global primary energy consumption and look for a function that will enable us to extrapolate energy consumption over the coming years.

- We import the sheet "Primary Energy - Consumption":

In[ ]:= **primaryenergyDataYear = Import[**
       "https://www.bp.com/content/dam/bp/business-sites/en/global/corporate/
          xlsx/energy-economics/statistical-review/bp-stats-review-2022-all
          -data.xlsx", {"Sheets", "Primary Energy Consumption"}];

- With TableView the imported data can be shown:

In[ ]:= **TableView[primaryenergyDataYear]**

	1	2	3	4	5	6	7	8
1	Primary							
2								
3	Exajoules	1965.	1966.	1967.	1968.	1969.	1.97×10^3	1971.
4								
5	Canada	5.00053	5.32365	5.56665	5.92412	6.28325	6.72683	6.93772
6	Mexico	1.05597	1.1181	1.13958	1.24115	1.36234	1.45451	1.52938
7	US	51.9829	54.8905	56.8027	60.1898	63.3003	65.541	66.9204
8	Total	58.0394	61.3323	63.509	67.3551	70.9459	73.7224	75.3875
9								

- Notice that the desired information is in row 3 (year) and row 111 (Total World, in Exajoules). We also eliminate not relevant rows and columns.

In[ ]:= **primaryevolution =**
       Drop[Drop[Transpose[primaryenergyDataYear[[{3, 112}]]], 1], -4];
       Shallow[primaryevolution]

Out[ ]//Shallow=
       {{1965., 110.01}, {1966., 115.411}, {1967., 119.987},
        {1968., 127.923}, {1969., 136.364}, {1970., 144.}, {1971., 147.92},
        {1972., 155.41}, {1973., 164.257}, {1974., 162.18}, ≪46≫ }

- We plot the data. Using Tooltip, if you click on a curve point, its value will be shown:

In[ ]:= `dataplot = ListLinePlot[Tooltip[primaryevolution],`
   `PlotStyle → PointSize[.01], Mesh → All, AxesLabel → {"Year", "Exajoules"},`
   `PlotLabel → Style["Energy Consumption Evolution", Bold]]`

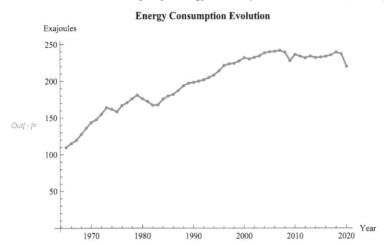

Out[ ]=

- Here, FindFormula is used to find a function that approximates the data:

In[ ]:= `fit = FindFormula[primaryevolution, t]`

Out[ ]=
$$\begin{cases} -1.12669 \times 10^7 - 861.742\, t + 1.70908 \times 10^6\, \text{Log}[t] & 1965. \le t < 1980.64 \\ -6393.34 + 3.31167\, t & 1980.64 \le t < 1995.29 \\ -3317.39 + 1.77334\, t & 1995.29 \le t < 2008.22 \\ -1038.06 + 0.631693\, t & 2008.22 \le t < 2020. \\ 0 & \text{True} \end{cases}$$

- After that, we extract the fit function for the last period to extrapolate beyond of 2021:

In[ ]:= `model = fit[[1, 4, 1]]`

Out[ ]= `-1038.06 + 0.631693 t`

In[ ]:= `Show[Plot[model, {t, 2006, 2025}, AxesOrigin → {1960, 100},`
   `PlotStyle → {Pink, Thick}, PlotRange → All], dataplot]`

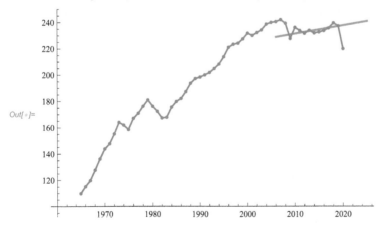

Out[ ]=

5.7 Time Series Analysis

In this section we are going to see an example of how to use time series to forecast the weather.

- We start collecting the daily average temperatures in the LESA weather station from 1980 until the end of 2021:

In[ ]:= `averagetempSalamanca =`
        `WeatherData["LESA", "MeanTemperature", {{1980}, {2021}, "Day"}]`

Out[ ]= TimeSeries[⊞ 〰 Time: 01 Jan 1980 to 31 Dec 2021 Data points: 15 215]

- The next function fits the data to a typical time series model:

In[ ]:= `tsm = TimeSeriesModelFit[averagetempSalamanca]`

Out[ ]= TimeSeriesModel[⊞ 〰 Family: ARMA Order: {2, 7}]

- We can now use the fitted model to forecast for example the average daily temperature today and one month from now:

In[ ]:= `today = DateValue[Today, {"Year", "Month"}]`

Out[ ]= {2022, 5}

In[ ]:= `tsm[today]`

Out[ ]= 10.7181 °C

In[ ]:= `tsm[DatePlus[today, {1, "Month"}]]`

Out[ ]= 11.0916 °C

- Next, we estimate average monthly temperatures using only the data from the last 10 years:

In[ ]:= `start = DateValue[DatePlus[Today, {-10, "Year"}], {"Year", "Month"}];`
        `tspec = {start, today, "Month"};`

In[ ]:= `temp = TimeSeries[WeatherData["Salamanca", "Temperature", tspec, "Value"],`
            `{start, Automatic, "Month"}];`

- Then display them graphically. Notice the seasonality.

In[ ]:= `DateListPlot[temp, Joined → True]`

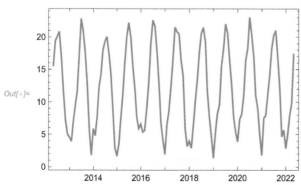

- We fit the data to a time series model:

In[ ]:=   **tsm = TimeSeriesModelFit[temp]**

Out[ ]=   TimeSeriesModel[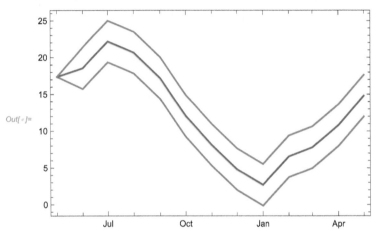 Family: SARIMA   Order: {{0, 0, 0}, {1, 1, 1}$_{12}$}]

- We visualize next year's forecast for the average monthly temperature using 95% confidence bands:

In[ ]:=   **fdates = DateRange[today, DatePlus[today, {1, "Year"}], "Month"];**

In[ ]:=   **forecast = tsm /@ fdates;**
         bands = tsm["PredictionLimits"][#] & /@ fdates;

In[ ]:=   **DateListPlot[{bands⟦All, 1⟧, forecast, bands⟦All, 2⟧},**
         {today, Automatic, "Month"}, Joined → True, PlotStyle → {Gray, Automatic}]

Out[ ]=

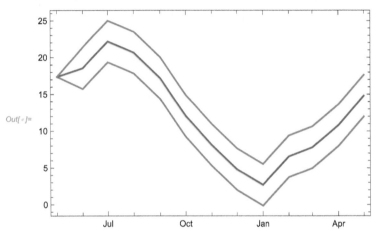

- The data records are highly correlated (The *p*-values are small when the data is autocorrelated):

In[ ]:=   **ListPlot[Table[AutocorrelationTest[temp, i], {i, 1, 6}], Filling → Axis]**

Out[ ]=

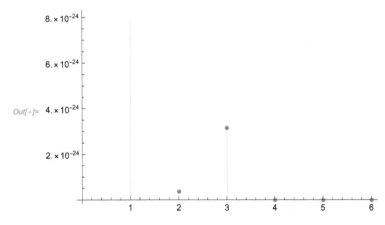

5.8 Spatial Statistics

The newest version of *Mathematica* has the capability to perform spatial statistics calculations:

https://reference.wolfram.com/language/guide/SpatialStatistics.html

We can now compute continuous values like temperature or elevation associated to spatial data points. Additionally, we can also get statistical measures such as center, density, and homogeneity when analyzing locations of things such as trees, earthquakes, or crimes.

In this example, we show how to estimate the temperature of points in a certain geographical area (Italy in this case) starting from a single point.

- First, we get a list of all the Italian weather stations accessible through Wolfram Resources and eliminate those for which there's no information. We extract the last recorded temperature for each one. Alternatively, we could have chosen data from a specific period of time.

```
In[ ]:= stations = GeoEntities[Entity["Country", "France"], "WeatherStation"];
```

```
In[ ]:= {temp1, locs1} = {DeleteMissing[AirTemperatureData /@ stations],
       DeleteMissing[Through[stations["Coordinates"]]]};
```

```
In[ ]:= sf = SpatialEstimate[locs1 → temp1]
```

Out[ ]= SpatialEstimatorFunction[⊞ ⋮⋮ Dimension: 2
⋮⋮⋮ SpatialTrendFunction: 1
SpatialNoiseLevel: Automatic]

- Now we can display the temperature at a specified location:

```
In[ ]:= UnitConvert[Quantity[sf[{45, 10}], "Kelvins"], "DegreeCelsius"]
```

Out[ ]= 14.3398 °C

- We can even generate a temperature map for the entire country:

```
In[ ]:= GeoDensityPlot[locs1 → temp1,
       ColorFunction → "Rainbow",
       RegionFunction → Entity["Country", "France"], GeoBackground → None,
       PlotLegends → Automatic]
```

Out[ ]=

5.9 Additional Resources

To access the following resources, if they refer to the help files, write their locations in a notebook (e.g., guide/ListManipulation), select them, and press <F1>. In the case of external links, copy the web addresses in a browser:

Importing and Exporting: guide/ImportingAndExporting

List Manipulation: guide/ListManipulation

How to Do Statistical Analysis: howto/DoStatisticalAnalysis

Descriptive Statistics: guide/DescriptiveStatistics

Parametric Statistical Distributions: guide/Parametric Statistical Distributions

Curve Fitting: guide/CurveFittingAndApproximateFunctions

Time Series: guide/TimeSeries

DatabaseLink Tutorial: DatabaseLink/tutorial/Overview

Spatial Statistics: guide/SpatialStatistics

Statistics Demonstrations: http://demonstrations.wolfram.com/topic.html?topic=Statistics

Probability Demonstrations: http://demonstrations.wolfram.com/topic.html?topic=Probability

6

Machine Learning and Neural Networks

*The number of applications in the real world that use machine learning continues to increase unabated. The Wolfram Language has both, general and domain-specific functions, that implement machine learning algorithms. After briefly discussing the concept of machine learning and showing some examples of Mathematica's capabilities in this area for identifying languages, estimating probability distributions and highlighting faces, this chapter focuses on the two most important functions for performing machine learning tasks with WL: **Classify** and **Predict**. Using data from the Titanic and Fisher's iris datasets among others, the reader will learn how to classify data into different clusters and predict numerical outcomes. The last section contains a short introduction to neural networks.*

6.1 What Is Machine Learning?

Nowadays, we know that many of our interactions with our mobile phones, tablets, computers, and other digital appliances are mediated by machine learning algorithms, but what is machine learning? Let's ask the relevant Wikipedia entry and retrieve its first five sentences:

In[ ]:= `TextSentences[WikipediaData["Machine Learning"]][[ ;; 5]]`

Out[ ]= {Machine learning (ML) is the study of computer algorithms that can improve automatically through experience and by the use of data., It is seen as a part of artificial intelligence., Machine learning algorithms build a model based on sample data, known as training data, in order to make predictions or decisions without being explicitly programmed to do so., Machine learning algorithms are used in a wide variety of applications, such as in medicine, email filtering, speech recognition, and computer vision, where it is difficult or unfeasible to develop conventional algorithms to perform the needed tasks., A subset of machine learning is closely related to computational statistics, which focuses on making predictions using computers; but not all machine learning is statistical learning.}

In this chapter, we focus on showing *Mathematica* functions for developing machine learning (ML) applications or functions that use ML algorithms for solving domain-specific problems. However, we are not going to explain how those algorithms actually work.

The two most important functions in WL for ML are Classify and Predict. They are used for either classifying inputs into categories or class labels or predicting numeric outputs. The next two sections will cover these functions in more detail.

Next, we show some toy examples of domain-specific functions that use ML:

- The function below, identifies the language used to write text:

In[ ]:= **LanguageIdentify[**
    { "thank you", "merci", "dar las gracias", "感謝", "благодарить"}]

Out[ ]= { English , French , Spanish , Chinese , Russian }

In this case, we analyze data from patients suffering from kidney failure to find the probability distribution that best approximates the time to infection after having the catheter replaced:

- We get the description of the data and store the time to kidney infection (months):

In[ ]:= **ExampleData[{"Statistics", "KidneyInfection"}, "LongDescription"]**

Out[ ]= Time to first exit–site infection in months
      in patients with renal insufficiency. The data are right censored.

In[ ]:= **KidneyInfection = ExampleData[{"Statistics", "KidneyInfection"}][[All, 1]];**

In[ ]:= **KidneyInfection // Short**

Out[ ]//Short=
      {1.5, 3.5, 4.5, 4.5, 5.5, 8.5, 8.5, 9.5, 10.5,
       ≪101≫, 19.5, 19.5, 20.5, 22.5, 24.5, 25.5, 26.5, 26.5, 28.5}

- Next, we use FindDistribution to find the best distribution that fits the data. This function uses ML algorithms when computing the answer.

In[ ]:= **e𝒟 = FindDistribution[KidneyInfection]**

Out[ ]= ExponentialDistribution[0.108901]

- Finally, we compare the histogram of the data to the PDF of the estimated distribution:

In[ ]:= **Show[Histogram[KidneyInfection, {0, 28, 2}, "ProbabilityDensity"],**
    Plot[PDF[e𝒟, x], {x, 0, 28}, PlotStyle → Thick, PlotRange → All]]

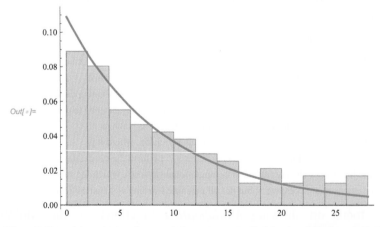

The following examples use images available in *Mathematica* help to show some functions related to image analysis.

- The image below is from the 5th Solvay International Conference on Electrons and Photons held in 1927, that included the world's most famous physicists at that time. Can you identify Niels Bohr? Albert Einstein?

In[ ]:= i =  ;

In[ ]:= `HighlightImage[i, FindFaces[#, "BoundingBox"] &]`

Out[ ]=

- We can use ImageCases to identify certain subimages:

In[ ]:= `ImageCases[`  `, ` **dog** SPECIES SPECIFICATION `]`

Out[ ]=

- The function ImageContents tries to identify the contents of an input image and returns a dataset of identified entities:

In[ ]:= **ImageContents**[ ]

Image	Concept	BoundingBox	Probability
	elephant	Rectangle[{434.504, 101	0.837853
	elephant	Rectangle[{334.952, 155	0.780911
	zebra	Rectangle[{35.9143, 18.	0.779942
	elephant	Rectangle[{133.816, 108	0.694424
	zebra	Rectangle[{3.92214, 152	0.631345
	zebra	Rectangle[{133.058, 96.	0.603996

We can also create a dataset containing information from a given text:

- First, we create the text and ask *Mathematica* to identify all possible entities:

In[ ]:= str = "Japan is an island country in East Asia. It is situated in the
        northwest Pacific Ocean, and is bordered on the west by the
        Sea of Japan, while extending from the Sea of Okhotsk in the
        north toward the East China Sea and Taiwan in the south.";
    content = TextContents[str]

String	Type	Position	Probability	HighlightedSnippet
Japan	Country	{1, 5}	0.780174	Japan is an island country in East Asia. It is situated in the
Japan	HistoricalCountry	{1, 5}	0.630876	Japan is an island country in East Asia. It is situated in the
island	SportObject	{13, 18}	0.63843	Japan is an island country in East Asia. It is situated in the
East Asia	GeographicRegion	{31, 39}	0.977164	Japan is an island country in East Asia . It is situated in the
Asia	Continent	{36, 39}	0.970012	is an island country in East Asia . It is situated in the
Asia	EthnicGroup	{36, 39}	0.966716	is an island country in East Asia . It is situated in the
Asia	GovernmentAgency	{36, 39}	0.994286	is an island country in East Asia . It is situated in the
Pacific Ocean	Ocean	{74, 86}	0.992036	is situated in the northwest Pacific Ocean , and is bordered on the west
Sea of Japan	Ocean	{124, 135}	0.975868	bordered on the west by the Sea of Japan , while extending from the
Sea of Okhotsk	Ocean	{163, 176}	0.989431	while extending from the Sea of Okhotsk in the north toward the East
East China Sea	Ocean	{202, 215}	0.970533	Okhotsk in the north toward the East China Sea and Taiwan in the south.
Taiwan	Country	{221, 226}	0.887452	in the north toward the East China Sea and Taiwan in the south.
Taiwan	HistoricalCountry	{221, 226}	0.763275	in the north toward the East China Sea and Taiwan in the south.
Taiwan	Island	{221, 226}	0.907025	in the north toward the East China Sea and Taiwan in the south.

- Then, we make use of the power of Interpreter to transform those entities into a dataset:

```
In[ ]:= TextCases[str,
         Union@Lookup[Normal[content], "Type"] → "Interpretation"] // Dataset
```

Continent	{ Asia }
Country	{ Japan , Taiwan }
EthnicGroup	{Asia}
GeographicRegion	{East Asia}
GovernmentAgency	{Asia}
HistoricalCountry	{Japan, Taiwan}
Island	{ Taiwan }
Ocean	{ Pacific Ocean , Sea of Japan , Okhotsk Sea , East China Sea }
SportObject	{island}

The ML algorithms implemented in *Mathematica* can also be applied to video processing as the example below shows:

- Find the number of faces in each video frame:

```
In[ ]:= v = Import["http://exampledata.wolfram.com/office.ts"];
```

```
In[ ]:= VideoFrameList[v, 3]
```

Out[ ]= {  }

```
In[ ]:= ts = VideoMapTimeSeries[Length[FindFaces[#Image]] &, v];
```

Plot the result:

In[ ]:= `ListLinePlot[ts]`

Out[ ]=

6.2 Classification

In ML there are two approaches for teaching computers to classify inputs: unsupervised learning and supervised learning.

Under the first one, an algorithm will try to identify patterns from inputs that are neither classified nor labeled. Functions such as ClusterClassify and FindClusters implement the cluster analysis unsupervised algorithm to classify data into clusters, groups sharing similar characteristics. For additional information please refer to the "Partitioning Data into Clusters" technical note available at:

> https://reference.wolfram.com/language/tutorial/NumericalOperationsOnData.html#2948.

6.2.1 Classify

The supervised learning approach uses inputs that have been already labeled to teach algorithms how to classify new data. One of the most important functions in *Mathematica* for supervised learning is Classify. Let's see it in action:

- In this example, we use a set of images captured either during the day or at night to train an algorithm to recognize when the image was taken:

In[ ]:= `daynight = Classify[`

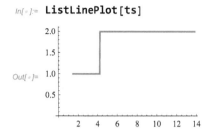

Out[ ]= `ClassifierFunction[` ⊞ ⬚ Input type: Image
Classes: Day, Night `]`

■ After generating a classifier function based on the training set, we can now ask the algorithm to classify new images:

In[ ]:= **daynight**[{  }]

Out[ ]= {Day, Night, Day, Night, Night, Night}

This function has numerous options that allow us to get additional information such as the accuracy of the classifier or the confusion matrix:

■ We first generate a training list and train a classifier with it:

In[ ]:= **traininglist = {1 → "A", 2 → "A", 3.5 → "B", 3 → "A", 5 → "B", 6 → "B"};**

In[ ]:= **c = Classify[traininglist]**

Out[ ]= ClassifierFunction[⊞ · Input type: Numerical / Classes: A, B]

■ Now, we can use test data to assess how good the classifier function is by measuring its accuracy and uncertainty:

In[ ]:= **testset = {1.2 → "A", 2.1 → "A", 3. → "B", 6.4 → "B"};**

In[ ]:= **cm = ClassifierMeasurements[c, testset];**

In[ ]:= **cm["Accuracy", ComputeUncertainty → True]**

Out[ ]= 0.75 ± 0.25

■ Finally, we can also visualize the confusion matrix to find out how many new inputs were misclassified:

In[ ]:= **ClassifierMeasurements[c, testset, "ConfusionMatrixPlot"]**

Out[ ]=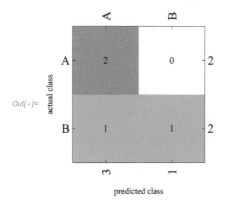

Let's imagine we have a wide variety of colors obtained by mixing red and blue and we would like to classify them into just two categories: Red or Blue.

- The first thing we need to do is to train a supervised learning algorithm with a labeled data set. In principle, the bigger the training set, the better the resulting classifier will be. For this step, we generate a list of colors mixing red and blue, with *1-x* and *x* being the proportion of red and blue, respectively. That is 100% red is $x = 0$, and 100% blue is x = 1.

In[ ]:= `Table[x, {x, 0, 1, 1/16}]`

Out[ ]= $\left\{0, \dfrac{1}{16}, \dfrac{1}{8}, \dfrac{3}{16}, \dfrac{1}{4}, \dfrac{5}{16}, \dfrac{3}{8}, \dfrac{7}{16}, \dfrac{1}{2}, \dfrac{9}{16}, \dfrac{5}{8}, \dfrac{11}{16}, \dfrac{3}{4}, \dfrac{13}{16}, \dfrac{7}{8}, \dfrac{15}{16}, 1\right\}$

In[ ]:= `colorsblend = Table[Blend[{Red, Blue}, x], {x, 0, 1, 1/16}]`

Out[ ]= {■, ■, ■, ■, ■, ■, ■, ■, ■, ■, ■, ■, ■, ■, ■, ■, ■}

- Next, we tell Classify which colors are red and which ones are blue . If the mix is 50%/50%, we don't assign any label to that color:

In[ ]:= `reddish = colorsblend[[1 ;; 8]]`

Out[ ]= {■, ■, ■, ■, ■, ■, ■, ■}

In[ ]:= `blueddish = colorsblend[[10 ;; 17]]`

Out[ ]= {■, ■, ■, ■, ■, ■, ■, ■}

In[ ]:= `c = Classify[<|Red → reddish, Blue → blueddish|>]`

Out[ ]= ClassifierFunction[⊞ ⣿ Input type: Color
                                    Classes: ■, ■]

- We can now test the classifier with mixtures close to 50%/50%:

In[ ]:= `GatherBy[Table[x, {x, 1/2 - 1/7, 1/2 + 1/7, 1/15}], # > 1/2 &] // N`

Out[ ]= {{0.357143, 0.42381, 0.490476}, {0.557143, 0.62381}}

In[ ]:= `colorsblendtest1 =`
            `Table[Blend[{Red, Blue}, x], {x, 1/2 - 1/7, 1/2 + 1/7, 1/15}]`

Out[ ]= {■, ■, ■, ■, ■}

- Just like people, the classifier may have difficulties assigning a color to mixtures close to 50%/50%. However, in this case it's doing a great job by correctly assigning the red color to the third element of the list (49.048% blue):

In[ ]:= `c[colorsblendtest1]`

Out[ ]= {■, ■, ■, ■, ■}

- Finally, we can check the associated probabilities for each possible class:

In[ ]:= `c[colorsblendtest1, "Probabilities"]`

Out[ ]= { <|■ → 0.000068972, ■ → 0.999931|>,
            <|■ → 0.0060001, ■ → 0.994|>, <|■ → 0.345658, ■ → 0.654342|>,
            <|■ → 0.978826, ■ → 0.0211736|>, <|■ → 0.999753, ■ → 0.000247122|> }

In this next example, we use **colorsblend**, the list of colors we have just created, as an input to LearnDistribution to generate a new classifier in the form of a probability distribution.

- We create the learned distribution first:

```
In[ ]:= ld = LearnDistribution[
        {■, ■, ■, ■, ■, ■, ■, ■, ■, ■, ■, ■, ■, ■, ■, ■, ■}]
```

```
Out[ ]= LearnedDistribution[ ⊞ 🌀  Input type: Color
                                   Method: Multinormal ]
```

- Then, we generate 100 random colors using it:

```
In[ ]:= RandomVariate[ld, 100]
```

```
Out[ ]= {■, ■, ■, ■, ■, ■, ■, ■, ■, ■, ■, ■, ■, ■, ■, ■,
         ■, ■, ■, ■, ■, ■, ■, ■, ■, ■, ■, ■, ■, ■, ■, ■,
         ■, ■, ■, ■, ■, ■, ■, ■, ■, ■, ■, ■, ■, ■, ■, ■,
         ■, ■, ■, ■, ■, ■, ■, ■, ■, ■, ■, ■, ■, ■, ■, ■,
         ■, ■, ■, ■, ■, ■, ■, ■, ■, {■, ■, ■, ■, ■, ■, ■, ■, ■}}
```

- Approximately half are red, half are blue, and none are green. This is the expected result given the initial mixtures.

```
In[ ]:= PDF[ld, {■, ■, ■}]
```

```
Out[ ]= {4371.41983563893, 4370.57360271242, 1.7340803597×10^{-52699}}
```

Let's see an example where multiple inputs are associated to just one output.

- We are going to mix colors again using, in this case, the three basic colors: Red (R), Green (G), and Blue (B). We can do this in the WL with the function RGBColor$[r, g, b]$

```
In[ ]:= Map[Graphics[{RGBColor[#], Disk[]}] &, {{1, 0, 0}, {0, 1, 0},
        {0, 0, 1}, {0.5, 0.5, 0}, {0.5, 0, 0.5}, {0, 0.5, 0.5}}]
```

- We generate 100 random lists of the form {r,g,b} to reproduce the colors.

```
In[ ]:= SeedRandom[1234]
```

```
Out[ ]= RandomGeneratorState[ Method: ExtendedCA
                              State hash: 3 259 750 555 964 806 719 ]
```

```
In[ ]:= list = RandomReal[1, {100, 3}];
```

```
In[ ]:= RGBColor[#] & /@ list
```

```
Out[ ]= {■, ■, ■, □, ■, ■, □, □, □, ■, ■, ■, ■, ■, ■, ■, ■, ■, ■, ■, □,
         □, ■, ■, ■, ■, ■, ■, ■, ■, □, □, ■, ■, ■, ■, ■, ■, □, ■, ■, ■,
         ■, ■, □, ■, ■, □, ■, ■, ■, □, ■, ■, ■, ■, ■, ■, □, ■, ■, ■, ■,
         ■, □, ■, □, ■, ■, ■, ■, □, ■, ■, ■, □, ■, ■, ■, ■, ■, ■, ■, ■,
         ■, □, ■, □, ■, ■, ■, ■, □, ■, ■, ■, ■, ■, ■, ■, ■, ■, □}
```

- We can now label each list either Red, Green, or Blue depending on the position of its largest element. For example, if the biggest value is the second element, we assign the color Green to the list.

```
In[ ]:= listcolor =
        TakeLargest[# → "Index", 1] & /@ list /. {1 → Red, 2 → Green, 3 → Blue };
```

```
In[ ]:= training = Thread[list → Flatten[listcolor]];
```

In[ ]:= **clcolor = Classify[training]**

Out[ ]:= ClassifierFunction[⊞ Input type: NumericalVector (length: 3)]
　　　　　　　　　　　　　　　　　　Classes: ■, ■, ■

Let's see how well the new classifier works.

- We generate a 5×5 image first:

In[ ]:= **img = Image[RandomReal[1, {5, 5, 3}], ColorSpace → "RGB"]**

Out[ ]:=

- Next, using the new classifier, we convert that image into a different one containing only the three basic colors:

In[ ]:= **Image[Map[clcolor[#] &, Table[ImageData[img]〚i, 1 ;; 5〛, {i, 5}]]]**

Out[ ]:=

In[ ]:= **imgexample = ExampleData[{"TestImage", "House"}]**

Out[ ]:=

In[ ]:= **ImageDimensions[imgexample]**

Out[ ]:= {256, 256}

In[ ]:= **Image[**
　　　　Map[clcolor[#] &, Table[ImageData[imgexample]〚i, 1 ;; 256〛, {i, 256}]]]

Out[ ]:=

In the previous example, Classify chose automatically the learning method. However, we can also tell the function what algorithm to use (for further details refer to the documentation pages).

- In this case, we use "GradientBoostedTrees" instead of "LogisticRegression", the one that Classify automatically chose:

In[ ]:= **clcolor1 = Classify[training, Method → "GradientBoostedTrees"]**

Out[ ]= ClassifierFunction[⊞ ⬚ Input type: **NumericalVector** (length: 3) Classes: ■, ■, ■]

In[ ]:= **Image[**
     Map[clcolor1[#] &, Table[ImageData[imgexample]⟦i, 1 ;; 256⟧, {i, 256}]]]

Out[ ]=

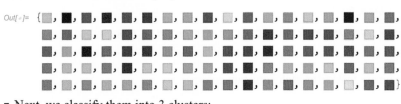

As mentioned previously, ClusterClassify can be very useful when our data is not labeled. Let's see an example.

- First, we randomly generate 90 colors:

In[ ]:= **colors = RandomColor[90]**

Out[ ]= {■, ■, ■, ■, ■, ■, ■, ■, ■, ■, ■, ■, ■, ■, ■, ■, ■, ■,
       ■, ■, ■, ■, ■, ■, ■, ■, ■, ■, ■, ■, ■, ■, ■, ■, ■, ■,
       ■, ■, ■, ■, ■, ■, ■, ■, ■, ■, ■, ■, ■, ■, ■, ■, ■, ■,
       ■, ■, ■, ■, ■, ■, ■, ■, ■, ■, ■, ■, ■, ■, ■, ■, ■, ■,
       ■, ■, ■, ■, ■, ■, ■, ■, ■, ■, ■, ■, ■, ■, ■, ■, ■, ■}

- Next, we classify them into 3 clusters:

In[ ]:= **c = ClusterClassify[colors, 3]**

Out[ ]= ClassifierFunction[⊞ ⬚ Input type: **Color** Classes: 1, 2, 3]

- And, finally, gather them by their class number:

In[ ]:= **GatherBy[colors, c] // TableForm**

Out[ ]//TableForm=

The example below shows how to find the dominant color in an image.

- We choose an image of a sunflower in this case:

In[ ]:= **image** =  ;

- Although we often represent colors using the RGB specification, sometimes it is more convenient to use a different color space such as LAB. The command below uses ColorConvert to change the color specification of the image .

In[ ]:= **ColorConvert[image, "LAB"]**

Out[ ]=

- To process it, we first extract the data:

In[ ]:= **imageData = Flatten[ImageData[ColorConvert[image, "LAB"]], 1];**

- Then, we cluster the array of pixels just obtained:

In[ ]:= **c = ClusterClassify[imageData, 4, Method → "KMedoids"];**

- Next, we use the classifier to assign a cluster to each pixel and find four dominant colors:

In[ ]:= **decision = c[imageData];**

In[ ]:= **LABColor @@@ (Mean /@ (Pick[imageData, decision, #] & /@ Range[4]))**

Out[ ]= {▢, ▢, ▢, ▢}

- Finally, we get binary masks for each dominant color:

In[ ]:= **Image /@**
   ComponentMeasurements[{image, Partition[decision, 426]}, "Mask"]〚All, 2〛

Out[ ]= {   }

6.2.2 Titanic Survival

In this example, we use the Titanic dataset that includes the class, age, gender, and survivor status for 1,309 passengers that were aboard the ship.

This dataset contains a mixture of numerical and textual data.

In[ ]:= **Short[data = ExampleData[{"MachineLearning", "Titanic"}, "Data"]]**

Out[ ]//Short=
   {{1st, 29., female} → survived, {1st, 0.9167, male} → survived,
   ≪1305≫, {3rd, 27., male} → died, {3rd, 29., male} → died}

In[ ]:= **titanic = Classify[data]**

Out[ ]= ClassifierFunction[⊞ ⬚ Input type: {Nominal, Numerical, Nominal}
Classes: died, survived]

- Now that we have our classifier, we can ask questions, such as the probability that a 50-year old male would have either died or survived:

In[ ]:= **titanic[{"1st", 50, "male"}, "Probabilities"]**

Out[ ]= <| died → 0.551655, survived → 0.448345 |>

- We can also visualize the probability of survival for first-class males based on their ages:

In[ ]:= **Plot[titanic[{"1st", age, "male"},**
**   {"Probability", "survived"}], {age, 0, 70}]**

Out[ ]=

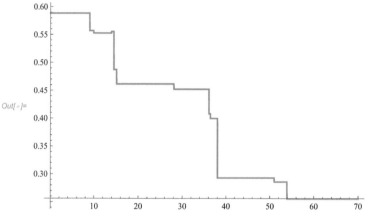

- We can even predict the age of a passenger using the rest of the features. For example, the predicted age of a female in 3rd class that survived is:

In[ ]:= **titanicAge = Predict[data /. HoldPattern[{class_, age_, sex_} → outcome_] :→**
**   {class, sex, outcome} → age];**
titanicAge[{"3rd", "female", "survived"}]

Out[ ]= 21.5452

For our last example in this section, we use Fisher's Iris dataset. The dataset consists of 150 samples from three species of Iris (setosa, versicolor, and virginica). Each sample contains four measurements: the length and the width of the sepals and petals. We want to be able to correctly classify new Iris flowers based on those four measurements:

- Let's download the data first:

In[ ]:= **irisData = ExampleData[{"MachineLearning", "FisherIris"}, "Data"];**

In[ ]:= **irisData // Short**

Out[ ]//Short=
    {{5.1, 3.5, 1.4, 0.2} → setosa, {4.9, 3., 1.4, 0.2} → setosa, ≪147≫, {5.9, 3., 5.1, 1.8} → virginica}

- Next, we train the algorithm using 100 randomly selected samples:

```
In[ ]:= SeedRandom[1234];
       irisTrainingData = RandomSample[irisData, 100];
       irisCf = Classify[irisTrainingData]
```

Out[ ]= ClassifierFunction[▦ ◹ Input type: **NumericalVector** (length: 4)
                                   Classes: setosa, versicolor, virginica]

- The remaining samples can then be used as test data:

```
In[ ]:= irisTestData = Complement[irisData, irisTrainingData];
```

- Finally, using ClassifierMeasurements on the test dataset, we calculate the accuracy of the algorithm, visualize the confusion matrix (2 samples of versicolor were misclassified as virginica) and see what examples had the highest actual-class probabilities:

```
In[ ]:= irisCm = ClassifierMeasurements[irisCf, irisTestData];
```

```
In[ ]:= irisCm["Accuracy"]
```

Out[ ]= 0.96

```
In[ ]:= irisCm["ConfusionMatrixPlot"]
```

Out[ ]=
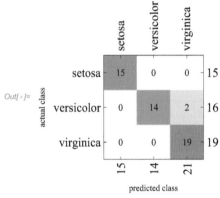

```
In[ ]:= irisCm["BestClassifiedExamples"]
```

Out[ ]= {{7.7, 2.6, 6.9, 2.3} → virginica, {7.7, 2.8, 6.7, 2.} → virginica,
       {7.7, 3.8, 6.7, 2.2} → virginica, {7.2, 3.6, 6.1, 2.5} → virginica,
       {6.7, 3.3, 5.7, 2.5} → virginica, {6.3, 3.4, 5.6, 2.4} → virginica,
       {4.4, 3.2, 1.3, 0.2} → setosa, {6.8, 3.2, 5.9, 2.3} → virginica,
       {5.5, 4.2, 1.4, 0.2} → setosa, {4.3, 3., 1.1, 0.1} → setosa}

- Evaluating the probabilities for the first element of the previous output, we can see the classifier is 100% certain that those measurements correspond to an Iris virginica flower:

```
In[ ]:= irisCf[{7.7, 2.6, 6.9, 2.3}, "Probabilities"]
```

Out[ ]= ⟨| setosa → 2.24715 × 10^{-21}, versicolor → 3.51601 × 10^{-8}, virginica → 1. |⟩

6.3 Prediction

6.3.1 The Predict function

Apart from Classify, the other main function for implementing supervised learning in *Mathematica* is Predict. The example below shows how to use it for fitting functions to data.

- We first simulate input-output data of the form: $x \rightarrow y$

In[ ]:= **trainingdata = Table[x → Sin[x] + RandomVariate[NormalDistribution[0, .3]],**
    {x, RandomReal[{-5, 5}, 50] }];

- Then, we predict any intermediate values using the "GaussianProcess" method. For further information about all the methods available for prediction, read the function documentation.

In[ ]:= **p = Predict[trainingdata, Method → "GaussianProcess"]**

Out[ ]= PredictorFunction[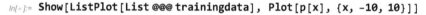]

- Finally, we compare visually the training data with values obtained with the predictor function:

In[ ]:= **Show[ListPlot[List @@@ trainingdata], Plot[p[x], {x, -10, 10}]]**

Out[ ]=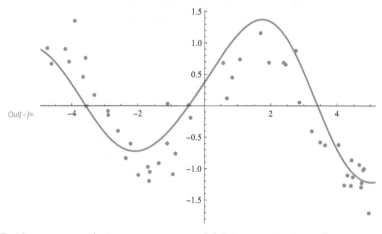

Let's suppose that a person was drinking water heavily contaminated with lead for 100 days. To find out the lead concentration after the exposure, urine samples were taken during the next 200 days. Our objective is to create a function to accurately predict the evolution of lead concentration after the first 200 days.

```
sampleslead = {{1, 2.58}, {2, 2.95}, {3, 2.66}, {4, 3.83}, {5, 4.18},
    {6, 4.18}, {7, 15.1}, {8, 12.65}, {9, 12.53}, {10, 11.57},
    {11, 11.69}, {12, 10.97}, {13, 11.97}, {14, 12.18}, {15, 16.56},
    {16, 16.2}, {17, 15.9}, {18, 21.02}, {19, 19.52}, {20, 18.63},
    {21, 18.2}, {22, 17.72}, {23, 18.63}, {24, 18.24}, {25, 19.18},
    {26, 18.37}, {27, 18.81}, {28, 18.41}, {29, 18.36}, {30, 18.31},
    {31, 19.87}, {32, 20.08}, {33, 32.34}, {34, 30.29}, {35, 30.77},
    {36, 29.19}, {37, 28.35}, {38, 30.07}, {39, 29.58}, {40, 30.7},
    {41, 29.19}, {42, 28.17}, {43, 32.65}, {44, 30.96}, {45, 30.6},
    {46, 31.97}, {47, 32.95}, {48, 32.04}, {49, 33.9}, {50, 32.59},
    {51, 31.74}, {52, 31.18}, {53, 33.22}, {54, 32.06}, {55, 31.64},
    {56, 33.04}, {57, 35.26}, {58, 34.64}, {59, 33.78}, {60, 36.14},
    {61, 34.41}, {62, 33.87}, {63, 33.02}, {64, 32.61}, {65, 32.58},
    {66, 31.76}, {67, 32.17}, {68, 35.32}, {69, 35.04},
    {70, 33.88}, {71, 33.14}, {72, 34.5}, {73, 34.54}, {74, 33.24},
    {75, 33.58}, {76, 33.75}, {77, 33.13}, {78, 45.23}, {79, 41.97},
    {80, 39.77}, {81, 42.11}, {82, 42.49}, {83, 41.58}, {84, 39.78},
    {85, 43.89}, {86, 42.41}, {87, 41.36}, {88, 40.3}, {89, 39.76},
    {90, 39.84}, {91, 39.77}, {92, 42.76}, {93, 41.29}, {94, 41.15},
    {95, 40.71}, {96, 39.84}, {97, 39.62}, {98, 39.89}, {99, 39.24},
    {100, 38.8}, {101, 37.56}, {102, 36.53}, {103, 35.64},
    {104, 34.83}, {105, 34.07}, {106, 33.36}, {107, 32.68},
  {108, 32.01}, {109, 31.37}, {110, 30.75}, {111, 30.13}, {112, 29.53},
    {113, 28.95}, {114, 28.37}, {115, 27.81}, {116, 27.26},
    {117, 26.72}, {118, 26.19}, {119, 25.67}, {120, 25.17},
    {121, 24.67}, {122, 24.18}, {123, 23.7}, {124, 23.23}, {125, 22.77},
    {126, 22.32}, {127, 21.88}, {128, 21.44}, {129, 21.02}, {130, 20.6},
    {131, 20.2}, {132, 19.8}, {133, 19.4}, {134, 19.02}, {135, 18.64},
    {136, 18.27}, {137, 17.91}, {138, 17.56}, {139, 17.21},
    {140, 16.87}, {141, 16.53}, {142, 16.21}, {143, 15.89},
    {144, 15.57}, {145, 15.26}, {146, 14.96}, {147, 14.66},
  {148, 14.37}, {149, 14.09}, {150, 13.81}, {151, 13.54}, {152, 13.27},
    {153, 13.01}, {154, 12.75}, {155, 12.5}, {156, 12.25}, {157, 12.01},
    {158, 11.77}, {159, 11.54}, {160, 11.31}, {161, 11.08},
    {162, 10.86}, {163, 10.65}, {164, 10.44}, {165, 10.23},
    {166, 10.03}, {167, 9.83}, {168, 9.64}, {169, 9.44}, {170, 9.26},
    {171, 9.07}, {172, 8.89}, {173, 8.72}, {174, 8.55}, {175, 8.38},
    {176, 8.21}, {177, 8.05}, {178, 7.89}, {179, 7.73}, {180, 7.58},
    {181, 7.43}, {182, 7.28}, {183, 7.14}, {184, 7.}, {185, 6.86},
    {186, 6.72}, {187, 6.59}, {188, 6.46}, {189, 6.33}, {190, 6.21},
    {191, 6.08}, {192, 5.96}, {193, 5.84}, {194, 5.73}, {195, 5.62},
    {196, 5.5}, {197, 5.39}, {198, 5.29}, {199, 5.18}, {200, 5.08}};
```

In[ ]:= `{inp, out} = Transpose[sampleslead];`

In[ ]:= `p = Predict[inp → out]`

Out[ ]= `PredictorFunction[`  `Input type: Numerical`
`                                 Method: NearestNeighbors      ]`

- Once the predictor function has been created, we can plot the predicted values against the training sample:

In[ ]:= **Show[Plot[p[x], {x, 1, 250}, Exclusions → None],**
**   ListPlot[sampleslead, PlotStyle → Green]]**

Out[ ]=

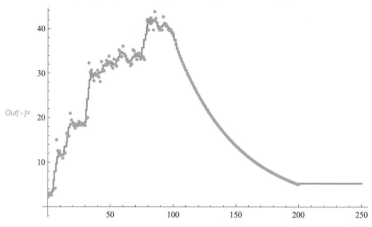

6.3.2 Wine Quality

In this example (available in the documentation), we want to predict subjective opinions regarding the quality of wine based on certain physical measurements.

- Learn additional details about the example:

In[ ]:= **ExampleData[{"MachineLearning", "WineQuality"}, "LongDescription"]**

Out[ ]=   Predict the subjectively reported quality of a white wine
         (on a scale of 1-10), given 11 physical features of the wine.
            These features include properties like the pH of
            the wine and its alcohol content. There are 4898 examples.

- Load the training dataset:

In[ ]:= **trainingset =**
**   ExampleData[{"MachineLearning", "WineQuality"}, "TrainingData"];**

- Visualize a few data points:

In[ ]:= **RandomSample[trainingset, 3] // TableForm**

Out[ ]//TableForm=
         {7.5, 0.3, 0.32, 1.4, 0.032, 31., 161., 0.99154, 2.95, 0.42, 10.5} → 5.
         {6.8, 0.19, 0.4, 9.85, 0.055, 41., 103., 0.99532, 2.98, 0.56, 10.5} → 6.
         {9.4, 0.28, 0.3, 1.6, 0.045, 36., 139., 0.99534, 3.11, 0.49, 9.3} → 5.

- Get a description of the variables:

In[ ]:= **ExampleData[{"MachineLearning", "WineQuality"}, "VariableDescriptions"]**

Out[ ]= {fixed acidity, volatile acidity, citric acid, residual sugar,
         chlorides, free sulfur dioxide, total sulfur dioxide, density,
         pH, sulphates, alcohol} → wine quality (score between 1-10)

- Visualize the distribution of the "alcohol" and "pH" variables:

```
In[ ]:= {Histogram[trainingset[[All, 1, 11]], PlotLabel → "alcohol"],
        Histogram[trainingset[[All, 1, 9]], PlotLabel → "pH"]}
```

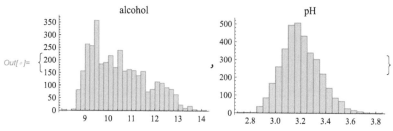

- Train a predictor on the training set:

```
In[ ]:= p = Predict[trainingset]
```

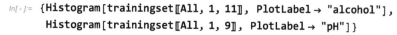

Out[ ]= PredictorFunction[Input type: NumericalVector (length: 11)
                           Method: RandomForest]

- Predict the quality of an unknown wine:

```
In[ ]:= unknownwine = {7.6`, 0.48`, 0.31`, 9.4`,
        0.046`, 6.`, 194.`, 0.99714`, 3.07`, 0.61`, 9.4`};
```

```
In[ ]:= p[unknownwine]
```

Out[ ]= 4.96642

- After using the "GradientBoostedTrees" method to create a new predictor function, we can create "SHAPPlots" to measure the impact of specific features on the final prediction. Note that in this case we download the data set using ResourceData.

```
In[ ]:= wine = ResourceData["Sample Data: Wine Quality"];
        p = Predict[wine → "WineQuality",
            Method → "GradientBoostedTrees", PerformanceGoal → "Speed"];
```

```
In[ ]:= pm = PredictorMeasurements[p, wine];
```

```
In[ ]:= pm["SHAPPlots"]
```

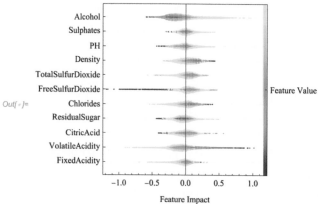

The three features that seem to have a significant influence in the predicted quality according to their SHAP (Shapley additive explanations) values are: Alcohol, FreeSulfurDioxide, and VolatileAcidity.

6.4 Working with Neural Networks

Many machine learning algorithms use neural networks. At their core, neural networks are a collection of regression layers that feed into each other in a way that mimics the architecture of the brain. These collection of layers can be represented as a symbolic graph where each layer holds fixed information about its type, inputs and outputs, and weights. The following video from the 3Blue1Brown YouTube channel provides an excellent overview of the topic:

https://www.youtube.com/watch?v=aircAruvnKk

WL features a wide selection of functions to build and train them. For example, most of the *Mathematica* commands that contain the word "Layer" or start with "Net" can be used for working with neural networks:

In[ ]:= **? *Layer***

In *Mathematica*, we can manipulate the neural network expressions created by the language like any other symbolic object and, for example, optimize their weights using training data. A more detailed overview of neural networks in the Wolfram Language is available at:

https://reference.wolfram.com/language/tutorial/NeuralNetworksOverview.html

The Wolfram Neural Net Repository contains a vast collection of trained and untrained models to build, train an deploy neural networks:

https://resources.wolframcloud.com/NeuralNetRepository/

The following examples show how to get started using them in *Mathematica*.

6.4.1 A Basic Introductory Example
- Given a list of data points $\{x_i, y_i\}$:

In[ ]:= **data = {{0, 3.1}, {1, 3.4}, {2, 4.2}, {3, 4.4}, {4, 4.9}, {5, 5}};**

- if we want to find the best linear fit, we can just use Fit to obtain the desired equation of the line:

In[ ]:= **model = Fit[data, {1, x}, x]**

Out[ ]= $0.405714 x + 3.15238$

However, let's use this toy example to show how easy it would be to plot the line of best fit and visualize the errors using LeastSquaresPlot, one of the functions available in the Wolfram Function Repository:

https://resources.wolframcloud.com/FunctionRepository/resources/LeastSquaresPlot/

In[ ]:=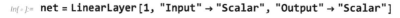

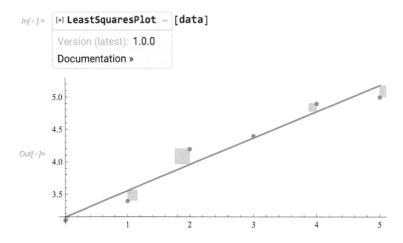

6.4.2 Using a Neural Network

■ An alternative approach would be to use a neural network to find the coefficients for the slope and the y-intercept:

In[ ]:= **net = LinearLayer[1, "Input" → "Scalar", "Output" → "Scalar"]**

Out[ ]=

In[ ]:= **inputOutputData = Rule @@@ data**

Out[ ]= {0 → 3.1, 1 → 3.4, 2 → 4.2, 3 → 4.4, 4 → 4.9, 5 → 5}

In[ ]:= **trained = NetTrain[net, inputOutputData]**

Out[ ]=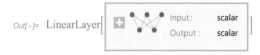

■ We can now extract the values of the slope and the intercept by using the options "Weights" and "Biases", respectively:

In[ ]:= **NetExtract[trained, "Weights"] // Normal**

Out[ ]= (0.407973)

In[ ]:= **NetExtract[trained, "Biases"] // Normal**

Out[ ]= {3.14473}

In[ ]:= **NetChain[{trained, trained}][1]**

Out[ ]= 4.59413

■ The resulting graph is very similar to the one obtained previously:

```
In[ ]:= Show[
    ListPlot[data],
    Plot[trained[x], {x, 0, 5}]
  ]
```

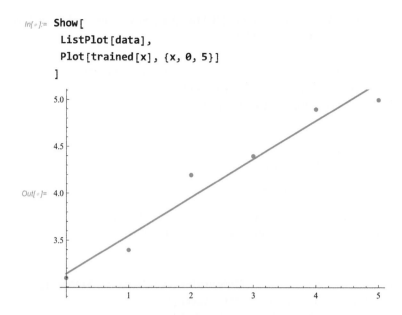

6.4.3 Revisiting the Titanic Dataset

In this example we build a neural network classifier to assign the labels "survived" or "died" to passengers, based on information about their ticket class, age, and sex:

▪ As usual, we download the dataset first:

```
In[ ]:= ResourceData["Sample Data: Titanic Survival"]
```

Class	Age	Sex	SurvivalStatus
1st	29. yr	female	survived
1st	0.9167 yr	male	survived
1st	2. yr	female	died
1st	30. yr	male	died
1st	25. yr	female	died
1st	48. yr	male	survived
1st	63. yr	female	survived
1st	39. yr	male	died
1st	53. yr	female	survived
1st	71. yr	male	died
1st	47. yr	male	died
1st	18. yr	female	survived
1st	24. yr	female	survived
1st	26. yr	female	survived
1st	80. yr	male	survived
1st	—	male	died
1st	24. yr	male	died
1st	50. yr	female	survived
1st	32. yr	female	survived
1st	36. yr	male	died

rows 1–20 of 1309

- We then modify the data to follow this pattern: {{Class, Age, Sex}→survived or {Class, Age, Sex}→died}. Missing data is represented by "—":

In[ ]:= **dataset =**
      Rule[Lookup[#, {"Class", "Age", "Sex"}], Lookup[#, "SurvivalStatus"]] & /@
        ResourceData["Sample Data: Titanic Survival"];
    {train, test} = Normal@TakeDrop[dataset, 1000];

    Short[train]

Out[ ]//Short=

  $\Big\{$ {3rd, Missing[Not Available], male} → survived,

     ≪307≫, $\Big\{$3rd, 29. yr, male$\Big\}$ → died$\Big\}$

- Next, we define an neural network using NetChain. In this case, we are creating a multilayer perceptron (MLP) classifier with a **"FeatureExtractor"** encoder to deal automatically with mixed and missing data:

In[ ]:= **net = NetChain[**
        {DropoutLayer[], LinearLayer[12], ElementwiseLayer[LogisticSigmoid],
         LinearLayer[4], LinearLayer[2], SoftmaxLayer[]},
        "Output" → NetDecoder[{"Class", {"survived", "died"}}],
        "Input" → NetEncoder["FeatureExtractor"]
       ];

- We now train the feature extractor and the classifier jointly, using NetTrain:

In[ ]:= **trainedNet = NetTrain[net, train]**

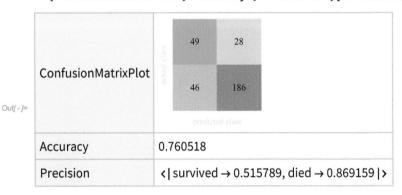

Out[ ]= NetChain[... Input port: feature extractor / Output port: class]

- Finally, we test the classifier on unseen data:

In[ ]:= **cm = ClassifierMeasurements[trainedNet, test];**
    AssociationMap[cm,
       {"ConfusionMatrixPlot", "Accuracy", "Precision"}] // Dataset

Out[ ]=

ConfusionMatrixPlot	49 28 / 46 186
Accuracy	0.760518
Precision	⟨| survived → 0.515789, died → 0.869159 |⟩

We leave it as an exercise for the reader to compare this MLP with the classifier we created previously using Classify. Which one generates better results?

6.4.4 Handwritten Numbers

This example shows how to download and train a neural network to recognize handwritten figures:

For this exercise NVIDIA GPU with supported compute capability should be installed, in other case skip it.

- Let's download the model first using NetModel:

In[•]:= `net = NetModel["LeNet Trained on MNIST Data"]`

Out[•]= NetChain[Input port: image; Output port: class; Number of layers: 11]

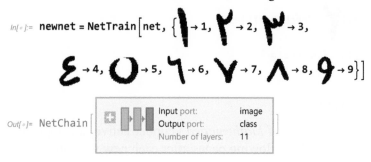

- Next, we test the net:

In[•]:= `net /@` 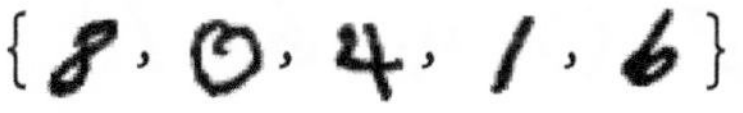

Out[•]= `{8, 0, 4, 1, 6}`

- We can also train the model with new cases using NetTrain and test it:

In[•]:= `newnet = NetTrain[net,` 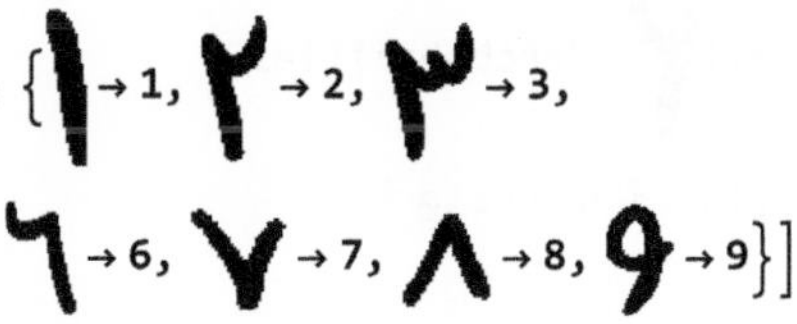

Out[•]= NetChain[Input port: image; Output port: class; Number of layers: 11]

In[•]:= `newnet[`  `]`

Out[•]= 7

- Based on the existing net, it's very easy to initialize a new model (NetInitialize) and train it (NetTrain) to classify handwritten inputs as either "Arabic" or "Latin":

```
In[•]:= languageNet = NetInitialize[
   NetJoin[
    NetTake[net, 9],
    NetChain[{2, SoftmaxLayer[]},
     "Output" → NetDecoder[{"Class", {"Arabic", "Latin"}}]]],
   All]
```

Out[•]= NetChain[Input port: image; Output port: class; Number of layers: 11]

In[∘]:= **trainedLanguageNet = NetTrain[languageNet,**

{٢ → "Arabic", ٣ → "Arabic", ٤ → "Arabic", ٦ → "Arabic",

٧ → "Arabic", ٨ → "Arabic", 8 → "Latin", ٥ → "Latin",

4 → "Latin", 6 → "Latin", 1 → "Latin"}, TargetDevice → "GPU"]

Out[∘]:= NetChain[

Input port:	image
Output port:	class
Number of layers:	11

]

In[∘]:= **trainedLanguageNet /@**

{٢, ٣, ٤, ٦, ٧, ٨, 8, ٥, 4, 6, 1}

Out[∘]:= {Arabic, Arabic, Arabic, Arabic,
        Arabic, Arabic, Latin, Latin, Latin, Latin, Latin}

In[∘]:= **trainedLanguageNet[٢, "Probabilities"]**

Out[∘]:= ⟨|Arabic → 1., Latin → 1.9582×10^{-7}|⟩

6.5 Additional Resources

The machine learning functionality of *Mathematica* is under active development so it would be a good idea to visit the following pages after each new version release:

https://reference.wolfram.com/language/guide/MachineLearning.html
https://reference.wolfram.com/language/workflowguide/MachineLearning.html
https://reference.wolfram.com/language/guide/NeuralNetworks.html
https://reference.wolfram.com/language/tutorial/NeuralNetworksOverview.html

Introduction to Machine Learning by Etienne Bernard is a very nice introduction to machine learning using *Mathematica* that can be read online at: https://wolfr.am/iml. A printed version is also available.

7

Calculating π and Other Mathematical Tales

The first six sections of this chapter are dedicated to π, one of the most famous mathematical constants. What looks like a simple geometric ratio between the length of a circle's circumference and its diameter has surprising properties and unexpected connections to seemingly unrelated fields. Many famous mathematicians throughout history have created their own formulas to calculate the largest number of decimals of π in the most efficient possible way, something that they currently still do. The last two sections discuss the Riemann hypothesis, one of the most important unsolved problems in mathematics, and its connection to the prime numbers.

7.1 The Origins of π

Approximately 4,000 years ago, Egyptians and Babylonians already knew that the ratio between the circle's circumference and its diameter, what we call π (pi) nowadays, had a constant value slightly higher than 3.

The first estimates were most likely based on experimental measurements. After drawing a circle and measuring its perimeter and diameter, you will notice that the perimeter is a bit more than three times the diameter.

The Old Testament includes an estimation of π in the First Book of Kings 7:23, referring to the construction of Solomon's temple. We can read: "And he made a molten sea, ten cubits from the one brim to the other: it was round all about, and his height was five cubits: and a line of thirty cubits did compass it round about"; that is, its circumference was 30 cubits and its diameter was 10 cubits, so the relationship between the circumference and the diameter is 3.

In the Great Pyramid of Giza there are ratios clearly showing that the Egyptians had obtained a good approximation to the value of π. The Rhind Papyrus (circa 1650 B.C.), one of the best sources about mathematics in Ancient Egypt, contains the following ratio as the value of π: $\frac{256}{81}$ =3.16049.

The Babylonians noticed that the length of a circle was slightly bigger than the one of an hexagon inscribed inside it getting a π value of $3\frac{1}{8} = 3.125$, quite close to the correct value of 3.14159....

The following example describes a function to inscribe a n-sided polygon inside a circle.

- The display of a polygon inscribed in a circle can be done using the `Circle` and `Polygon` functions inside the `Graphics` command. `Circle[]` without arguments indicates a circle of radius $r = 1$ and center at $\{x, y\} = \{0, 0\}$. `Polygon[ {x_1, y_1}, ..., {x_n, y_n}]` represents a polygon whose vertices are given by the coordinates $\{x_1, y_1\}, ..., \{x_n, y_n\}$. We use `Inset` to place text inside the graphical object. In the function definition we include `?EvenQ` and `?Positive` to execute the function only if n is an even positive integer.

```
In[ ]:= polig1[(n_ ?EvenQ) ?Positive] := Graphics[{Circle[],
        {LightGray, Polygon[Table[{Cos[2 π k /n], Sin[2 π k /n]}, {k, n}]]},
        Inset["Inscribed Polygon", {0, 0}]}, ImageSize → 130]
```

- We can add a warning message that will be displayed if the requirements for n are not met. This function must be defined immediately after the previous one. This way, **polig1** will check first whether n is a positive even number and if not, it will display the warning message. If we reverse the order in which we define the functions, it will not work.

```
In[ ]:= polig1[n_] := "n must be an even positive number."
```

```
In[ ]:= polig1[6]
```

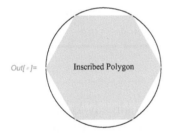

Out[ ]=

```
In[ ]:= polig1[5]
```

Out[ ]= n must be an even positive number.

```
In[ ]:= polig1[-2]
```

Out[ ]= n must be an even positive number.

An alternative way to display a 2D figure without using *Mathematica* functions is to use the Drawing Tools, available in the **Graphics** menu or by pressing CTRL+D in Windows or CMD+T in OS X.

7.2 Archimedes' Approximation

Archimedes (287–212 B.C.) is one of the most famous individuals in the history of science. Even today some of his discoveries are still being studied in schools and his famous palimpsest, the oldest surviving copy of his works, has been the subject of extensive research since 1999. He lived at the end of the Ancient Greece era, the same period as other famous mathematicians such as Euclid (ca. 325–ca. 270 B.C.) and Eratosthenes (ca. 284–ca. 192 B.C.). They were all located in a small geographical region consisting of several cities around the Aegean sea, a lucky coincidence that has not happened again in the history of humankind. Archimedes' life is the subject of many legends of which the best-known one is probably the story of how he was able to set on fire a large fleet of Roman galleys that were trying to attack Syracuse. His extraordinary intelligence was acknowledged even by his own enemies. Plutarch tells us that during the siege of Syracuse a Roman soldier found the old man so absorbed in his geometric drawings that he ignored his orders and was killed as a result. Marcellus, the person in charge of the Roman troops, had given express orders to spare his life and when he heard the news he punished the soldier, mourned Archimedes' death, and buried him with honors.

About the The Archimedes Palimpsest: http://www.archimedespalimpsest.org/

In Archimedes' tomb there is a cylinder enclosing a sphere, a reference to one of his greatest findings, that the sphere has 2/3 the volume of its circumscribing cylinder. The same 2/3 ratio applies to all surfaces.

Next we are going to show how to inscribe a sphere inside a cylinder.

- To show a cylinder enclosing a sphere, we use the syntax Graphics3D[{{*properties*, Cylinder}, {*properties*, Sphere}}, options]. We use Opacity to define the degree of transparency in the cylinder. We can also choose the color and other relevant details using options. Since we are not using any arguments for either Cylinder[] or Sphere[], the commands assume by default that $r = 1$, $\{x, y, z\} = \{0, 0, 0\}$ and, in the case of the cylinder, $h = 1$.

In[ ]:= **Graphics3D[{{Opacity[0.2], Blue, Cylinder[]}, { Green, Sphere[]}},**
   Boxed → False, ImageSize → 100]

Out[ ]=

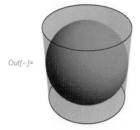

- The same graph can be done with ContourPlot3D, a *Mathematica* command used for any 3D function for which we know its analytic expression. PlotLabel, a general graphics option, adds a legend to the graph. In this example we use the equations of both geometric figures as legends and display them on separate lines with TableForm.

```
In[•]:= ContourPlot3D[{x^2 + y^2 + z^2 == 1, x^2 + y^2 == 1}, {x, -1, 1}, {y, -1, 1},
        {z, -1, 1}, ContourStyle → {Green, {Blue, Opacity[0.2]}}, Mesh → None,
        Axes → False, Boxed → False, ImageSize → 200, PlotLabel → TableForm[
        {"Sphere:" x^2 + y^2 + z^2 == 1, "Cylinder:" x^2 + y^2 == 1 ∧ 0 ≤ z ≤ 1}]]
```

Sphere: $x^2 + y^2 + z^2 = 1$
Cylinder: $x^2 + y^2 = 1 \wedge 0 \leq z \leq 1$

Out[•]=
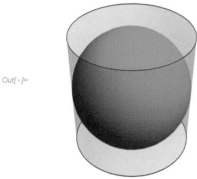

The method that follows enables us to verify that the ratio between the volume of a sphere and that of its circumscribing cylinder is 2/3 (without using the formulas for their respective volumes).

We first compute the volume of the sphere, V_{sphe}, and of the cylinder, V_{cyl}, with $r = 1$ by integrating their analytical expressions and then calculate V_{sphe}/V_{cil}. Archimedes obviously did not use this method since he didn't know integral calculus.

- We use Boole to indicate that the integral of the points that do not satisfy the condition is 0.

```
In[•]:= volsphe = Integrate[ Boole[x^2 + y^2 + z^2 ≤ 1],
        {x, -1, 1}, {y, -1, 1}, {z, -1, 1}]
```

Out[•]= $\dfrac{4\pi}{3}$

```
In[•]:= volcyl =
        Integrate[ Boole[x^2 + y^2 ≤ 1], {x, -1, 1}, {y, -1, 1}, {z, -1, 1}]
```

Out[•]= 2π

```
In[•]:= volsphe / volcyl
```

Out[•]= $\dfrac{2}{3}$

One of Archimedes' biggest discoveries was a new method for computing the value of π. He realized that by dividing a circle into identical sectors, the sum of their areas would resemble that of a sum of triangles whose total area would approximate the area of the circle. Geometrically, the previous method is equivalent to placing the sectors together inverting every other one. If each sector is replaced by its inscribed triangle, that Archimedes knew how to calculate, the resulting area would be very close to that of the circle. Furthermore, when $r = 1$, since the area of the circle is $A_c = \pi r^2$, then $A_c = \pi$, and from that formula we can deduce that the value of π is going to be between the value area of the inscribed polygon, A_{ins} and that of the circumscribed polygon A_{cir}, that is, $A_{\text{ins}} \le A_c = \pi \le A_{\text{cir}}$.

- The previous procedure can be easily modeled in *Mathematica*. We use GraphicsRow to display all the figures in the same row. The higher the number of triangles, the closer the resemblance of the resulting graphical object to a rectangle.

```
In[ ]:= Manipulate[
    GraphicsRow[Table[Graphics[{EdgeForm[Opacity[1]], Table[{Hue[i/n, .5],
        Disk[{i Sin[π/n], ((-1)^(i + 1) + 1) Cos[π/n]/2}, 1,
          π {(-1)^i/2 - 1/n, (-1)^i/2 + 1/n}]}, {i, n}]}], {n, 4, m}]],
    {{m, 6, "number of sectors"}, 4, 20, 2, Appearance → "Labeled"}]
```

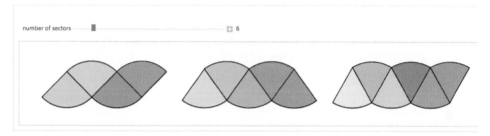

7.3 π with More Than One Trillion Decimals

A 16-digit approximation of π, like the one available in any calculator, is more than enough for practical computations.

For example, the calculation error when measuring the earth circumference with such a number would be much smaller than 1 mm.

- The average distance from the earth to the sun is:

```
In[ ]:= sunDist = ⊟ Average Sun distance in meters
```

Out[ ]= 1.49598×10^{11} m

- The difference between using the exact value of π and its approximation with a 10^{-16} precision is insignificant.

```
In[ ]:= 2 π 10^-16 sunDist
```

Out[ ]= 0.0000939951143117 m

Despite this fact, the number of functions created to compute π with the largest number of decimals in the fastest possible way since Archimedes is remarkable. Some of the best mathematicians in history have developed their own approaches. In the following paragraphs, we'll show some of them.

Although Archimedes' method is inefficient, little progress was made until Newton in 1666 described a radically new method.

(Ve)The Discovery That Transformed Pi: https://www.youtube.com/watch?v=gMlf1ELvRzc

He found a pi approximation by evaluating the first 22 terms of this sum:

$$\pi = \frac{3\sqrt{3}}{4} + 24\left(\frac{1}{12} - \frac{1}{5\times 2^5} - \frac{1}{28\times 2^7} - \frac{1}{72\times 2^9}\cdots\right)$$

Newton generated this expression for π by considering the shape below. His approach to calculate the famous constant is based on computing the area of a semicircle using an integral.

```
In[ ]:= Graphics[{{LightBlue, Disk[{0.5, 0}, 0.5, {2 Pi/3., Pi}]},
         {LightGreen, Polygon[{{0.25, 0}, {0.5, 0}, {0.25, Sqrt[3]/4.}}]},
         Circle[{0.5, 0}, 0.5, {0, Pi}], Line[{{0, 0}, {0, 0.6}}]],
         Line[{{0.5, 0}, {0.25, 0.5 Sqrt[3]/2}, {0.25, 0}}], ,
         Text[Style["A", 16], {0., 0.}, -{1, 1}],
         Text[Style["B", 16], {0.26, 0}, -{1, 1}],
         Text[Style["C", 16], {0.5, 0}, -{1, 1}],
         Text[Style["D", 16], {0.25, 0.5 Sqrt[3.]/2}, -{1, 1}]],
       Axes → {True, False}, Ticks → {{1/4, 1/2, 1}, {1/2}},
       PlotRange → {{-0.1, 1.1}, {-.1, 0.6}}]
```

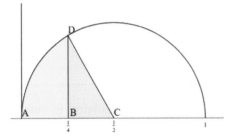

Here is a more detailed explanation of his solution:

- The area α under the arc AD (light blue) is equal to the area of the sector ADC ($\frac{\pi(1/2)^2}{6} = \frac{\pi}{24}$), minus the area of triangle BCD (light green) ($\frac{1}{2}(\text{BC}\times\text{BD}) = \frac{1}{2}(\frac{1}{4}\times\sqrt{\frac{1}{4}-\frac{1}{4^2}}) = \frac{\sqrt{3}}{32}$). That is, $\frac{\pi}{24} - \frac{\sqrt{3}}{32}$.

```
In[ ]:= sol1 = Solve[α == pi/24 - Sqrt[3]/32, pi]
```

$$Out[ ]= \left\{\left\{pi \to \frac{3}{4}\left(32\,\alpha + \sqrt{3}\right)\right\}\right\}$$

- The circle has center (1/2, 0) and radius = 1/2. Therefore:

```
In[ ]:= sol2 = SolveValues[(x - 1/2)^2 + y^2 == 1/4, y]
```

$$Out[ ]= \left\{-\sqrt{x-x^2}, \sqrt{x-x^2}\right\}$$

- Now the integration can be computed directly

```
In[ ]:= Integrate[sol2[[2]], {x, 0, 1/4}]
```

$$Out[ ]= \frac{1}{96}\left(4\pi - 3\sqrt{3}\right)$$

$In[\circ]:=$ $\mathbf{Simplify}\left[\dfrac{3}{4}\left(\sqrt{3}+32\times\%\right)\right]$

$Out[\circ]=$ π

Obviously, in the 17th century, these calculations took much longer.

- Since $y = \sqrt{x - x^2} = \sqrt{x}\,\sqrt{1-x}$, we can now find the value of $\sqrt{x}\,\sqrt{1-x}$ using the binomial expansion:

$In[\circ]:=$ $\mathbf{bexpansion} = \sqrt{\mathbf{x}}\ \mathbf{Series}\left[\sqrt{\mathbf{1-x}},\ \{\mathbf{x},\ \mathbf{0},\ \mathbf{4}\}\right]$

$Out[\circ]=$ $\sqrt{x} - \dfrac{x^{3/2}}{2} - \dfrac{x^{5/2}}{8} - \dfrac{x^{7/2}}{16} + O\!\left(x^{9/2}\right)$

$In[\circ]:=$ $\mathbf{Integrate}\left[\left\{\sqrt{\mathbf{x}},\ \dfrac{\mathbf{x}^{3/2}}{2},\ \dfrac{\mathbf{x}^{5/2}}{8},\ \dfrac{\mathbf{x}^{7/2}}{16}\right\},\ \{\mathbf{x},\ \mathbf{0},\ \mathbf{1/4}\}\right]$

$Out[\circ]=$ $\left\{\dfrac{1}{12},\ \dfrac{1}{160},\ \dfrac{1}{3584},\ \dfrac{1}{36\,864}\right\}$

So $\alpha = \displaystyle\int_{0}^{1/4}\sqrt{x} - \dfrac{x^{3/2}}{2} - \dfrac{x^{5/2}}{8} - \dfrac{x^{7/2}}{16} - \dfrac{5\,x^{9/2}}{128} + O\!\left(x^{11/2}\right)\,dx = \dfrac{1}{12} - \dfrac{1}{160} - \dfrac{1}{3584} - \dfrac{1}{36\,864} - \ldots =$

$\left(\dfrac{1}{12} - \dfrac{1}{5\times 2^5} - \dfrac{1}{28\times 2^7} - \dfrac{1}{72\times 2^9}\ \ldots\right).$

- Therefore:

$In[\circ]:=$ $\mathbf{sol1[\![1,\ 1]\!]}$

$Out[\circ]=$ $pi \to \dfrac{3}{4}\left(32\,\alpha + \sqrt{3}\right)$

$In[\circ]:=$ $\mathbf{\%\ /.\ \alpha} \to \dfrac{1}{12} - \dfrac{1}{160} - \dfrac{1}{3584} - \dfrac{1}{36\,864}\ \mathbf{//\ N}$

$Out[\circ]=$ $pi \to 3.14169$

Contrary to expectations given the original definition of π as the ratio between the perimeter of a circumference and its diameter, many of its later definitions do not imply a geometric relationships. The original one was of a practical nature. Other non-geometrical definitions are also valid. As a matter of fact, we could take any of them as the definition of π. Next we present some examples.

- The formula below is probably the easiest way (but not the most effective one) to calculate π. Notice that *Mathematica* is able to obtain an exact result for infinite series.

$In[\circ]:=$ $4\displaystyle\sum_{j=0}^{\infty}\dfrac{(-1)^{j}}{2\,j+1}$

$Out[\circ]=$ π

- The next one shows π as the result of an integral computation. It's due to Leibniz, whose integration methods won over those from his contemporary fellow mathematician Newton.

$In[\circ]:=$ $\displaystyle\int_{0}^{\infty}\dfrac{1}{t^2+1}\,dt$

$Out[\circ]=$ $\dfrac{\pi}{2}$

- There are very curious methods such as the ones that compute π using a series of continuous fractions. Here's one example:

$ln[\circ]:=$ $4/\pi == 1 + 1/\left(3 + 2^2/\left(5 + 3^2/\left(7 + 4^2/\left(9 + 5^2/\left(11 + 6^2/(13 + ...)\right)\right)\right)\right)\right)$

$Out[\circ]=$ $\dfrac{4}{\pi} = \cfrac{1}{3 + \cfrac{4}{5 + \cfrac{9}{\cfrac{16}{\cfrac{25}{\cfrac{36}{...+13} + 11} + 9} + 7}}} + 1$

One of the most surprising formulas to calculate π comes from Ramanujan (1887–1920), the famous Indian mathematician.

Born to a poor family, he didn't have a formal education, learning mathematics completely on his own (The movie *The Man Who Knew Infinity* is based in his life). When he was in his teens, he read a book containing only theorems and a list of problems with their respective solutions. Since he could mentally arrived at the answer without the need to write down intermediate calculations, he thought that would be the ideal way to proceed: to figure out mentally the solution and how to get it and just write down the final result. Accordingly, he made a list of several identities that according to him were self-evident and sent them to several mathematicians. One of them, Hardy, quickly realized that the author of the list was a genius, and he invited him to Oxford. From his time in England, we have the following famous anecdote: One day Hardy was telling Ramanujan that he had taken a cab with a very boring license plate number 1729, to which Ramanujan replied that on the contrary that number was very interesting since it was the smallest one that could be decomposed into the sum of two different cubes.

- In fact:

$ln[\circ]:=$ $9^3 + 10^3 == 1^3 + 12^3 == 1729.$

$Out[\circ]=$ True

- It's not by chance then that in 1914 Ramanujan found the following surprising formula:

$ln[\circ]:=$ $\mathtt{ramapi[n\_]} := \dfrac{2\sqrt{2}}{9801} \sum_{j=0}^{n} \dfrac{(4j)!\,(26390j + 1103)}{\left(4^{4j} j!^4\right) 99^{4j}}$

$ln[\circ]:=$ $\mathtt{ramapi[\infty]}$

$Out[\circ]=$ $\dfrac{1}{\pi}$

- To check how fast the above formula converges we proceed as follows: Since $\pi = 1/\mathrm{ramapi}[\infty]$, we increase the number of digits in the expression $\pi - 1/\mathrm{ramapi}[n]$, using N[] to force *Mathematica* to display the numeric approximation. We can see that with just the first 3 terms, the approximation is accurate to 30 decimal places:

$In[\circ]:=$ $\mathtt{TableForm}\Big[\mathtt{Table}\Big[\Big\{\mathtt{n}, \mathtt{N}\Big[\pi - \dfrac{1}{\mathtt{ramapi}[\mathtt{n}]}, \mathtt{10}\Big]\Big\}, \{\mathtt{n}, \mathtt{1}, \mathtt{5}\}\Big]\Big]$

Out[◦]//TableForm=

1	$-6.395362624 \times 10^{-16}$
2	$-5.682423256 \times 10^{-24}$
3	$-5.238896280 \times 10^{-32}$
4	$-4.944187579 \times 10^{-40}$
5	$-4.741011769 \times 10^{-48}$

- If we add a 6th term we'll get an error message. The reason is that the series convergence is so fast that we need to substantially increase the computation precision. To do that we change the value of the variable $MaxExtraPrecision and limit the scope of the new assignment with Block to the previous expression only; otherwise, the new value would be applied to all calculations done after the latest definition.

$In[\circ]:=$ $\mathtt{Block}\Big[\{\mathtt{\$MaxExtraPrecision = 10\,000}\},$

$\qquad \mathtt{TableForm}\Big[\mathtt{Table}\Big[\Big\{\mathtt{n}, \mathtt{N}\Big[\pi - \dfrac{1}{\mathtt{ramapi}[\mathtt{n}]}, \mathtt{10}\Big]\Big\}, \{\mathtt{n}, \mathtt{1}, \mathtt{10}\}\Big]\Big]\Big]$

Out[◦]//TableForm=

1	$-6.395362624 \times 10^{-16}$
2	$-5.682423256 \times 10^{-24}$
3	$-5.238896280 \times 10^{-32}$
4	$-4.944187579 \times 10^{-40}$
5	$-4.741011769 \times 10^{-48}$
6	$-4.598865016 \times 10^{-56}$
7	$-4.499979208 \times 10^{-64}$
8	$-4.433276097 \times 10^{-72}$
9	$-4.391477319 \times 10^{-80}$
10	$-4.369600809 \times 10^{-88}$

By March 14, 2019 (the Pi day) the approximation reached 31.4 trillions digits ($31.4 \cdot 10^{12}$) using the Chudnovsky algorithm (it is the used by *Mathematica*). To put that figure in perspective, if we wanted to print that many digits we would need a library of 30 million volumes (500 km worth of shelves) .

$In[\circ]:=$ $\mathtt{chudnovsky}[\mathtt{n\_}] := 12 \displaystyle\sum_{\mathtt{k=0}}^{\mathtt{n}} \dfrac{(-1)^{\mathtt{k}} * (6*\mathtt{k})\,! * (13\,591\,409 + 545\,140\,134 * \mathtt{k})}{(3*\mathtt{k})\,! * (\mathtt{k}\,!)^{3} * 640\,320^{3*\mathtt{k}+3/2}}$

$In[\circ]:=$ $\mathtt{chudnovsky}[\infty]$

Out[◦]= $\dfrac{1}{\pi}$

To verify that a sequence is correct, we compare part of it with the ones obtained by different methods. For example:

- We can combine RealDigits and [[...]] to show the digits located in positions *n* to *m + n* after the decimal point. We extract the digit *n+1* since 3, the first digit, is not decimal. We then use $n + m + 5$ to increase the precision with 5 more digits than the required ones to calculate $m + n$.

$In[\circ]:=$ $\mathtt{pinth}[\mathtt{n\_}, \mathtt{m\_}] := \mathtt{RealDigits}[\mathtt{N}[\pi - 3, \mathtt{n} + \mathtt{m} + 5]][[1, \mathtt{n} ;; \mathtt{n} + \mathtt{m}]]$

- For example, we can check that after position 762, the digit 9 appears six consecutive times.

In[ ]:= **pinth[760, 10]**

Out[ ]= {3, 4, 9, 9, 9, 9, 9, 9, 8, 3, 7}

But, how can a record be tested to see if it is still valid? Usually the Bailey–Borwein–Plouffe's formula (1996) is used.

In[ ]:= **bayley1[n_] :=** $\sum_{k=0}^{n} \frac{1}{16^k} \left(\frac{4}{8k+1} - \frac{2}{8k+4} - \frac{1}{8k+5} - \frac{1}{8k+6} \right)$

"On the rapid computation of various polylogarithm constants" (*Mathematics of Computation* 66–218, 1997) the authors describe a method in which by using the previous formula it can calculate directly in base 16 the nth digit of π. We can use this fact to verify whether a record is valid.

- The formulas that we have seen so far represent just a small fraction of the existing ones. We can take a look at some of those formulas by using *Mathematica*'s free-form input. Press = and type Pi. To see the results, expand the output by clicking on the + symbol in the top right corner.

7.4 Buffon's Method

The Frenchman Georges-Louis Leclerc (1707–1788), Comte de Buffon, was a naturalist whose encyclopedia *Histoire Naturelle* had a great influence on his 19th century colleagues. Charles Darwin even cites him in his book *On the Origin of Species*. In mathematics, he's famous for his experimental method to calculate π.

The method, as shown in Figure 7.1, consists of randomly throwing a needle of length l onto a plane containing parallel lines at a fixed distance $d > l$ from each other.

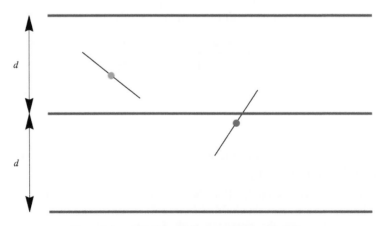

Figure 7.1   Visualizing Buffon's method for calculating π.

When throwing the needle, there are two possibilities: either the needle crosses one of the lines or it doesn't. The probabilities associated to each of those outcomes are, respectively, $p_0 = 1 - 2\,r\,\theta$, and $p_1 = 2\,r\,\theta$, with $r = l\,/\,d$ and $\theta = 1\,/\,\pi$.

Let N_1 be the number of times that the needle crosses a line out of n throws. Then we can estimate p_1 using $\hat{p}_1 = N_1\,/\,n$ and consequently obtain θ. Therefore, $\pi = 1/\theta$.

$$\hat{\theta} = \frac{\hat{p}_1}{2\,r} = \frac{N_1}{2\,r\,n}.$$

The previous experiment has been performed on several occasions, and it has always required thousands of throws to achieve a precision better than two digits. Nowadays we use simulations where the throws are replaced by computer-generated (pseudo)random numbers.

Simulation-based procedures, usually known as Monte Carlo methods, enable us to calculate π probabilistically in many different ways.

A very simple example to approximate the value of π is to first draw a circle of radius 1 circumscribed by a square with sides of length 2 and then to drop a ball m times onto the square counting how many times (n) the ball falls inside the circle. Assuming that the ball can land with equal probability on any point in the square, the probability that it will land inside the circle is given by the ratio between the area of the circle and that of the square, that is, $\frac{n}{m} \approx \pi\,r^2\,/\,l^2$, since $l = d = 2$, then $4\,n/m \to \pi$.

- The above experiment can be simulated with *Mathematica*. We just need to generate m pairs of real numbers between -1 and 1 and count how many of them "land" inside a circle with radius $r = 1$ (we can use Cases to check that the pairs verify $x^2 + y^2 < 1$ and count with Length how many they are).

```
In[ ]:= pib[m_] := Module[{data, n},
          data = RandomReal[{-1, 1}, {m, 2}];
          n = Length[Cases[data, {x_, y_} /; x^2 + y^2 < 1]];
          4. n / m]
```

- The method converges very slowly.

```
In[ ]:= pib[100 000]
```

```
Out[ ]= 3.14408
```

- The previous function can be extended to display the results graphically. Notice that text needs to go between quotation marks and that inside a string of text we can use \t or \n to insert a tab or a new line, respectively.

```
In[ ]:= Manipulate[Module[{data, n},
          data = RandomReal[{-1, 1}, {m, 2}];
          n = Length[Cases[data, {x_, y_} /; x^2 + y^2 < 1]];
          Text@Style[Column[{Graphics[
                {PointSize[0.006], Orange, Disk[{0, 0}, 1], Black, Point[data]}],
              Row[{"inside: ", n, "\toutside: ", m - n, "\ntotal:   ", m,
                "\tπ ≈ 4 × ", n, "/", m, " = ", 4. n / m}]}], "Label"]],
        {{m, 100, "Random pairs generated"}, 1, 10 000,
          1, Appearance → "Labeled"}]
```

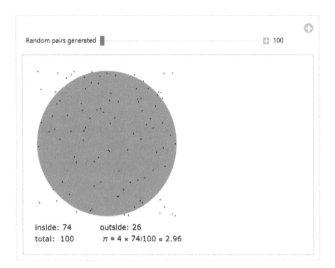

To find out more about Buffon's method you can visit:

http://demonstrations.wolfram.com/BuffonsNeedleProblem

Throwing Buffon's Needle with Mathematica by Enis Siniksaran: https://www.mathematica-journal.com/2009/01/12/throwing-buffons-needle-with-mathematica/

7.5 Application: Are the Decimal Digits of π Random?

It doesn't matter what method we use, if it's correct, we will always get the same sequence of π decimals even though such a sequence will never exhibit any periodicity since π is an irrational number (actually a transcendental one) and, therefore, it cannot be expressed as the ratio of two integers. However, if we explore a long enough sequence of its decimals, we will discover an interesting behavior: the frequency in which the digits appear is similar, that is, the numbers 0, 1, 2, ..., 9 will appear approximately with the same probability (this characteristic doesn't depend on expressing π in base 10. The same behavior happens regardless of the chosen base). Since this property is the same that random numbers originated from a discrete uniform distribution have, unless we knew in advance that certain digits were π decimals, we wouldn't be able to differentiate them from random numbers generated by a roulette with 10 positions, all equally likely. Numbers that have this property are called normal numbers. In the case of π, at least considering only the digits computed until now, it's been checked that it is normal but it has not been proved that it would be normal for an infinite number of digits although everything suggests that it would.

- To convert the digits of π to list elements, we use the function RealDigits to decompose the decimal approximation of π into a list of numbers. For example: 3.14159... becomes:

In[ ]:= **RealDigits[N[π]]**

Out[ ]= {{3, 1, 4, 1, 5, 9, 2, 6, 5, 3, 5, 8, 9, 7, 9, 3}, 1}

- The 1 at the end indicates that the first element corresponds to the integer part. If we just want the decimal part we can subtract 3 from 3.14159... : $3.14159... - 3 = 0.14159$. For example, the first 10 decimal digits of π are:

In[ ]:= `RealDigits[N[π - 3, 10]][[1]]`

Out[ ]= {1, 4, 1, 5, 9, 2, 6, 5, 3, 5}

- The following expression enables us to find a specific π digit in a short time. In the example below we find the 10,000,000th digit.

In[ ]:= `Timing[First[RealDigits[Pi, 10, 1, -10^7]]]`

Out[ ]= {9.875, {7}}

- This short expression tells us how often numbers 0–9 appear in the first 100,000 decimals of π.

In[ ]:= `Sort[Tally[RealDigits[N[π - 3, 100000]][[1]]]]`

Out[ ]=

$$\begin{pmatrix} 0 & 9999 \\ 1 & 10\,137 \\ 2 & 9908 \\ 3 & 10\,025 \\ 4 & 9971 \\ 5 & 10\,026 \\ 6 & 10\,029 \\ 7 & 10\,025 \\ 8 & 9978 \\ 9 & 9902 \end{pmatrix}$$

The previous output seems to confirm that in base 10 all digits are present with approximately the same probability. This indicates that they are distributed following a random discrete uniform distribution.

- To test this hypothesis for the first 10,000 π decimals we use a combination of the functions DistributionFitTest and DiscreteUniformDistribution.

In[ ]:= `vals = RealDigits[N[π - 3, 10000]][[1]];`

In[ ]:= `h = DistributionFitTest[vals,`
`        DiscreteUniformDistribution[{μ, ω}], "HypothesisTestData"];`
`    h[{"TestDataTable", All}]`

Out[ ]=

	Statistic	P-Value
Pearson χ^2	9.318	0.408453

P-Values above 0.05 (as in this case) suggest that the data come from a discrete uniform distribution. If this were true then π would be normal.

- The command below returns the frequency of numbers 0–9 as a function of the total number of digits used in the calculation.

In[ ]:= `Manipulate[`
`    BarChart[BinCounts[RealDigits[N[π - 3, 10000]][[1, 1 ;; u]], {0, 10, 1}],`
`        ChartLabels → Range[0, 9]],`
`    {{u, 100, "Total number of digits"},`
`        1, 10000, 1, Appearance → "Labeled"}]`

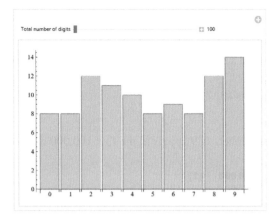

In a discrete uniform random distribution, all number combinations are possible. However, the probability of some of them happening is very small. For example: the probability that the same number, in this case any digit between 0 and 9, will repeat itself is very small, although sometimes it happens.

Let's create a command to identify when that happens with the digits of π.

- Split is very useful to divide lists into sublists consisting of identical elements.

In[ ]:= `Split[RealDigits[Pi, 10, 100][[1]]]`

Out[ ]= {{3}, {1}, {4}, {1}, {5}, {9}, {2}, {6}, {5}, {3}, {5}, {8}, {9}, {7}, {9}, {3}, {2}, {3}, {8},
{4}, {6}, {2}, {6}, {4}, {3, 3}, {8}, {3}, {2}, {7}, {9}, {5}, {0}, {2}, {8, 8}, {4}, {1}, {9},
{7}, {1}, {6}, {9}, {3}, {9, 9}, {3}, {7}, {5}, {1}, {0}, {5}, {8}, {2}, {0}, {9}, {7}, {4}, {9},
{4, 4}, {5}, {9}, {2}, {3}, {0}, {7}, {8}, {1}, {6}, {4}, {0}, {6}, {2}, {8}, {6}, {2}, {0}, {8},
{9, 9}, {8}, {6}, {2}, {8}, {0}, {3}, {4}, {8}, {2}, {5}, {3}, {4}, {2}, {1, 1}, {7}, {0}, {6}, {7}}

- Notice that some sublists contain more than one element. This happens when a digit repeats itself once or more consecutively. With the next function we can count the size of each sublist:

In[ ]:= `Length /@ Split[RealDigits[Pi, 10, 100][[1]]]`

Out[ ]= {1, 1, 1, 1, 1, 1, 1, 1, 1, 1, 1, 1, 1, 1, 1, 1, 1, 1, 1, 1, 1, 1, 1, 2, 1, 1, 1, 1, 1, 1, 1,
1, 2, 1, 1, 1, 1, 1, 1, 1, 2, 1, 1, 1, 1, 1, 1, 1, 1, 1, 1, 1, 1, 1, 1, 2, 1, 1, 1, 1, 1, 1, 1,
1, 1, 1, 1, 1, 1, 1, 1, 1, 1, 1, 2, 1, 1, 1, 1, 1, 1, 1, 1, 1, 1, 1, 1, 1, 2, 1, 1, 1, 1}

We see that in 6 cases a digit repeated itself once, and there were no instances of a number appearing consecutively 3 or more times.

- The next function counts the number of sublists with 1, 2, ..., n terms. Using Drop we eliminate the first element since it corresponds to the number of empty sublists. The function is then applied to the first 10^6 digits of π.

In[ ]:= `data1 =`
`    Drop[BinCounts[Length /@ Split[RealDigits[Pi, 10, 1000000][[1]]]], 1];`

- We present the results in table format: 1 indicates that a digit does not appear in consecutive positions, 2 that the same digit appears in two consecutive positions and so on.

In[ ]:= `TableForm[{Table[i, {i, Length[data1]}], data1},`
`    TableHeadings → {{"repetitions", "frequency"}, None}]`

Out[ ]//TableForm=

repetitions	1	2	3	4	5	6	7
frequency	809 776	81 256	7976	830	77	12	1

- By adding Position we can identify the order in which a certain repetition happens. For example, to check the positions in which the sequence {9,9} appears within a 100 decimal digits approximation of π, we would type:

In[ ]:= `Position[Split[First@RealDigits[Pi - 3, 10, 100]], {9, 9}]`

Out[ ]= $\begin{pmatrix} 42 \\ 75 \end{pmatrix}$

- To relate the sublist position to the number of digits, we proceed as follows:

In[ ]:= `Part[Accumulate[Length /@ Split[RealDigits[Pi - 3, 10, 100][[1]]]],`
        `{42, 75} - 1] + 1`

Out[ ]= `{44, 79}`

- This result indicates that the sequence 99 appears first in the 44th and 45th digits and the next one in the 79th and 80th ones. We can check it using the table below. In it, π decimals are grouped by rows containing 10 digits each.

In[ ]:= `TableForm[Partition[First@RealDigits[Pi - 3, 10, 100], 10],`
        `TableHeadings → {Range[10], Range[10]}]`

Out[ ]//TableForm=

	1	2	3	4	5	6	7	8	9	10
1	1	4	1	5	9	2	6	5	3	5
2	8	9	7	9	3	2	3	8	4	6
3	2	6	4	3	3	8	3	2	7	9
4	5	0	2	8	8	4	1	9	7	1
5	6	9	3	9	9	3	7	5	1	0
6	5	8	2	0	9	7	4	9	4	4
7	5	9	2	3	0	7	8	1	6	4
8	0	6	2	8	6	2	0	8	9	9
9	8	6	2	8	0	3	4	8	2	5
10	3	4	2	1	1	7	0	6	7	9

- The function SequencePosition can be used to find the position where the digit n will be repeated r consecutive times.
  In the example below, using the first 10,000,000 digits of π, we find the cases in which the number 9 appears consecutively exactly 7 times.

In[ ]:= `digitsPi = First[RealDigits[N[Pi, 10^7]]];`

In[ ]:= `SequencePosition[digitsPi, PadRight[{9}, 7, 9]]`

Out[ ]= $\begin{pmatrix} 1\,722\,777 & 1\,722\,783 \\ 3\,389\,381 & 3\,389\,387 \\ 4\,313\,728 & 4\,313\,734 \\ 5\,466\,170 & 5\,466\,176 \end{pmatrix}$

According to the output in the first 10,000,000 digits of π in 4 occasions the number 9 appears consecutively 7 times. The first occurrence starts at the 1,722,777-th digit (or 1,722,776-th digit after the decimal point) and finishes at position 1,722,783-th.

We may be tempted to believe that any possible combination of numbers, no matter how unlikely, will eventually happen as long as we use enough digits. For example, the most recent π approximations have shown that there's a sequence in which 0 is repeated 12 consecutive times and the same happens with the numbers 1–9. There's even a position in which 8 appears repeated 13 consecutive times. The frequency of occurrence is consistent with a discrete uniform distribution: the probabilities, in base 10, that 0 will appear consecutively once, twice, or 12 times are: $1/10$, $1/10^2$

and $1/10^{12}$, respectively. This means that, using a π approximation of 10^{12}digits, we'd most likely see one sequence of 12 consecutive 0s. The same could be said for other sequences.

- In the example below, we look for the pattern 3131313 in the first then million digits of Pi:

In[ ]:= `SequencePosition[digitsPi, {3, 1, 3, 1, 3, 1, 3}]`

Out[ ]= (3 662 426   3 662 432)

These types of patterns can lead us to a variant of what Borges proposed in his famous work *The Library of Babel*. The text describes a library containing all the books ever written and the ones that may be written in the future (actually, Borges mentions the number of pages and the number of lines per page but this is irrelevant for the purpose of our argument). Let's suppose that we could write a book where all its letters were *a*, followed by one containing only *b's* and so on until writing a final book with all *z's*. This way we would cover all the possible combinations of letters. The resulting number would be huge, immensely bigger than the number of atoms in the universe. Nevertheless it would be a finite number.

A variant of the previous story consists of replacing the order of the letters with probabilities. Let's supposed that we have a roulette with 35 positions representing the 26 letters of the English alphabet plus certain symbols such as punctuation marks, whitespace characters, etc. If we played such roulette enough times, we would end up not only with every possible word, sentence, and paragraph but also with every possible book! The only problem would be the time and the space to store the results. We can speculate that the same applies to π. For example, if we express it in base 35 and assign the digit 1 to the letter *a*, the digit 2 to *b* and so on, with a number of digits big enough we will find the library of Babel (it doesn't matter how big the number needs to be since π has an infinite number of digits). This means that all books: past, present, and future, have already been written and they are in the decimals of π.

Look at: https://community.wolfram.com/groups/-/m/t/1623981

Surprisingly, in contemporary physics when we speculate about the existence of parallel universes, we arrive at what we might consider an extreme version of the previous speculation. If there is an infinite number of universes, all the possibilities that don't violate the laws of physics occur. Therefore, right now there's someone identical to you doing exactly the same as you're doing. According to this theory all the variants of your life will happen although in different universes.

7.6 The Strange Connection

π is a transcendental number. This means that a finite polynomial of the form $a_0 + a_1 x + a_2 x^2 + ... + a_n x^n = 0$, with the coefficients $a_0 ... a_n$ being rational numbers (or integers), in which π is one of its roots (solutions) doesn't exist. As a result, it can be proven that constructing a square with the same area of a given circle by a finite number of steps only using a compass and a straightedge, also known as squaring the circle, is not possible. There are other transcendental numbers with e being probably the second best-known one after π.

In[ ]:= **N[e, 100]**

Out[ ]= 2.7182818284590452353602874713526624977572470936999595749669676277240766303535475˙.
94571382178525166427

- As with π, the number e can be computed using several different methods (Click on + in the next cell).

> ▤ **Number E representations** ⊞
> ↳ Alternative representation
>   **E == E^z /; z == 1**

- When $e^{\pi\sqrt{163}}$ is expressed in decimal form, something funny happens (we use NumberForm combined with ExponentFunction to avoid the output being displayed in scientific notation).

In[ ]:= **NumberForm$\left[\text{N}\left[e^{Pi\ \sqrt{163}},\ 30\right],\ \text{ExponentFunction} \rightarrow (\text{Null \&})\right]$**

Out[ ]//NumberForm=
     262537412640768743.999999999999

We can see that the first 12 decimal digits are all 9s. This could give the false impression that all the decimal digits are 9, but that is not the case. It would be enough to increase the precision by one more digit to check that it is only a lucky coincidence.

In a separate category we include imaginary numbers, those whose square is a negative number as in the solution to the equation $x^2 + 1 = 0$.

- Notice that to express imaginary numbers, *Mathematica* uses the symbol i and not the Latin letter i.

In[ ]:= **Solve[x^2 + 1 == 0, x]**

Out[ ]= $\{\{x \rightarrow -i\}, \{x \rightarrow i\}\}$

Numbers consisting of a real part and an imaginary one are called complex numbers. As a matter of fact, all real numbers can be considered a special case of complex numbers where the imaginary element is 0. Sometimes strange connections exist among different numbers as in the case of the following equation discovered by Euler:

$e^{ix} = \cos(x) + i\sin(x) \wedge x \in \mathbb{R}$

This connection between different fundamental constants cannot be random. If we assume $x = \pi$, the previous expression becomes:

$e^{i\pi} = -1$

We can see that in just one expression there are two transcendental constants and the imaginary unit connected to each other and, the most surprising thing, the result is the integer -1. It's hard to believe that all these strange connections between numbers are random.

Pi is also connected to one of Riemann's Zeta functions: $\zeta(s) = \sum_{i=1}^{\infty} \frac{1}{k^s}$, which we will discuss in further detail in the next section. We can see that Zeta [2] converges to:

In[ ]:= **Zeta[2]**

Out[ ]= $\dfrac{\pi^2}{6}$

π also appears in fundamental notions in physics such as Heisenberg's uncertainty principle or Einstein's general relativity theory. It seems that π is much more than the ratio between the perimeter of a circumference and its diameter!

7.7 The Riemann Hypothesis

We've seen in the previous section that π, at least according to its latest approximations using billions of digits, is probably a normal number. However, there's no proof that this is the case if we consider an infinite number of digits. This is an open problem in mathematics although not the only or most important one. David Hilbert, one of the most influential mathematicians of the 19th and 20th centuries, at the Paris conference of the International Congress of Mathematicians in 1900 gave a task to mathematicians for the 20th century: the resolution of 23 problems (ten given in the conference and the rest published later). Although most of them have been already solved, the most famous unsolved one is the Riemann's hypothesis about which Bernhard Riemann himself stated that if he were to resuscitate after 500 years, the first question he would ask would be whether the "damn hypothesis" had been proven or not. This problem remains unsolved and is one of the seven "Millennium Problems", a list of problems named by the Clay Mathematics Institute that carry a prize of 1 million dollars for their complete mathematical solution.

The Riemann's Zeta function definition is:

Let $s = a + b\,i \in \mathbb{C}$, for $\mathrm{Re}\,(s) > 1$, $\zeta(s) = \sum_{k=1}^{\infty} \dfrac{1}{k^s}$.

http://mathworld.wolfram.com/RiemannZetaFunction.html

In[ ]:= **ClearAll["Global`*"]**

- Given that s is a complex variable, $\zeta(s)$ is also complex. We can visualize its real component Re (s) and its imaginary one Im (s) with the following *Mathematica* command:

In[ ]:= **{Plot3D[Re[Zeta[a + b I]], {a, –10, 10}, {b, –10, 10}, Mesh → None],**
Plot3D[Im[Zeta[a + b I]], {a, –10, 10}, {b, –10, 10}, Mesh → None]}

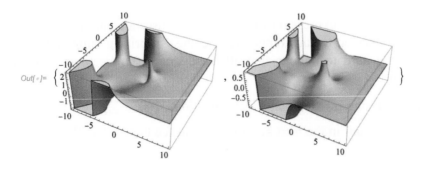

■ It can be shown (using expansions of analytic functions, a complex analysis concept), that for negative even values, that is when $s = -2n$, $\zeta(s) = 0$. We can check it for some cases (in the example below when n ranges from 1 to 100 in steps of 10).

In[ ]:= **Table[Zeta[-2 n], {n, 1, 100, 10}]**

Out[ ]= {0, 0, 0, 0, 0, 0, 0, 0, 0, 0}

■ Graphically we can see that by plotting $s = a + bi$ with $b = 0$, $s = 0$ for $a = \{-2, ..., -2n\}$:

In[ ]:= **Plot[{Re[Zeta[a]], Im[Zeta[a]]}, {a, -30, -2}, PlotLegends → "Expressions"]**

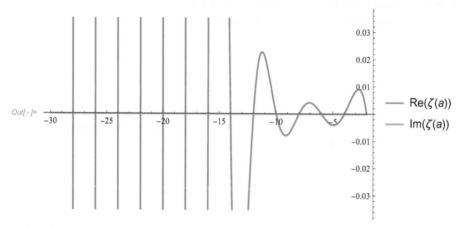

■ Another way to visualize this is by plotting $1/|\zeta(s)|$, with $|\zeta(s)| = |x + yi|$. Remember that $|x+y\,i| = \sqrt{x^2 + y^2}$. Using the inverse expression, the zeros become asymptotes that look like columns. In the graph below, we can see these "columns" appearing when $s = -2n$.
(This function and several others that follow are based on the ones described by Stan Wagon in his book *Mathematica in Action*).

In[ ]:= **Plot3D[1 / Abs[Zeta[x + i y]], {x, -20, -2}, {y, -1, 1},**
    MeshFunctions → (#3 &), MeshStyle → Blue, Mesh → 10,
    PlotPoints → 60, MaxRecursion → 3 ,
    Boxed → False, BoxRatios → {5, 2, 2}, AxesLabel → {"x", "y", None},
    Ticks → {Automatic, Range[-22, -2], Range[0, 4]}]

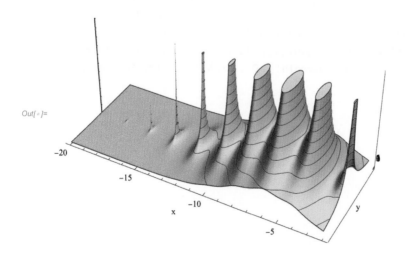

The $\zeta(-2\,n) = 0$ zeros, are named **trivial zeros**.

- Besides the trivial zeros, there are other complex values of s, with $0 < \mathrm{Re}(s) < 1$, for which the zeta function becomes also zero. These zeros are called **non-trivial zeros**. The Riemann's hypothesis states that all these non-trivial zeroes are located in the line $s = \frac{1}{2} + b\,i$ (known as the critical line) or put it in other words, that the real part of every non-trivial zero in the Riemann's Zeta function is 1/2. The graph below shows that this is true for the chosen interval.

In[ ]:= `Plot[{Re[Zeta[1 / 2 + b I ]], Im[Zeta[1 / 2 + b I ]]},`
　　　　`{b, -30, 30}, PlotLegends → "Expressions"]`

Out[ ]=

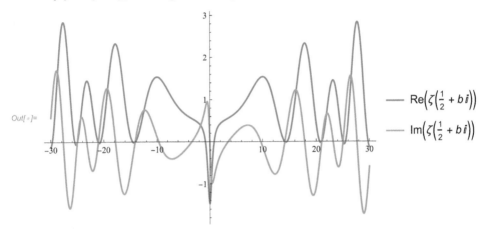

- In *Mathematica* the function **ZetaZero[k]** represents the *k-th zero* of $\zeta(s)$ in the critical line $s = \frac{1}{2} + i\,b$ with the smallest imaginary part. For example the first zero corresponds to:

In[ ]:= `s = N[ZetaZero[1], 15]`

Out[ ]= `0.5000000000000 + 14.1347251417347 i`

- We check that for that value $\zeta(s) = 0$.

In[ ]:= `Zeta[s] // Chop`

Out[ ]= `0`

- We've seen that the first zero is approximately at 14.1347251417347 i. If we would like to visualize all the existing zeros for a given interval, we can use the following function in terms of the imaginary component (remember that the real component is always 1/2).

In[ ]:= `Manipulate[Plot[RiemannSiegelZ[t], {t, 0, k}, Epilog → {PointSize[.02], Red,`
　　　　`Point[Table[{Im[ZetaZero[n]], 0}, {n, k}]]}], {k, 50, 100}]`

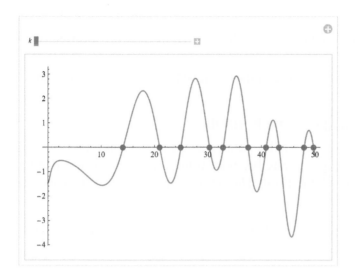

- Next, we plot $1/|\zeta(s)|$ below. We can see that all the zeros (columns) are clearly aligned.

```
In[ ]:= Plot3D[1/Abs[Zeta[x + ⅈy]], {x, -4, 5}, {y, -10, 40},
    MeshFunctions → (#3 &), MeshStyle → Blue, Mesh → 10,
    PlotPoints → 60, MaxRecursion → 3, ViewPoint → {8, 1, 3},
    Boxed → False, BoxRatios → {5, 10, 2}, AxesLabel → {"x", "y", None},
    AxesEdge → {{1, -1}, Automatic, Automatic},
    Ticks → {Automatic, Range[0, 30, 10], Range[0, 4]},
    ClippingStyle → None, PlotRange → {0, 5}]
```

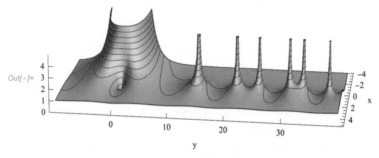

As of the end of 2021, all the non-trivial 10^{13} zeroes that have been found lie on the critical line. However, in mathematics, this doesn't count as a proof. Of all the approaches that have been proposed to prove the Riemann hypothesis, one of the most curious ones comes from D. Hilbert and G. Pólya (proposed independently). They suggested that the zeros of the Riemann Zeta correspond to the eigenvalues of a linear operator given by an infinite matrix. In 1972 the mathematician Hugh Montgomery and the physicist Freeman Dyson went even further and proposed a connection between the zeroes and heavy nuclei energy levels.

We can model these energy levels using what is known as the Spectrum of Riemannium:

https://www.americanscientist.org/article/the-spectrum-of-riemannium

In *Mathematica*, we can explore these two theories using random Hermitian matrices as described in:

https://www.wolfram.com/language/11/random-matrices/zeros-of-the-riemann-zeta-function.html

However, few mathematicians and scientists believe that there's a connection.

7.8 Looking for the Magic Prime Formula

One of the main characteristics of the Zeta function is its connection with prime numbers. Remember that a prime number p is an integer greater than 1 that has no positive divisors other than 1 and itself. The set of prime numbers (that since Euclid we know is infinite) is called $\wp$.

It can be shown that $\zeta(s) = \sum_{k=1}^{\infty} \frac{1}{k^s}$, can also be computed using Euler's product.

Euler's product

```
In[ ]:= WolframAlpha["Euler's product",
         {{"NamedMathematicalFormulas", 1}, "ComputableData"}]
```

$$Out[\circ]= \left\{ \text{Hold}\left[\prod_{k=1}^{\infty}(1-(p_k)^{-s}) = \frac{1}{\zeta(s)} \;/;\; \text{Re}(s) > 1\right],\ \text{Hold}\left[\zeta(s) = \prod_{k=1}^{\infty}\frac{1}{1-p_k^{-s}} \;/;\; \text{Re}(s) > 1 \wedge p_k = p_k\right]\right\}$$

- That is:

$$\zeta(s) = \left(1 - \frac{1}{2^s}\right)\left(1 - \frac{1}{3^s}\right)\left(1 - \frac{1}{5^s}\right) + \ldots = \prod_{p \in \wp}\frac{1}{1-p^{-s}}$$

- Prime numbers seem to be randomly distributed. For a given prime number, there's no general formula that will give us the next one.

```
In[ ]:= Table[Prime[n], {n, 45, 60}]
```

$Out[\circ]= \{197, 199, 211, 223, 227, 229, 233, 239, 241, 251, 257, 263, 269, 271, 277, 281\}$

However, there are algorithms to obtain good approximations of the prime counting function. The prime counting function is the function, notated $\pi(n)$, that returns the number of primes less than or equal to a natural number n. In *Mathematica*, we refer to it as PrimePi[n].

Prime Counting Function: https://mathworld.wolfram.com/PrimeCountingFunction.html

- The graph below shows how many prime numbers are less than or equal to an arbitrary number n. We denote this function $\pi(n)$ (in the example $n = 200$).

In[ ]:= `Plot[PrimePi[n], {n, 1, 200}, AxesLabel → {"n", "π(n)"}]`

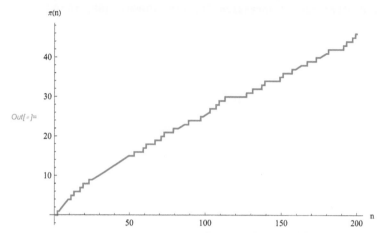

Out[ ]=

- PrimePi[n] gives the option to choose an specific algorithms. The time of computation for some of then are shown below

In[ ]:= `BarChart[Table[{First[AbsoluteTiming[PrimePi[10^n]]],`
`        First[AbsoluteTiming[PrimePi[10^n, Method → "LMO"]]], First[`
`            AbsoluteTiming[PrimePi[10^n, Method → "Meissel"]]]}, {n, 7, 11, .5}],`
`        ChartLegends → {"Automatic", "Lagarias-Miller-Odlyzko", "Meissel" }]`

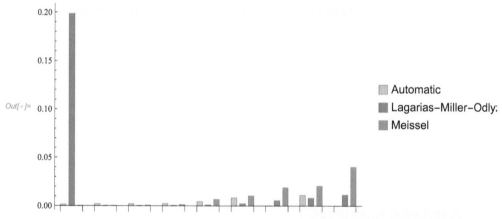

Out[ ]=

The trend is that the frequency of prime numbers decreases as *n* gets bigger.

There are several approximations to calculate $\pi(n)$, such as: $t/\mathrm{Log}(t)$, or $\mathrm{li}(n) = \int_0^z dt/\log t$ (in *Mathematica* LogIntegral[z]), but the best one is done by using Riemann's Zeta function. Specifically, the program uses the function $R(n) =$ RiemannR[*x*] .

For $x > 0$, $R(x) = \sum_n^\infty \mu(n)\,\mathrm{li}\left(x^{1/n}\right)/n$.

- The $t/\mathrm{Log}(t)$ approximation can be found using the asymptotic term for PrimePi at Infinity:

In[ ]:= `Asymptotic[PrimePi[t], t → Infinity]`

Out[ ]= $\dfrac{t}{\log(t)}$

- In the following plot, we compare the real value of π(n) with all these approximations.

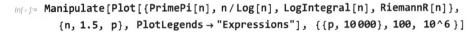

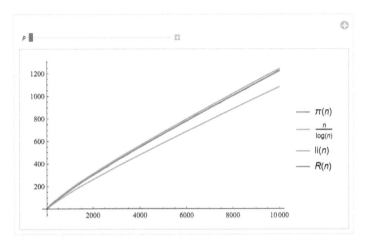

- We can visualize the comparison in an even better way by representing the difference between each of the alternative approximations and $\pi(n)$. In the case of $R(n)$, the approximation is very accurate since, at least in the chosen interval, the error never gets too big (actually it never exceeds the real value by more than ± 3).

In[ ]:= `Plot[{n / Log[n] - PrimePi[n], LogIntegral[n] - PrimePi[n],`
`        RiemannR[n] - PrimePi[n]}, {n, 200, 1000}, PlotLabels → "Expressions"]`

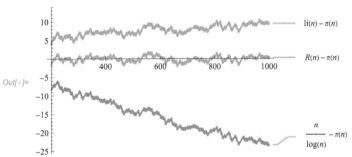

7.9 Additional Resources

A compilation of different methods used to compute π, their *Mathematica* implementations:

http://mathworld.wolfram.com/Pi.html and http://functions.wolfram.com/Constants/Pi

The Wikipedia article https://en.wikipedia.org/wiki/Pi is an excellent resource in terms of both its contents and its links.

Regarding the Riemann hypothesis see for example: (3Blue1Brown) https://www.youtube.com/watch?v=sD0NjbwqlYw

8

Looking at the Sky

This chapter gives a brief introduction to astronomy using some of the functions available in Mathematica. It shows how to get information about planets using entities, generate solar analemmas, and even create sky charts for a given location among other things. It also explains the differences between the Julian and the Gregorian calendars and why, contrary to popular belief, Miguel de Cervantes and William Shakespeare did not die on the same day. Applications related to determining the color of the stars and measuring distances across the universe are also covered. The last section discusses how to model the orbits of eclipsing binary stars by creating and analyzing light curves, very useful when searching for exoplanets.

8.1 A Short Astronomical Walk

Contemplating the sky on a starry night, away from light pollution, is one of the most fascinating shows that we can enjoy. It's not strange then that for millennia humans have looked up at the sky and realized that the sun and the moon follow regular cycles. The majority of ancient cultures came up with mythological or religious explanations about celestial objects, but they also used those objects for practical purposes such as the creation of calendars to schedule harvests. For example, the Egyptians associated the appearance of Sirius, the brightest star in the norther hemisphere (now we know it's a binary system), with the annual flooding of the Nile. Despite the reduced number of celestial bodies that can be seen with the naked eye, it's surprising how much knowledge of the sky that astronomers already had before the invention of the telescope.

The arrival of the telescope about 400 years ago (1608) heralded a revolution: the number of known stars went from thousands to millions, the moon, far from being a perfect sphere, was found to be covered with mountains and craters, the wandering stars (the planets) were more than mere bright dots, they actually had satellites, and Saturn had rings (it's not the only planet). Later on we learned to analyze the composition of the light coming from space. Many of the objects that looked like blurry spots under the telescope became galaxies with myriads of stars. It was discovered that those galaxies were moving apart from each other: the universe was expanding, an expansion that started around 14 billion years ago. It turned out that the visible stars and galaxies contain less than 5% of the mass of the Universe. The rest is dark energy and matter, we know it exists (at least the dark matter) but we don't really know what it is. The idea of the existence of dark energy ("a riddle wrapped inside a mystery") started with an amazing discovery (1998): The expansion of the universe is accelerating, a discovery that was worth a Nobel Prize in 2011. And the discovery race continues, promising new surprises: the important thing is still the journey.

For astronomical calculations, *Mathematica* has several functions: Comet (or CometData) , Exoplanet (or ExoplanetData), Galaxy (or GalaxyData), MinorPlanet (or MinorPlanetData), MoonPosition, Sunset, Nebula (or NebulaData), Planet (or PlanetData), Pulsar (or PulsarData), SiderealTime, StarCluster (or StarClusterData), Star (or StarData), SunPosition, Supernova (or SupernovaData), Sunrise and many more. These functions replace and extend AstronomicalData although you can still use it.

The syntax of all these functions is similar to that of other computable functions. Example: **Function** [*"name"*, *"property"*] where *"name"* or entity (Entity) is an object and *"property"* is a characteristic or parameter of the object.

- For example: to refer to the solar system planets we can type either PlanetData[] or:

In[ ]:= **EntityValue ["Planet", "Entities"]**

Out[ ]= { Mercury , Venus , Earth , Mars , Jupiter , Saturn , Uranus , Neptune }

If we place the cursor on one of the list elements in the output, for example "Mercury", *Mathematica* displays Entity["Planet", "Mercury"]. This tells us that "Mercury" belongs to the "Planet" entity.

- We can get the available properties for planets using the free-form input: **Insert ▶ Inline Free-format input** and then type "planet" or "planets". Also, in this example and the following one, we use Shallow or Short to avoid getting the complete output. If you want to see all the properties just remove Shallow when executing the command:

In[ ]:= ▦ **planets** PLANETS [···] ✓ ["Properties"] // Shallow

Out[ ]//Shallow=

  { absolute magnitude H , age , albedo , alphanumeric name ,

  alternate names , altitude , next maximum altitude , angular diameter ,

  angular radius , largest distance from the Sun , ≪125≫}

- The next output informs us about the property "OrbitPeriod" for each one of the planets. Notice the use of "EntityAssociation" to link names with properties.

In[ ]:= **planets** PLANETS ⬚ ✓ ["OrbitPeriod", "EntityAssociation"]

Out[ ]= ⟨| Mercury → 87.96926 days, Venus → 224.70080 days,

Earth → 365.25636 days, Mars → 1.8808476 a, Jupiter → 11.862615 a,

Saturn → 29.447498 a, Uranus → 84.016846 a, Neptune → 164.79132 a |⟩

- Here we show Mercury's average orbit velocity writing ▤ Mercury average velocity   or

  EntityValue[Entity["Planet","Mercury"],"AverageOrbitVelocity"]

In[ ]:= **Mercury** PLANET [*average orbit velocity*]

Out[ ]= 47.4 km/s

- Sometimes, we may want to get the value without its corresponding unit. In those cases, we can use QuantityMagnitude.

In[ ]:= **QuantityMagnitude** [**Mercury** PLANET [*average orbit velocity*]]

Out[ ]= 47.4

The result is given in light-years (ly), a common unit to express distances between stars.

- The parsec (pc), and its multiples (kiloparsec or kpc, and megaparsec or Mpc), are used to measure very long distances such as distances between galaxies. Let's see the relationship between a parsec and a light year.

In[ ]:=

convert parsec to light year » ⬚

↳ Result

  3.262 ly

Out[ ]= 3.262 ly

- The table showing the equivalence to 1 light-year in different units:

In[ ]:= **Pane[TableForm[Map[{#, N@UnitConvert[Quantity["LightYears"], #]} &,**
        {"Meters", "Parsecs", "SIBase", "SI", "Imperial", "Metric"}]] , 200]

Out[ ]=

Meters	9.46073×10^{15} m
Parsecs	0.306601 pc
SIBase	9.46073×10^{15} m
SI	63 241.1 au
Imperial	5.87863×10^{12} mi
Metric	63 241.1 au

- For some properties there may be no information available. Remember that with DeleteMissing we can remove those properties whose headings contain the word "Missing".

In[ ]:= **DeleteMissing** [▤ Plutoids ["Density", "EntityAssociation"]]

Out[ ]= ⟨| 136199 Eris (2003 UB313) → 2.53 g/cm³, Pluto → 1.860 g/cm³ |⟩

Some of the functions related to astronomy need the precise location for which the information is desired. By default they will use our current time and position.

- Example: The following function returns the sunrise time on February 28, 2023, at a place located 40° N and 10° W, a location that is one hour ahead of GMT as we can see.

In[ ]:= `Sunrise[GeoPosition[{40, 10}], DateObject[{2023, 03, 01, 0, 0}]]`

Out[ ]=   Minute: Wed 1 Mar 2023 07:54 GMT+2

Dates and hours are of crucial importance in astronomy. The existence throughout history, even nowadays, of different calendars makes it difficult to compare dates of astronomical and historical events. To avoid this problem we use Julian dates. A Julian date (not to be confused with the Julian calendar: Julian year) indicates the days that have gone by since January 1, 4713 B.C. at 12:00 PM UT (November 24, 4714 B.C., in the proleptic Gregorian calendar). The name of the Julian date comes from Julius Scaliger (not Julius Caesar) who proposed it in 1583 after realizing that such a day was the least common multiple of 3 calendar cycles used by the Julian calendar. Using 12:00 PM as the starting time instead of 12:00 AM has the advantage that astronomical observations, usually done at night, are recorded on the same day. Starting with *Mathematica* 10.2 the function JulianDate is available. The function converts dates from our calendar, the Gregorian one, to Julian dates. It takes into account the modifications to the Julian calendar done by Gregorio XIII in 1582 in most Christian countries that improved the handling of leap years, but in which the year 0 doesn't exist.

- In England and other countries, the Gregorian calendar was introduced at a later date. This fact makes statements such as that Shakespeare and Miguel de Cervantes died on the same day incorrect.

In[ ]:= `dateSh =` | William Shakespeare PERSON | [date of death]

Out[ ]=   Tue 23 Apr 1616 (Julian calendar)

In[ ]:= `dateCe =` | Miguel de Cervantes PERSON ⋯ ✓ | [date of death]

Out[ ]=   Fri 22 Apr 1616

- Actually, it is not clear when Cervantes died. Although April 22 is the most frequently used date, the parish register shows April 23. In any case, Cervantes's and Shakespeare's deaths happened 10 days apart. The reason for the confusion is that in England they were still using the Julian calendar. To compare both dates, we need to convert Shakespeare's death date to the Gregorian calendar.

In[ ]:= `CalendarConvert[dateSh, "Gregorian"]`

Out[ ]=   Tue 3 May 1616

- We can alternatively find out the Spanish writer's death date using the Julian calendar and calculate the difference in days between both deaths.

In[ ]:= `JulianDate[dateSh] - JulianDate[dateCe]`

Out[ ]= 11.

- We can now use the above mentioned approach to convert the current date to a Julian date. We have added Dynamic to dynamically update the calculation with an interval defined with UpdateInterval, in this case 1 second.

In[ ]:= `Dynamic[AccountingForm[JulianDate[Date[]], 12], UpdateInterval → 1]`

Out[ ]= 2459870.19238

We can stop the dynamic updating by unchecking it in the Evaluation menu: **Evaluation ▶ Dynamic Updating Enabled**.

Star positions in the sky don't follow an annual cycle exactly. They are subject to perturbations due to their own motions and other celestial movements, such as the precession of equinoxes. Because of this, it's necessary to regularly set up dates to use when referring to celestial coordinates. These standard reference dates are known as epochs. The last one was set on January 1, 2000 at 11:58:55.816 AM. When referring to this epoch we use **J2000.0**. StarData and other astronomical functions refer to the position of stars using this latest epoch without taking into account either the precession of equinoxes or the stars' own motions. For practical purposes, these adjustments are irrelevant, but we need to consider them when making historical projections.

8.2 Solar Analemma

The curve describing the sun and the visible planets in the sky every day of the year always observed at the same location and time of the day is called analemma.

A solar analemma can be obtained with a camera and patience, taking pictures of the horizon every day or almost every day of the year from the same position, with the same exposure and at the same solar time. The result will be a composition similar to the one below. It shows an analemma taken by Jack Fishburn between 1998—1999 from Bell Laboratories in Murray Hill, New Jersey:

In[ ]:= `WikipediaData["Analemma", "ImageList"][[3]]`

Out[ ]=

You can also make analemma for other celestial objects such as Venus or Mercury.

With less patience and without having to wait for a year, we can simulate an analemma with *Mathematica* from our own computer.

- We need to define the location (latitude and longitude).

In[ ]:= **salamanca = GeoPosition[Salamanca CITY ⋯ ✓]**

Out[ ]= GeoPosition[{40.97, −5.67}]

SunPosition returns the sun position from our position (azimuth and latitude) at 9:00 AM in GMT (choosing TimeZone→0) on January 1, 2016.

In[ ]:= **SunPosition[salamanca, DateObject[{2016, 1, 1, 9}, TimeZone → 0]]**

Out[ ]= {132.99°, 10.15°}

We apply the same function to get the position every 5 days, at the same time, to complete a year:

In[ ]:= **Graphics[{Orange, PointSize[Large],**
**   Point@Map[QuantityMagnitude, SunPosition[salamanca,**
**   DateRange[DateObject[{2016, 1, 1, 9, 0}, TimeZone → 0],**
**   DateObject[{2016, 12, 31, 9, 0}, TimeZone → 0], 5]][**
**   "Values"], {2}]}, Frame → True]**

Out[ ]=

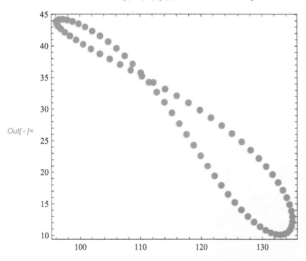

8.3 Stargazing

In[ ]:= **Clear["Global`*"]**

An incredible experience is to look at the sky away from light-polluted cities, trying to imagine the hidden celestial mechanism generating the incredible view. We will soon realize the greatness of those night owl geniuses (Hipparchus, Ptolemy, Copernicus, Brahe, Kepler, and others) that established the foundations of modern astronomy without any other means than their plain sight and basic instruments such as the astrolabe to measure the stars' positions. Everything changed in 1609 when Galileo used the telescope for astronomical purposes for the first time. Another equally important invention was the pendulum clock that enabled the measurement of time with great precision. Centuries later, telescopes mounted on satellites and atomic clocks heralded a new revolution.

8.3.1 Naked-Eye Astronomy

When looking at the sky, almost all of the objects that can be seen with the naked eye are stars. Throughout history, different cultures imagined that some stars resembled figures, named constellations. The same phenomenon occurred in civilizations that were not connected in any way. For example: The seven brightest stars from the Ursa Major were interpreted the same way in very different places. The inhabitants of the British islands said that those stars were the legendary King Arthur's cart. For the Germans, they represented a wagon pulled by three horses. The Greeks came up with a more imaginative story: In a Greek legend, the god Zeus and the mortal Callisto had a son named Arcas. Hera, Zeus's jealous wife, turned Callisto into a female bear. Arcas, unaware that the bear was actually her mother almost killed her. Afterward, Zeus also turned Arcas into another female bear. Callisto is Ursa Major and Arcas is Ursa Minor. Greek mythology is full of such stories to explain almost all the stars that appear near each other in the sky. Many of those myths have actually given names to the constellations. Over the centuries, many constellations were added until they eventually covered the entire visible sky from any place on Earth. There are 88 constellations in total. In 1930, the International Astronomical Union established the limits of each of them. Obviously, they are imaginary lines covering the entire firmament, including both hemispheres. As in maps, we don't need to keep all the constellations' details. They and the names of their most important stars are represented in a planisphere, a celestial map that shows us the sky each night for a given date and time.

- The function below returns the sky chart for our location at the time given in the input. We can get *Mathematica* to create it for us by typing "Sky chart at 21 GMT 2022-02-15" in free-form input.

In[ ]:= ▪ **SkyChart** [

       DateObject[{2022, 2, 15}, TimeObject[{21, 0}, TimeZone → "GMT"]]]

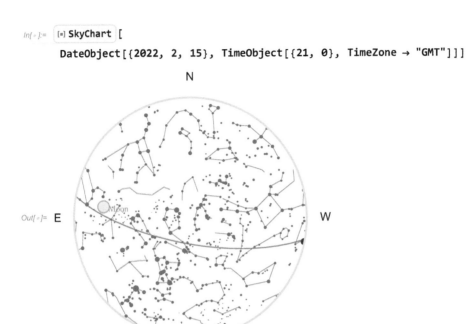

Out[ ]= E   W

- The program uses "SkyChart", a function from the repository. To find out more details about its functionality, we type:

In[ ]:= **ResourceObject["SkyChart"]**

Out[ ]= ResourceObject[**f[▪]** Name: SkyChart »
                          Type: Function
                          Description: Visualize the sky for a specified location, time and date]

- With the instruments and methods available to astronomers nowadays, the information that we can obtain about stars is very extensive. In the next example, we show the properties most commonly used when describing stars:

In[ ]:= **properties = {"AbsoluteMagnitude", "ApparentMagnitude",**
       "RightAscension", "Declination", "Altitude", "Azimuth",
       "Constellation", "DistanceFromSun", "SpectralClass"};

- Let's find those properties for Polaris, the North Star:

In[ ]:= **Polaris** STAR ••• ✓ **[properties, "PropertyAssociation"]**

Out[ ]= ⟨| AbsoluteMagnitude → −3.64, ApparentMagnitude → 1.97, RightAscension → $2^h\,31^m\,47.06^s$,
    Declination → $89°\,15'\,51.''$, Altitude → $40°\,23'\,43.''$, Azimuth → $359°\,23'\,2.\times10^1\,''$,
    Constellation → Ursa Minor , DistanceFromSun → 431.217 ly, SpectralClass → F7:Ib–IIvSB |⟩

The apparent magnitude tells us how bright an astronomical object appears to an observer regardless of its intrinsic brightness:

http://scienceworld.wolfram.com/astronomy/ApparentMagnitude.html

The smaller the magnitude, the bigger its luminosity. For the most luminous stars, we use negative values. Most of the people looking at the sky with the naked eye and with little light pollution just see stars with apparent magnitude of less than 5.

- Here are the 10 closest stars to Earth (the Sun is excluded), many of them not visible to the naked eye:

In[ ]:= `Through[EntityList[EntityClass["Star", "StarNearest10"]][`
      `{"Name", "DistanceFromEarth", "ApparentMagnitude"}]]`

Out[ ]=

$$\begin{pmatrix}
\text{Barnard's Star} & 5.93796\ \text{ly} & 9.54 \\
\text{Lalande 21185} & 8.30785\ \text{ly} & 7.49 \\
\text{Luyten 726--8 A} & 8.56318\ \text{ly} & 12.57 \\
\text{Luyten 726--8 B} & 8.56318\ \text{ly} & 12.7 \\
\text{Proxima Centauri} & 4.2465\ \text{ly} & 11.01 \\
\text{Rigil Kentaurus} & 4.39282\ \text{ly} & -0.01 \\
\text{Toliman} & 4.40311\ \text{ly} & 1.35 \\
\text{Sirius} & 8.59682\ \text{ly} & -1.44 \\
\text{Sun} & 1.01234\ \text{au} & -26.72 \\
\text{Wolf 359} & 7.79346\ \text{ly} & 13.45
\end{pmatrix}$$

We can immediately notice that most of them are not visible since their apparent magnitude is bigger than 5. With the exception of the Sun, the star with the biggest apparent magnitude is Sirius, a star that we can see during winter nights (in the Northern hemisphere) next to the Orion constellation. The closest one is Proxima Centauri that is only visible in the Southern hemisphere (Proxima Centauri rotates around the double system Centaurus A and B in a 500,000-year period. By chance, it's currently in its orbital position closest to the solar system).

- The number of stars visible without the use of technology is very small compared to the number of stars in our galaxy. The command below generates a plot representing the naked-eye stars and their distances to Earth in light years.

In[ ]:= `ListPlot[Sort[EntityClass["Star", "NakedEyeStar"]["DistanceFromSun"]],`
      `AxesLabel → {"stars", "ly"}]`

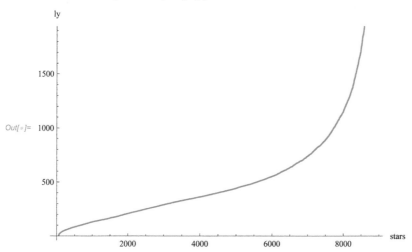

From the graph above we can see that almost all of the visible stars are less than 1,500 ly away. Our galaxy, the Milky Way, has a width of approximately 100,000 ly. Therefore, with our bare eyes we only see a very small fraction of the stars in our own galaxy, most of them the ones closest to Earth. The previous figure refers to the number of potentially visible stars without light pollution or moonlight. In practice, the actual number is 2,000 or 3,000, since we only see part of the sky depending on our position.

Besides stars with apparent magnitude 5 or less, the only other celestial bodies visible using only our eyesight are: the Moon, the 5 nearest planets and, sporadically, some comets. In the Northern hemisphere, the only object visible that doesn't belong to our galaxy is the Andromeda galaxy. In short, we see an insignificant fraction of our own galaxy, which in turn is just one among billions of galaxies.

We've seen that among the properties of Star (or StarData) and other astronomical functions are: "RightAscension", "Declination", "Altitude", and "Azimuth". These properties are commonly used to indicate the position of a celestial body.   In equatorial coordinates, "RightAscension" and "Declination" are similar to latitude and longitude, but they refer to the celestial sphere so they are independent of the observer . Another type of coordinates widely used are the alt-azimuthal (alt/az) that depend on the observer: The "Altitude" (alt) is the height of the star over the horizon. It goes from 0° to 90° and has a positive sign for stars located above the horizon and a negative sign for the ones located below it. The "Azimuth" (az) is the arc in the horizon measured counterclockwise from the South point until the object's vertical. Its value ranges from 0° to 360°.

For further details see: *The Celestial Sphere*, http://demonstrations.wolfram.com/TheCelestial-Sphere, by Jeff Bryant.

- The next example shows Sirius's displacement on the first of each month at 22:00 h using alt/az coordinates in local time.

Planet

```
In[ ]:= sirius = Table[QuantityMagnitude[EntityValue[
          Dated[Entity["Star", "Sirius"], DateObject[{2022, i, 1, 22, 0, 0}]]],
            {"Azimuth", "Altitude"}], "AngularDegrees"], {i, 1, 12}]
```

$$
Out[ ]= \begin{pmatrix}
118.713 & 6.38653 \\
143.138 & 23.8375 \\
171.697 & 31.9130 \\
205.412 & 28.4294 \\
231.945 & 14.4112 \\
253.045 & -6.00471 \\
271.772 & -28.0331 \\
296.751 & -50.3112 \\
346.221 & -65.1931 \\
46.9011 & -58.3531 \\
78.9981 & -37.7519 \\
98.8455 & -15.5399
\end{pmatrix}
$$

In[●]:= `Graphics[{Orange, Point[sirius]}, Frame → True,`
`    FrameLabel → {"azimuth", "altitude"}, AspectRatio → 1/2]`

Out[●]=

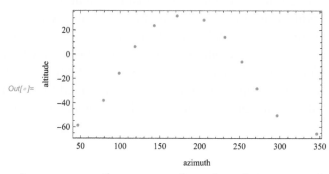

As we can see, there are negative values that correspond to the period of the year in which Sirius falls below the horizon and as a consequence is not visible. As a matter of fact, the Egyptians considered that the appearance of the star marked the beginning of a new year.

- In the previous example, we didn't specify the observer's location and the local time zone. They can be shown:

In[●]:= `{FindGeoLocation[], $TimeZone}`

Out[●]= {GeoPosition[{40.97, −5.66}], 2.}

- We can display the time in zone 0, using the Coordinated Universal Time (UTC)

In[●]:= `DateObject[Now, TimeSystem → "UTC", TimeZone → 0]`

Out[●]= Sun 22 May 2022 20:15:48 UTC

- However, it's important to keep in mind that the function actually gives us the position of the server that we are using to connect to the Internet based on the IP address. Sometimes it may be far away from our real position. This would be the case if, for example, we were using a phone to surf the net. We can check it by showing the obtained position in a map:

In[●]:= `GeoGraphics[`
`    GeoVisibleRegionBoundary[FindGeoLocation[]], GeoZoomLevel → 12]`

Out[●]=

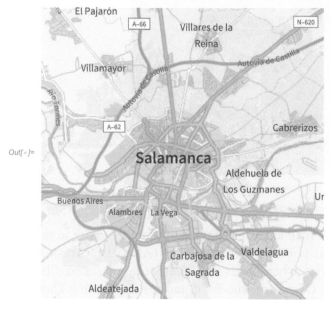

Naked-eye stars display an amazing regularity. It's possible to forecast the location of each star without prior knowledge of celestial mechanics. If we observe a star at a certain hour, for example at 24:00 h, ignoring the displacement due to the inclination of the Earth's axis with respect to the ecliptic, the next day it will be on the same spot 4 minutes earlier, that is at 23:56 h, completing a cycle in a year. The explanation that the Greek civilization and others provided for this fact was to assume that all the stars (which they called fixed, to distinguish them from the planets or wandering stars that did not exhibit such behavior) were glued to a dome that rotated with a daily cycle of 23 h and 56 min, the sidereal day (or stellar day).

In[ ]:= ⊟ `sidereal days` ⸬ ✓

Out[ ]= 1 sidereal day

- We can use UnitConvert to find out the actual duration. When executing the command above, the predictive interface will do the conversion automatically.

In[ ]:= `UnitConvert[Quantity[1, "SiderealDays"],`
       `MixedUnit[{"Hours", "Minutes", "Seconds"}]]`

Out[ ]= 23 h 56 min 4.09054 s

However, over very long periods we can notice discrepancies that at first sight may seem insignificant during a person's lifetime but become quite obvious after several generations. Hipparchus of Nice (c. 190 BC–c. 120 BC) compared the stellar maps available at the time (celestial cartography is over 2,500 years old!) and realized that there was a relative movement of the stars with respect to the ecliptic. This movement is known today as the precession of the equinoxes. This is probably his most famous discovery. The displacement happens when the Earth's axis, moving along a circumference with respect to the ecliptic, rotates with a period of 25,771 years.

Figure 8.1   Precession of the Equinoxes demonstration.

http://demonstrations.wolfram.com/PrecessionOfTheEquinoxes

As a result, the North celestial pole keeps on moving. Today it is close to the North Star but 4,800 years ago it was pointing toward Thuban (Alpha Draconis). William Shakespeare did not realize that, when in his play *Julius Caesar* stated: "But I am constant as the northern star, Of whose true-fix'd and resting quality There is no fellow in the firmament".

In reality, all the stars seen from the Earth have a slow displacement motion as a result of the precession and their own orbits in the galaxy. This last type of movement is known as proper motion.

- We can find the proper motion of the North Star as follows:

Out[ ]= 11.75 mas/yr

> "mas/yr" means milliarc seconds per year. It's a value that cannot be observed during a lifetime. We need to keep in mind that a telescope on the ground can rarely see details smaller than 1 arc second. However, during long periods even the shapes of the constellations change.

Astrology (a pseudoscience) attributed predictive powers to the constellations. For example, they were supposed to determine the future of people born under them. The firmament was divided into 12 signs corresponding approximately to the number of lunar cycles in a year. Each division was assigned a symbol called a zodiac sign. The names of the zodiac are associated with the first constellations named by the Greeks during the 5th century BC, even though the Babylonians were the first ones to mention them 4,000 years ago. The starting point was the constellation pointing toward Aries (the moment when the Sun moves from the south celestial hemisphere to the north one coinciding with the spring equinox) in the 5th century BC. Today, we still use the same division although the constellations are now in a different location to where they were 2,500 years ago. The real astronomical zodiac dates correspond to the constellation located behind the solar disk during the spring equinox, in the direction opposite to the Earth's direction (the Sun actually passes through 13 constellations and not 12, the 13th one is Ophiuchus). The zodiac signs must be adjusted by a month to adapt to the current astronomical reality, so chances are you may have to revise your zodiac sign (although it wouldn't be very useful).

- The natural language input below returns the location of the Ursa Minor stars in the year 100,000 compared to their current position (dashed line). The rest of the stars experience similar changes.

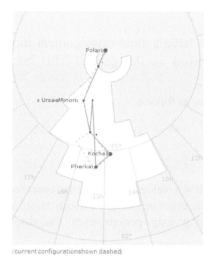

(current configuration shown dashed)

- The solar day (24 hours) is defined as the time it takes for the Sun to return to its culmination point (its highest point in the sky). That is, the Earth in 24 hours spins more than 360 °. In one average tropical year, the number of rotations is:

In[ ]:= `1 + QuantityMagnitude[UnitConvert[Quantity[1, "TropicalYears"], "Days"]]`

Out[ ]= 366.2421887920

However, the periodicity of the Earth's rotation is not constant, but it's slowing down. Because of that, the rotation is no longer associated to the definition of seconds, as the base time unit. Now, the official definition is (https://www.bipm.org/en/si-base-units/second):

"The unperturbed ground−state hyperfine transition

frequency of the caesium−133 atom, to be 9 192 631

770 when expressed in the unit Hz, which is equal to s^{-1}".

To keep time standards such as UT1, based on the Earth's rotation, accurate it's necessary to add or subtract discrete leap seconds (normally once or twice a year).

- Even though the UTC system, which is the standard "civil time", is not based on tracking the precise rotation speed of the Earth, there was still a discrepancy between it and the International Atomic Time (TAI) standard on 2021-12-31:

In[ ]:= | Wed 1 Dec 2021 00:00:00 UTC | − | Wed 1 Dec 2021 00:00:00 TAI |

Out[ ]= 37. s

- GeoOrientationData takes into account the measured rotation speed of the Earth. The following code shows the deviation of the duration of the day over 86,400 seconds (24 h) for the last 20 years:

In[ ]:= DateListPlot[GeoOrientationData[DateInterval[{Now - ▤ 20 yr ✓ , Now}],

"DayDurationExcess"], TargetUnits → "Seconds"]

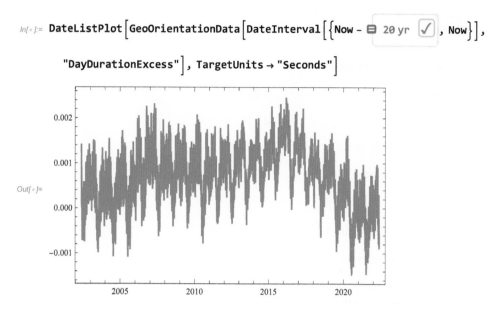

8.3.2 Wandering Planets

Since antiquity, people have realized that there are a small number of stars exhibiting a behavior different from the one of the fixed stars and their annual cycles. These stars were once called wandering stars. Nowadays we know that they correspond to the 5 visible planets: Mercury, Venus, Mars, Jupiter, and Saturn (Uranus was not added to the list of planets until 1783 given how difficult it is to observe it with the naked eye). To get information about planets, we use the Planet entity (or the function PlanetData).

- Here all available properties for Jupiter are shown (Click in "atmospheric composition".)

In[ ]:= **Jupiter** PLANET [···] [✓] ["Dataset"]

Out[ ]=

absolute magnitude H	−9.4
age	4.5×10^9 yr
albedo	0.52
alphanumeric name	Jupiter
alternate names	{the giant planet}
altitude	$-49°-3'-28."$
next maximum altitude	$48°\,52.1247'$
angular diameter	35.69"
angular radius	17.85"
largest distance from the Sun	8.1608146×10^8 km
largest distance from Sun	8.1608146×10^8 km
next apoapsis time	Sat 30 Dec 2028
last apoapsis time	Sat 18 Feb 2017
apparent altitude	$-49°-4'-21."$
apparent magnitude	−2.21
longitude of ascending node Ω	100.55615°
atmospheric composition	‹| ···20 |›
atmospheric pressure	0.3 bar
atmospheric scale height	27. km
authalic diameter	1.399×10^5 km

⊼ ∧ rows 1-20 of **135** ∨ ⊻

↪

- This function shows the mass and density for each planet:

In[ ]:= **Text[Grid[Prepend[EntityValue["Planet", {"Name", "Mass", "Radius"}],**
     {"Name", "Mass", "Radius"}], Frame → All,   Background → {None,
      {Lighter[Yellow, .9], {White, Lighter[Blend[{Blue, Green}], .8]}}}]]

Name	Mass	Radius
Mercury	3.301×10^{23} kg	2.44×10^3 km
Venus	4.867×10^{24} kg	6.05×10^3 km
Earth	5.97×10^{24} kg	6371.009 km
Mars	6.417×10^{23} kg	$3390.$ km
Jupiter	1.898×10^{27} kg	6.995×10^4 km
Saturn	5.683×10^{26} kg	5.830×10^4 km
Uranus	8.681×10^{25} kg	2.536×10^4 km
Neptune	1.0243×10^{26} kg	2.46×10^4 km

Out[•]=

- The next function displays the number of moons for each planet:

In[•]:= `Length /@ EntityValue["Planet", "Satellites", "EntityAssociation"]`

Out[•]= ⟨| Mercury → 0, Venus → 0, Earth → 1, Mars → 2,

Jupiter → 80, Saturn → 83, Uranus → 27, Neptune → 14 |⟩

As we mentioned previously, if we periodically observe a planet with the naked eye at the same time and from the same place and we record (sometimes we can take pictures, as we saw with the example of the Sun) its location with respect to the horizon, the resulting figure is called an analemma. In Chapter 5, we built the one corresponding to the Sun. Here we are going to do the same for Mars and Venus.

- We need first to define our position (latitude and longitude).

In[•]:= `zone = "Location" → GeoPosition[{40.96, -5.66}];`

- Next, we display the Mars and Venus positions at 20:00 h each 5 days for the first 90 days of 2023.

In[•]:=
```
analemmamars = Table[{
    QuantityMagnitude[PlanetData["Mars", EntityProperty["Planet", "Azimuth",
        {"Date" → DateObject[DateList[{2023, 1, i, 20}]], zone}]],
    "AngularDegrees"], QuantityMagnitude[
    PlanetData["Mars", EntityProperty["Planet", "Altitude",
        {"Date" → DateObject[DateList[{2023, 1, i, 20}]], zone}]],
    "AngularDegrees"]}, {i, 1, 90, 5}];
```

In[•]:=
```
analemmavenus = Table[{
    QuantityMagnitude[PlanetData["Venus", EntityProperty["Planet", "Azimuth"
        {"Date" → DateObject[DateList[{2023, 1, i, 20}]], zone}]],
    "AngularDegrees"], QuantityMagnitude[
    PlanetData["Venus", EntityProperty["Planet", "Altitude",
        {"Date" → DateObject[DateList[{2023, 1, i, 20}]], zone}]],
    "AngularDegrees"]}, {i, 1, 90, 5}];
```

- With GraphicsRow we can display both graphs in the same row. The graph for Mars has been modified using AspectRatio to make its visualization easier.

```
In[ ]:= GraphicsRow[{
        Graphics[{Orange, Point[analemmamars]},
          Frame → True, FrameLabel → {"azimuth", "altitude"},
          AspectRatio → 1/2, PlotLabel → Style["Mars Analemma", Bold]],
        Graphics[{Orange, Point[analemmavenus]}, Frame → True,
          FrameLabel → {"azimuth", "altitude"},
          PlotLabel → Style["Venus Analemma", Bold]]}, Frame → All]
```

Out[ ]=

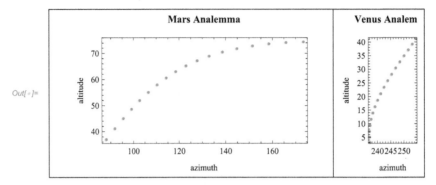

The points with negative altitude correspond to those times when the planet is below the horizon and therefore invisible. We cannot see the planet when it's behind the Sun either.

We can also see that in the case of Venus, the planet is only visible a few degrees over the horizon. This is because Venus is an interior planet and, as such, when observed from the Earth, it will never be very high above the horizon. This means that we will never be able to see it next to the zenith since the sunlight would blind us. Obviously, this doesn't happen with exterior planets such as Mars.

- The graph below shows the orbits of the terrestrial planets or inner planets: Mercury, Venus, the Earth and Mars, with the Sun (not in scale) in the center.

```
In[ ]:= Graphics3D[{{Yellow, Sphere[{0, 0, 0}, 0.02]}, EntityValue[
        EntityClass["Planet", "InnerPlanet"], "OrbitPath"]}, Boxed → False]
```

Out[ ]=

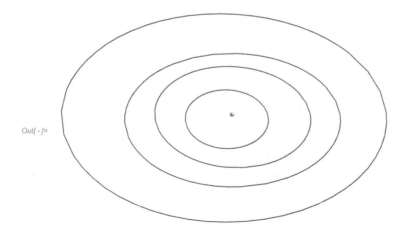

■ The next four functions, based on examples included in the Wolfram Documentation, enable us to calculate and display graphically the distance from Earth and the apparent magnitude of the visible planets over a year:

```
In[ ]:= visibleplanets[planet_, d1_, d2_ , property_] :=
          PlanetData[planet, Table[EntityProperty["Planet", property,
              {"Date" -> DateObject[date]}], {date, DateRange[d1, d2, "Week"]}]];
      ts[planet_, d1_, d2_, property_ ] :=
          TimeSeries[Transpose[{DateRange[d1, d2, "Week"],
              visibleplanets[planet, d1, d2, property]}]];
```

```
In[ ]:= DateListPlot[
          AssociationMap[ts[#, {2021, 1, 1}, {2022, 12, 31}, "DistanceFromEarth"] &,
              {"Mercury", "Venus", "Mars", "Jupiter", "Saturn"}],
          PlotLegends -> "Expressions", FrameLabel -> Automatic]
```

```
In[ ]:= DateListPlot[
          AssociationMap[ts[#, {2021, 1, 1}, {2022, 12, 31}, "ApparentMagnitude"] &,
              {"Mercury", "Venus", "Mars", "Jupiter", "Saturn"}],
          PlotLegends -> "Expressions", FrameLabel -> Automatic]
```

Even though naked-eye observation and the use of angle-measuring instruments (sextant, astrolabe, etc.) advanced the science of astronomy, the arrival of the telescope and other instruments such as precision clocks was no less important. They opened new possibilities for exploring the firmament. Later on, in the 20th century, radio telescopes and spacial astronomy would arrive taking the science to a whole new level.

8.3.3. Dwarf Planets

We've seen that PlanetData doesn't include Pluto, demoted to the "Dwarf Planet" category in a controversial meeting of the International Astronomical Union on August 24, 2006. This action was motivated mainly by the discovery of planets beyond Pluto, in an area known as the Kuiper Belt that probably includes thousands of planetoids. Pluto is considered to be part of this belt. All the planets that didn't fit the new definition were categorized as dwarf planets. There are reasons to justify that Pluto doesn't belong to the same category as the classical planets but the dwarf label doesn't seem the most adequate one since there are dwarf planets that are most likely bigger than the classical planet Mercury.

- The celestial bodies that orbit around the Sun but are not planets have been included in the MinorPlanet entity (or MinorPlanetData).

In[ ]:= **EntityValue["MinorPlanet", "EntityClasses"]**

Out[ ]= { minor planets , Amor asteroids , Apollo asteroids ,

asteroid belt , Aten asteroids , dwarf planets , Kuiper Belt objects ,

scattered disk objects , near-Earth asteroids , Plutoids , trans-Neptunian objects ,

Jupiter Trojan asteroids , centaur asteroids , inner main belt asteroids ,

main belt asteroids , Mars crossing asteroids , outer main belt asteroids }

- Next, we use a function to know how far away in astronomical units (au) it would be from the Earth each year starting in 1970-01-01 until 2300

In[ ]:= **DateListPlot[{{{#, 01, 01}, EntityValue[**
      Entity["MinorPlanet", "Pluto"], EntityProperty["MinorPlanet",
      "DistanceFromEarth", {"Date" → #}]]} & /@ Range[1970, 2300, 10]]

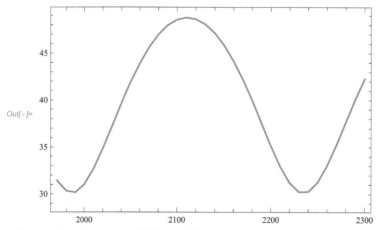

Notice that during our lifetimes Pluto will be further away each year. In July 2015, a probe, New Horizons, visited it for the first time.

- One of the most interesting characteristics of Pluto and other celestial bodies in the Kuiper Belt is that their orbits are normally slanted with respect to the ecliptic plane, next to the ones from the classical planets, as we can see with the following function:

In[•]:= ▣ Plutoids ["Inclination", "EntityAssociation"]

Out[•]:= ⟨| 136199 Eris (2003 UB313) → 44.143°, 136108 Haumea (2003 EL61) → 28.1950°,

136472 Makemake (2005 FY9) → 28.99899°, Pluto → 17.1°|⟩

In the coming years, the number of newly discovered dwarf planets will most likely increase substantially.

- We dynamically simulate the movement of the dwarf planets starting on January 1, 2020 using MinorPlanetData. Notice the use of Tooltip to see the name when placing the mouse over a planet. The function is slow.

In[•]:=
```
minorplanets = MinorPlanetData[EntityClass["MinorPlanet", "Plutoid"]];
Manipulate[
 Graphics3D[{{(Tooltip[Sphere[#⟦2⟧, 1], #⟦1⟧]) & /@
   Transpose[{minorplanets,
    QuantityMagnitude[
      MinorPlanetData[EntityClass["MinorPlanet", "Plutoid"],
     EntityProperty["MinorPlanet", "HelioCoordinates", {"Date" →
       DateObject[{2017, 1, date}]}]]]}]}, ColorData[1, 1],
    MinorPlanetData[#, "OrbitPath"] & /@ minorplanets},
  PlotLabel → DateString[DateList[{2017, 1, date}]]],
  {date, 1, 300 × 365.25}, SaveDefinitions → True]
```

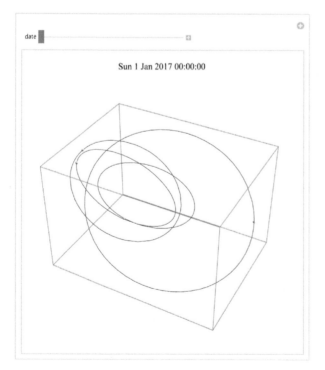

8.3.4 Exoplanets

To obtain information about planets outside of the solar system we can use the Exoplanet entity (or ExoplanetData).

- According to a paper published in *Nature* on August 25, 2016, there's an Earth-sized planet orbiting our nearest neighboring star other than the sun, Proxima Centauri.

In[ ]:= **EntityClass["Exoplanet",**
{"DistanceFromEarth" → TakeSmallest[1]}] // EntityList

Out[ ]= { Proxima Centauri b }

- Next all the available information regarding this planet:

In[ ]:= **DeleteMissing[Proxima Centauri b ["Dataset"]]**

alphanumeric name	Proxima Centauri b
alternate names	{}
altitude	$-18°0'-39.0"$
apparent altitude	$-18°-3'-38.6"$
average orbit distance	0.051 au
azimuth	$162°44'48."$
above the horizon	False
classes	{ ⊞ super-Earths }
component	b
constellation	Centaurus
declination	$-62°-40'-46.15"$
discovery date	2016
discovery method	radial velocity
distance from Earth	4.21 ly
distance from Sun	4.21 ly
distance interval	(4.20742 to 4.20742) ly
orbital eccentricity	0.35
entity classes	{ ⊞ super-Earths }
galactic latitude	$-2°-17'-37.2701"$
galactic longitude	$314°15'58.2097"$
↗ ⋀ rows 1-20 of 48 ⋁ ↘	

Although the planet orbits (0.051 au) closer to its star than Mercury does to the Sun, the star itself is far fainter than the Sun. The surface temperature would allow the presence of liquid water. However, the conditions on the surface may be strongly affected by the ultraviolet and X-ray flares from the star, far more intense than the Earth experiences from the Sun.

8.3.5 Galaxies and Nebulae

Today, we know that the Sun is just one of the stars in a galaxy known as the Milky Way. However, the notion of a galaxy was not introduced until the 1920s. Before then, the difference between galaxies and nebulae was not clear. Seen through the telescope both appeared like blurry objects with shapes resembling clouds. That's why the term nebulae was used to refer to both. Galaxies were considered a type of nebula, and our galaxy was the entire Universe.

- Let's see the definition of galaxy:

In[ ]:=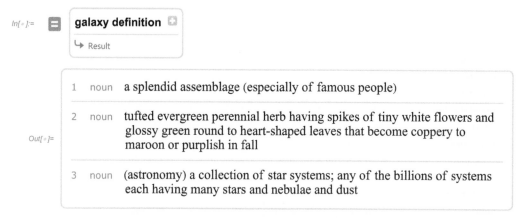

Based on the 3rd definition, nebulae are just part of galaxies. With telescopes we can only see the ones located in our own galaxy. The generic name nebulae actually includes two types of very different structures: the planetary nebulae ("PlanetaryNebula") generated by the accumulation of dust left after a supernova explosion, and nebulae ("Nebula") star nurseries, such as the Orion Nebula. As we can see the name is confusing as planetary nebulae have nothing to do with the formation of planets. Other very important galaxy elements are *clusters*, or globular clusters (groups of stars). Information about galaxies, clusters, and nebulae can be found using the Galaxy entity (or GalaxyData), the StarCluster entity (or StarClusterData), and the Nebula entity (or NebulaData).

- Here is the Milky Way diameter in light-year

In[ ]:=

Out[ ]:= 1.0×10^5 ly

- A very interesting nebula is Orion (M42), visible specially during winter in the northern hemisphere. It is an enormous gas cloud where new stars are constantly being born.

In[ ]:= ⊟ Orion Nebula

Out[ ]:= M42

In[ ]:= `Entity["Nebula", "M42"]["Dataset"]`

alphanumeric name	M42
alternate names	{NGC 1976, LBN 974, the Orion Nebula}
altitude	-7° -51' -6.8"
next transit altitude	43° 38.39'
angular diameter	1.3°
angular radius	39.'
apparent altitude	-7° -56' -5.1"
apparent magnitude	—
azimuth	269° 39' 56."
azimuth at rise	96° 39.26'
azimuth at set	263° 21.4'
blue band magnitude	4.0
above the horizon	False
classes	{deep sky objects, diffuse nebulae, Messier objects, NGC objects}
constellation	Orion
daily time above horizon	11 h 27. min
declination	-5° -23' -25.00"
average diameter	30. ly
distance from Earth	1.3×10^3 ly
distance from Sun	1.3×10^3 ly

rows 1–20 of 49

- We've seen that big distances, such as the ones between galaxies, are usually measured in parsec (pc), the equivalent of the distance corresponding to one arc second. The closest galaxies to our own are shown below, note the use of Dataset. They all form a group of galaxies known as the local group. Many of them are quite small.

```
In[ ]:= lggaxies = Dataset[ Local Group of galaxies GALAXIES [
         "DistanceFromSun", "EntityAssociation"]];
```

- We select the galaxies closer than 100 kpc:

In[ ]:= `lggaxies[Select[# < 100 kpc &]]`

Boötes Dwarf Galaxy	0.0604 Mpc
Canis Major Dwarf Galaxy	2.5×10^4 ly
Draco Dwarf Galaxy	0.082 Mpc
Large Magellanic Cloud	0.0500 Mpc
Milky Way	2.6×10^4 ly
Sagittarius Dwarf Elliptical Galaxy	0.020 Mpc
Sculptor Dwarf Galaxy	0.079 Mpc
Sextans Dwarf Galaxy	0.086 Mpc
Small Magellanic Cloud	0.0606 Mpc
Ursa Minor Dwarf Galaxy	0.066 Mpc
Willman I	0.037 Mpc

Out[ ]=

- To check which galaxies are visible to the naked eye (Apparent Magnitude < 5), we sort the apparent magnitudes of the local group members (remember that the smaller the magnitude the higher the luminosity).

In[ ]:= `Dataset[[`⊞ **Local Group of galaxies** GALAXIES `][`

    `"ApparentMagnitude", "EntityAssociation"]][Select[# < 5 &]]`

Large Magellanic Cloud	0.9
M31	3.5
Sagittarius Dwarf Elliptical Galaxy	4.5
Small Magellanic Cloud	2.2

Out[ ]=

As mentioned earlier, with our own eyes it is very difficult to see magnitudes greater than 5. This means that in the northern hemisphere the only visible galaxy is M31 (Andromeda), the rest of the galaxies with high luminosity (and correspondingly low apparent magnitude) are only visible from the southern hemisphere.

8.4 Application: Determining the Color of the Stars

In[ ]:= `ClearAll["Global`*"]`

The French philosopher Auguste Comte (1798–1857) believed that the composition of stars was going to be beyond human knowledge. In his *Cours de Philosophie* he used to say that stars would only be known as specks of light in the sky since they were very far away. However, the analysis of stellar light has enabled us to know reasonably well stars' composition, temperature, and evolution.

A star is what in physics is known as a black body and therefore has a light spectrum whose characteristics are related to its surface temperature following Planck's law.

- Using the free-from input we can access Planck's law formula. First we type "Planck law" using natural language, press ⊞ once Mathematica has understood our input, select the equation's pod, and after right-clicking on the subpod and choosing "Copy subpod content", paste the information on the cell below.

In[ ]:=

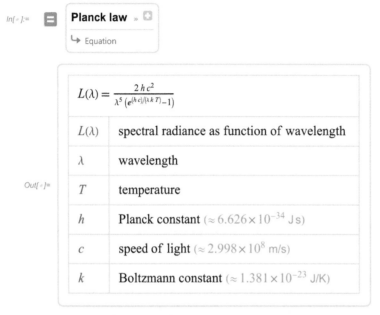

Planck law » ⊞

↳ Equation

Out[ ]=

$L(\lambda) = \dfrac{2\,h\,c^2}{\lambda^5\left(e^{(h\,c)/(\lambda\,k\,T)}-1\right)}$	
$L(\lambda)$	spectral radiance as function of wavelength
λ	wavelength
T	temperature
h	Planck constant ($\approx 6.626\times 10^{-34}$ J s)
c	speed of light ($\approx 2.998\times 10^{8}$ m/s)
k	Boltzmann constant ($\approx 1.381\times 10^{-23}$ J/K)

- The spectrum for a specific temperature, and other properties, can be represented using the PlanckRadiationLaw.

In[ ]:= `PlanckRadiationLaw[Quantity[8000, "Kelvins"], "SpectralPlot"]`

Out[ ]=

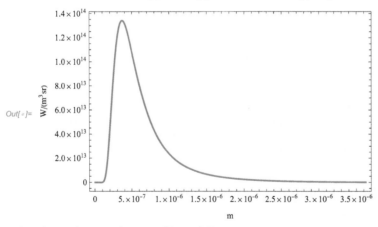

- An alternative way is to use FormulaData.

In[ ]:= `plancklaw1 = FormulaData[{"PlanckRadiationLaw", "Wavelength"}]`

Out[ ]= $L(\lambda) = \dfrac{2\,h\,c^2}{\lambda^5\left(\exp\left(\frac{h\,c/k}{T\,\lambda}\right)-1\right)}$

- Let's have a look at the inner formula composition:

In[ ]:= **InputForm[plancklaw1[[2]]]**

Out[ ]//InputForm=
```
Quantity[2, "PlanckConstant"*"SpeedOfLight"^2] /
    ((-1 + E^(Quantity[None, ("PlanckConstant"*"SpeedOfLight") /"BoltzmannConsta
        (QuantityVariable["T", "Temperature"]*QuantityVariable["λ", "Wavelength
    QuantityVariable["λ", "Wavelength"]^5)
```

- To get the constant values in the International System (SI):

In[ ]:= **plancklaw = UnitConvert[plancklaw1[[2]], "SI"]**

Out[ ]= $\mathrm{UnitConvert}\left[\dfrac{2\,h\,c^2}{\lambda^5\left(\exp\left(\frac{h\,c/k}{T\,\lambda}\right)-1\right)},\ \mathrm{SI}\right]$

- Next, we build a function where the wavelength is computed as function of the temperature.

In[ ]:= **blackbody[T_] := plancklaw /.**
    {QuantityVariable["λ", "Wavelength"] → Quantity[k , "Meters"],
    QuantityVariable["T", "Temperature"] → Quantity[T, "Kelvins"]}

In[ ]:= **Plot[{Callout[blackbody[8000], "8000 K", Above],**
    Callout[blackbody[5800], "5800 K", Above],
    Callout[blackbody[4000], "4000 K", Above]}, {k, 10^-7, 3×10^-6 },
    AxesOrigin → {0, 0}, PlotRange → All, MaxRecursion → 0]

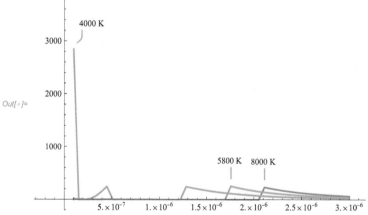

We are going to build a dynamic representation of Planck's law that shows the relationship between color and temperature.

- The next function associates a black body spectrum to its temperature (in degrees Kelvin).

In[ ]:= **BlackBody[T_] :=**
    With[{h = 6.62607×10^-34 , c = 2.99792458×10^8, k = 1.38065×10^-23},
    Plot[(2 h c^2/λ^5) / (Exp[h c / (λ k T)] - 1),
    {λ, 10^-9, 1000×10^-9}, MaxRecursion → 0, ColorFunction →
    (ColorData["VisibleSpectrum"][10^9 #] &), ColorFunctionScaling → False,
    Filling → Axis, Ticks → None, PlotLabel → Row[{T, " K"}]]]

- With Manipulate we create the dynamic representation:

Before executing the function, we need to turn off the message indicating machine underflows:

In[ ]:= **Off[General::munfl]**

```
In[ ]:= Manipulate[
        BlackBody[T], {{T, 5800, "temperature"}, 3000, 10 000},
        SaveDefinitions → True]
```

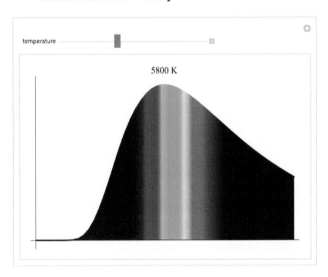

A more complete modelBlackbody Spectrum by Jeff Bryant can be found in:

http://demonstrations.wolfram.com/BlackbodySpectrum

- The next function simulates the color of any given star. We use *"EffectiveTemperature"* to get the superficial temperature and then relate it to its color:

```
In[ ]:= starColorPlot[star_] :=
        Graphics3D[{ColorData["BlackBodySpectrum"][QuantityMagnitude[
            star["EffectiveTemperature"]]], Sphere[]}, Boxed → False, Lighting →
           {{"Ambient", Gray}, {"Directional", LightGray, ImageScaled[{0, 0, 1}]}},
          PlotLabel → star["Name"]]
```

- Let's apply it to several known stars:

```
In[ ]:= stars = Entity["Star", #] & /@
        {"Rigel", "Sirius", "Sun", "Capella", "Betelgeuse"}
```

Out[ ]= { Rigel , Sirius , Sun , Capella , Betelgeuse }

```
In[ ]:= Through[stars[{"Name", "EffectiveTemperature"}]]
```

$$
Out[ ]= \begin{pmatrix}
\text{Rigel} & 1.21 \times 10^4 \, \text{K} \\
\text{Sirius} & 9940. \, \text{K} \\
\text{Sun} & 5772. \, \text{K} \\
\text{Capella} & 5.1 \times 10^3 \, \text{K} \\
\text{Betelgeuse} & 3.6 \times 10^3 \, \text{K}
\end{pmatrix}
$$

- Blue corresponds to hotter stars and orange to colder ones:

In[ ]:= **starColorPlot /@ stars**

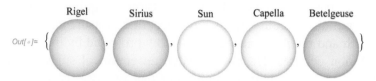

Out[ ]= { Rigel, Sirius, Sun, Capella, Betelgeuse }

8.5 The Measurement of Distances Across the Universe

In[ ]:= **ClearAll["Global`*"]**

Historically, the measurement of distances across the Universe was one of the fundamental problems in astronomy. The first step toward solving it was to measure the Earth's size but before even that, someone had to realize that our planet is a sphere. It is believed that it was the Pythagoreans, mysterious people that didn't leave any written works behind, who for the first time, circa 430 B.C., came up with the idea that the Earth was round, based probably on two observations: a) Sailors traveling from Greece to Northern Africa had noticed that several constellations, such as Ursa Major, would appear higher in the horizon in Egypt than in Athens for a given date and time (although there are other geometric shapes that may explain this fact, the sphere is the one that provides the simplest explanation), b) The way the Moon hides during lunar eclipses can be easily explained if one assumes that it's a shadow projected by a sphere, the Earth, upon the Moon.

Once it was clear that the Earth was round, the next step was its measurement. It was Eratosthenes, who eventually became the head of the library of Alexandria, the first person to measure the Earth's size. The method he used is legendary and well-known among astronomy enthusiasts, but it's worth describing it in the next few lines. He started measuring the distance between Alexandria and Syene (nowadays Aswan), that were located approximately on the same meridian (they really differ by 3°). He also assumed that the Sun was distant enough that its rays would hit the Earth in a parallel manner. He knew that during the summer solstice in Syene the light would illuminate the bottom of the wells in the city. At that moment, the sunbeams would hit the ground perpendicularly while at the same time in Alexandria they would form an angle of approximately 1/50th of the length of a circumference (number deduced by the projection of the shadows). He then used a basic trigonometric identity to figure out that the distance between Alexandria and Syene was 1/50th of the Earth's circumference. Since that distance was 5,000 stadia, the Earth's circumference was 250,000 stadia. There's no agreement about the exact conversion rate to meters. Depending on the reference, if we take a value ranging from 185 m to 157.2 m, the error is between 17% and 1% of the actual size, quite an extraordinary approximation in any case. In 1492, Christopher Columbus assumed a considerably smaller figure while trying to convince Queen Isabella of Spain to finance his trip against the opinion of her advisors, who knew Eratosthenes's estimate. He persisted in his error and ended up discovering America.

Shortly before Eratosthenes's death, Hipparchus of Nice was born (c. 190 BC). He's not only famous due to his discovery of the precession of equinoxes, as we have already mentioned, but also because of his quite accurate measurement of the distance between the Earth and the Moon using the shadow projected by the former onto the latter during Moon eclipses. He also initiated an empirical study of the apparent magnitude of the stars and their positions. For that purpose, he invented the ecliptic coordinates (different from the elliptic ones) and used the astrolabe (invention attributed to him). Many centuries later (from 1989 to 1993) a satellite, carrying its name, repeated his work with technology from the late-20th century. Its successor: Gaia (http://www.esa.int/Our_Activities/Space_Science/Gaia_overview),   currently   in mission, will measure the position of the objects in our galaxy with extreme precision.

We can compute the distance to the nearest stars using geometric methods (Figure 8.2). We measure the position of the star seen from a certain location, and 6 months later we will observe its apparent displacement. This phenomenon is known as parallax, and its explanation is similar to what happens when we place our thumb between and object and our eyes. When closing either eye, we will notice the apparent movement of the object. In the case of a stellar parallax, each eye corresponds to the vertices of the base of an isosceles triangle created when joining the Earth, position in two vertices A and B, separated by 6 months along the orbit of the star, with vertex C, the position of the star, considered fixed. The parallax is half the angle ACB.

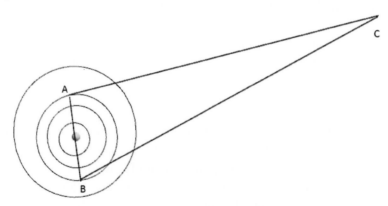

Figure 8.2   Calculating the distance to the nearest stars.

The distance, d, from the Sun to the star with parallax, p, in radians, can be computed as follows:

$$d = \frac{1\,\text{AU}}{\tan p} = \{\text{given that the angle is very small, } \tan p \simeq p \} \simeq \frac{1}{p}\,\text{AU}.$$

- Since the values of p are very small, it is convenient to express them in arcseconds, p_s. The parallax of Procyon is:

In[ ]:= $1 \Big/$ `QuantityMagnitude`$\Big[$ ⊟ Procyon parallax $\Big]$ `"pc"`

Out[ ]= 3.50 pc

- Or directly in parsecs:

In[ ]:= `UnitConvert`$\Big[$ **Procyon** STAR $\Big[$ *distance from Earth* $\Big]$, `"Parsecs"`$\Big]$

Out[ ]= 3.49568 pc

A big leap forward in the measurement of astronomical distances happened when Henrietta Swan Leavitt (1868–1921) noticed that certain stars, now known as Cepheid variables, exhibited regular changes in their luminosity from which their absolute magnitude could be calculated.

The relationship period/luminosity for the Cepheid variables has been revised several times since Henrietta Leavitt's original measurements. Currently, the following function is usually applied:

$$M = -2.78 \log (P) - 1.35$$

where M is the absolute magnitude of the star (the luminosity of the star if it was 1 pc away) and P is the average period in days.

- The previous expression can be written in *Mathematica* as:

In[ ]:= `abm[pd_] = -2.78 Log [10, pd] - 1.35;`

Applying a similar criterion to the one used to determine the distance between a bulb with known brightness and the brightness experienced by an observer, we can deduct the following relationship between M, the apparent magnitude m (magnitude seen from the Earth, calculated as the average between the maximum and minimum observed magnitudes), and the distance D (in pc):

$$m - M = 5 \log (D/10) = 5 \log(D) - 5$$

- From the previous formulas, we can establish the following equation to compute the distance using the period (pd) and the apparent magnitude (apm), both of which can be measured empirically:

In[ ]:= `distceph[pd_, apm_] := Solve[apm - abm[pd] - 5 Log[10, dist] + 5 == 0, dist]`

This method can be applied to stars in our galaxy with the help of big ground telescopes or space ones. We can also use it to measure the distances to our nearest galaxies assuming we can find Cepheid variable stars in them.

In https://www.eso.org/public/products/education/ we can find a set of basic astronomy exercises using real data. In one of them, we are given the results of the calculations done by Hubble in several Cepheid variables located in the M100 galaxy. According to the data, one star has a period of 53.5 days and an apparent magnitude of 24.5. Stars of such magnitude can only be observed with large telescopes.

- The absolute magnitude and distance, in pc, are:

In[ ]:= `{abm[53.5], dist /. distceph[53.5, 24.90]}`

Out[ ]= $\{-6.15482, \{1.62542 \times 10^7\}\}$

It's one of the most distant objects ever measured using this method.

For remote galaxies the method used is based on measuring the apparent luminosity of a type Ia supernovae and relating it to the real magnitude (that can be deduced using theory).

Another key parameter related to the previous one is the redshift experienced by light. This is an example of the Doppler effect caused by galaxy movements, mostly associated with the expansion of the universe.

- You can find information about redshift typing in a *free-form input cell* "**Cosmological redshift**"). After expanding the cell and telling *Mathematica* to interpret it as a formula, you can even do some interactive calculations (click +) .

In[ ]:=   ▦   **Cosmological redshift** » ▣

            ↳ Basic dimensions

In[ ]:= **UniverseModelData[Quantity[5×10^9, "LightYears"], "Redshift"]**

Out[ ]= 0.485872

In[ ]:= **QuantityMagnitude@**
         UniverseModelData[Quantity[5×10^9, "LightYears"], "Redshift"];

Astronomer Edwin Powell Hubble (1889–1953) (the famous telescope was named after him) realized that galaxies frequently presented redshift. The amount changed depending on the galaxy. Later on, this astronomical phenomenon became known as Hubble's law.

$$c\,z = H_0\,D$$

with:

z (redshift) = $(\lambda_{ob} - \lambda_e)/\lambda_e$, where λ_{ob} corresponds to the observed wavelength and λ_e to the emitted one; z is therefore a dimensionless number.
c is the speed of light.
D is the actual distance to the galaxy (in Mpc).
H_0 is the Hubble constant at the moment of observation (during our lifetime and for the duration of our civilization, this value will be practically constant).

Once z is known, we can calculate D using basic algebra:

$$D = \frac{c\,z}{H_0}$$

Hubble's law is considered a fundamental relation between recessional velocity and distance. However, the relation between recessional velocity and redshift depends on the cosmological model adopted, and is not established except for small redshifts.

A modified version of Hubble's law and other properties of the universe are now included in UniverseModelData.

- The Hubble constant, H_0 (HubbleH0) is regularly expressed in m/s/Mpc.

In[ ]:= **UniverseModelData["LambdaCDM"]**

Out[ ]= <|OmegaLambda → 68.89%, OmegaMatter → 31.11%, OmegaRadiation → 0.07%, HubbleH0 :→ H_0|>

- Determine the redshift of a comoving object 5×10^9 light years away:

In[ ]:= **UniverseModelData[Quantity[5×10^9, "LightYears"], "Redshift"]**

Out[ ]= 0.485872

- Determine the distance of a comoving object, in light years, for a redshift of 0.075:

In[ ]:= **f[t_?NumericQ] := QuantityMagnitude@
    UniverseModelData[Quantity[t, "LightYears"], "Redshift"];**

In[ ]:= **FindRoot[0.075 == f[dist], {dist, 10^9}]**

> ⋯ FindRoot: The line search decreased the step size to within tolerance specified by AccuracyGoal
>   and PrecisionGoal but was unable to find a sufficient decrease in the merit function. You may
>   need more than MachinePrecision digits of working precision to meet these tolerances.

Out[ ]= $\{\text{dist} \rightarrow 1.00056 \times 10^9\}$

- Let's apply this method to the galaxy IC1783.

In[ ]:= **zIC1783 = GalaxyData["IC1783", "Redshift"]**

Out[ ]= 0.011

In[ ]:= **FindRoot[zIC1783 == f[dist], {dist, 10^8}]**

Out[ ]= $\{\text{dist} \rightarrow 1.55537 \times 10^8\}$

- We compare this value with the best estimate available in the Galaxy entity.

In[ ]:= **UnitConvert[IC 1783 GALAXY [distance from Earth], "ly"]**

Out[ ]= 1.256×10^8 ly

We can see that there are some discrepancies. This is because galaxies are also affected by the gravitational forces of their galactic neighbors, e.g., Andromeda influencing the Milky Way. In fact, Andromeda and our galaxy are approaching each other and will collide in approximately 4 billion years from now.

The possibilities of the astronomical functions in *Mathematica* don't end here. Feel free to continue exploring.

8.6 Application: Binary Systems and the Search for Exoplanets

In[ ]:= **Quit[]**

In this section, we will show how we can use *Mathematica* for other Astronomy related matters. We will describe, in particular, several features developed by the author for the study of binary systems and, by extension, the search for exoplanets.

Since 1995, when the first extrasolar planet was found, more than 400 have been detected. NASA's Kepler spacecraft launched on March 6, 2009, was specifically designed to search for extrasolar planets. Probably, by the time you read these lines, their total number will have increased significantly.

Kepler site: http://www.nasa.gov/kepler

To find exoplanets, we mainly use two methods:

1) Radial velocity: A star with a planet will move in its own small orbit in response to the planet's gravity leading to variations in the radial velocity of the star with respect to Earth. These movements are only detectable for giant planets next to stars.

2) Transit photometry: When a planet crosses in front of its parent star's disk the observed brightness of the star drops by a small amount. This last method is becoming the most effective one. The principle is similar to the one used when studying binary star systems.

More than half of the stars that we see are actually systems of two or more stars. Normally it's not possible to distinguish between different stars in multiple systems (not to confuse them with naked-eye binaries that are usually stars that seem close to each other when seen from Earth but are actually located in different systems altogether). The detection is usually done with the transit photometry method. This approach is very effective when dealing with eclipsing binaries, systems of two stars with coplanar orbits when observed from Earth.

Next we are going to describe how to build this kind of system using *Mathematica* and study its brightness. In the example that follows, we use as reference *Eclipsing Binary Stars* by Dan Bruton:

http://www.physics.sfasu.edu/astro/ebstar/ebstar.html

Figure 8.3 represents a binary system (although we discuss the case of two stars, the method is similar if we consider a star and a planet). The coordinates of the centers are (x1, y1, z1) for star 1 and (x1, y1, z1) for star 2 (assuming homogeneous spheres, the centers of mass coincide with the geometric centers). We consider {0, 0, 0} as the coordinates for the center of mass of the system. These coordinates have been chosen so that an observer, located on Earth, is on the axis OZ. That is: the one that goes from the center of mass to Earth. For the OY axis, we take the perpendicular to OZ in the star 1 plane when it forms an angle $\theta = 0$. This happens when the star is closest to the observer.

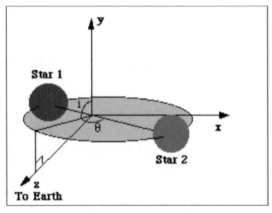

Figure 8.3   A binary system.

With:

i = orbital inclination: angle, in radians, between the line of vision and the orbital plane axis.

R = distance between the centers of both stars. It's expressed in arbitrary units, usually the average distance between the stars. In the case of a constant orbit, R = 1.

m1, m2 = stars' masses (we can choose arbitrary units, but it's important that they keep the equivalence q = m2/m1).

r1, r2 = radii. For convenient purposes we choose r1 > r2. In practice, we can choose stars 1 and 2 so that the condition is always met. It's convenient to express the radii in fractions of the major semi-axis. In the case of a circular orbit, obviously the major semi-axis is greater than or equal to the minor one and identical to the radius R.

θ = Phase, corresponding to the angle between the star and OZ, using θ = 0, the angle where the center of star 2 is the closest to the observer. In a binary system with circular orbit, θ is a variable parameter (it's time-dependent) while the previous parameters are constant (although that is not always the case, for example: There might be some intrinsic changes to the brightness or even the mass could change as in the case of a mass transfer between stars. However, in the example that follows we'll use the simplest case, which happens to be the most common one). θ is known as the azimuthal angle and follows $\theta = \frac{2\pi(t-t0)}{P}$, with P being the orbital period.

In a binary system, the coordinates {x1, y1, z1} and {x2, y2, z2} for the location of the stars are given by:

$$x1 = \frac{-x}{1 + (1/q)} \, , \quad y1 = \frac{-y}{1 + (1/q)} \, , \quad z1 = \frac{-z}{1 + (1/q)} \, ;$$

$$x2 = \frac{x}{1 + q} \, , \quad y2 = \frac{y}{1 + q} \, , \quad z2 = \frac{z}{1 + q} \, ;$$

- We can convert them to spherical coordinates using the formulas:

```
In[ ]:= x = R Sin[θ];
        y = R Cos[i] Cos[θ];
        z = R Sin[i] Cos[θ];
```

- Then, the position of each star as a function of R, i, θ, q with q = m2/m1 can be calculated with the following functions:

```
In[ ]:= star1[R_, i_, θ_, q_] = {─────────, ─────────, ─────────};
                                 -x           -y           -z
                              1 + (1/q)    1 + (1/q)    1 + (1/q)

        star2[R_, i_, θ_, q_] = {─────, ─────, ─────};
                                   x       y       z
                                 1 + q   1 + q   1 + q
```

- We have used the previous equations to define a *Mathematica* function to visually display the movement of both stars as a function of R, i, θ, m1, and m2:

`eclbin[R, i, θ, m2/m1, r1, r2]` (The condition m2 < m1 is necessary).

```
In[ ]:= eclbin[rr_, ii_, rho_, qq_, rr1_, rr2_] :=
      Module[{R, i, θ, q, r1, r2},

        Manipulate[Graphics3D[{ControlActive[Opacity[1], Opacity[.6]],

            Sphere[st1[R, i, θ, q], r1], Sphere[st2[R, i, θ, q], r2],
            Opacity[1], RGBColor[.6, .74, .36], PointSize[.02],
            Point[{0, 0, 0}], Black}, PlotRange → {{-2, 2}, {-2, 2}, {-2, 2}},
          SphericalRegion → True, Boxed → False, Background → Black,
          ImageSize → {500, 340}, ViewPoint → {0, 0, 20}],
          {{R, rr, "R"}, 0.1, 5, Appearance → "Labeled"},
          {{i, ii, "i"}, 0, π, Appearance → "Labeled"},
          {{q, qq, "mass 2 / mass 1"}, 0.1, 1, Appearance → "Labeled"},
          {{r1, rr1, "radius 1"}, 0.1, .5, Appearance → "Labeled"},
          {{r2, rr2, "radius 2"}, 0.1, .5, Appearance → "Labeled"},
          {{θ, rho, "θ (rotation)"}, 0, 4π, Appearance → "Labeled"},
```
$$
\text{Initialization} \Rightarrow \left\{ st1[R\_, i\_, θ\_, q\_] := \left\{ -\frac{R\,\mathrm{Sin}[θ]}{1+\frac{1}{q}}, \right. \right.
$$
$$
\left. -\frac{R\,\mathrm{Cos}[i]\,\mathrm{Cos}[θ]}{1+\frac{1}{q}}, -\frac{R\,\mathrm{Cos}[θ]\,\mathrm{Sin}[i]}{1+\frac{1}{q}} \right\},\ st2[R\_, i\_, θ\_, q\_] :=
$$
$$
\left. \left\{ \frac{R\,\mathrm{Sin}[θ]}{1+q}, \frac{R\,\mathrm{Cos}[i]\,\mathrm{Cos}[θ]}{1+q}, \frac{R\,\mathrm{Cos}[θ]\,\mathrm{Sin}[i]}{1+q} \right\} \right\} \right]\,]
$$

The initial moment $θ = 0$ is the time when both stars are aligned and can be seen from the OZ axis (where the observer is located), that is when the interposition of the stars is at its maximum. In a real system, all the parameters would be fixed and only $θ$ would change as a function of time.

- Let's consider the binary system defined by the following parameters R = 2, i = 2 Pi 85 /360, m1 = 0.6, m2 = 0.1, R1 = 0.4, R2 = 0.3.

```
In[ ]:= eclbin[2, 2 Pi 85 / 360, 0, 0.1/0.6, 0.4, 0.3]
```

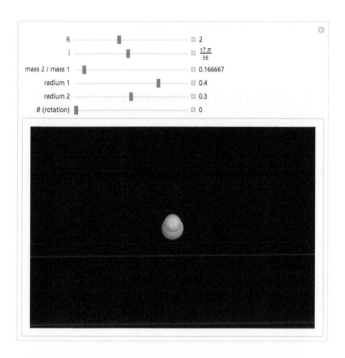

8.7 Light Curves

8.7.1 Light Curves in Binary Systems

In[ ]:= **Quit[]**

The transit photometry method discussed in the previous section is based on the study of the changes in brightness when a star or a planet interposes itself between another star and the observer. As mentioned before, since stars are not visually distinguishable, even less a star and a planet, what we will see is a variation in the apparent brightness.

The method described here, with the arrival of astronomical CCD cameras, is within the reach of Astronomy fans. There have been instances where fans have even been able to detect exoplanets using this technique. This is a great example of how science can be done without the need for major resources.

The luminosity l of a star is defined as the amount of energy emitted by the star per unit of time. The flow F is the energy emitted per unit of surface and per unit of time, that is $F1 = \dfrac{l_1}{4\pi R1^2}$ and $F2 = \dfrac{l_2}{4\pi R2^2}$ where l_1, l_2 = luminosities (in arbitrary units, as long as they keep the ratio l_2/l_1).

Then an observer will see the system with a luminosity given by:

$l = K\,(F1\,A1\ +\ F2\,A2)$

where A1 and A2 represent each of the areas of the stars' disks as seen by the observer, and K can be determined based on the observer's detector area, on the OZ axis, and the distance between the Earth and the binary system.

To find these areas, we need to know the apparent distance ρ as seen by the observer (located on the OZ axis). That distance can be calculated as follows:

$$\rho = \sqrt{(x_1 - x_2)^2 + (y_1 - y_2)^2}$$

- The next function computes $\rho(R, i, \theta, q)$:

In[ ]:= `rho[R_, i_, ɵ_, q_] = Module[{x, y, x1, x2, y1, y2}, x = R Sin[ɵ];`

`y = R Cos[i] Cos[ɵ];`

$$\{x1, y1\} = \left\{\frac{-x}{1 + (1/q)}, \frac{-y}{1 + (1/q)}\right\};$$

$$\{x2, y2\} = \left\{\frac{x}{1 + q}, \frac{y}{1 + q}\right\};$$

`Sqrt[ (x2 - x1)^2 + (y2 - y1)^2]];`

In[ ]:= `rho[2, 2 Pi 85 / 360, ɵ, 0.1/0.6]`

Out[ ]= $\sqrt{0.0303845 \, Cos[ɵ]^2 + 4. \, Sin[ɵ]^2}$

- With the example data (R = 2, i = 2 Pi 85 /360, m_1 = 0.6, m_2 = 0.1), the apparent distance ρ changes according to the graph below. Based on our problem formulation, the lowest apparent size corresponds to the maximum interposition that happens when $\theta = 0$, Pi, 2 Pi, ..., n Pi.

In[ ]:= `Plot[rho[2, 2 Pi 85 / 360, ɵ, 0.1/0.6], {ɵ, 0, 4 Pi}, AxesLabel → {ɵ, ρ}]`

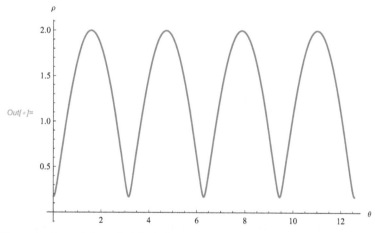

To calculate A1 and A2, it is necessary to take into account the current stage in the cycle.

- The two figures below show the star with the bigger radius (r_1) in red, and the one with the smaller radius (r_2) in blue.

In[ ]:= `Graphics[{Red, Disk[{0, 0}, 2], Blue, Disk[{4, 0}, 1]}]`

Out[ ]=

Without eclipse: Corresponds to the period in which all the light emitted by both stars in the direction of the observer reaches him. Therefore, this is when the luminosity is at its maximum. In this stage, the following relationship holds: $\rho > r_1 + r_2$ with:

$A_1 = \pi \, r_1^2$ and $A_2 = \pi \, r_2^2$

Partial eclipse: This happens when one star partially covers the other one. During this stage, $r_1 + r_2 > \rho > r_1 - r_2$.

- Figure 8.4 was created using the following function:

```
Graphics[{{Red, Circle[{0, 0}, 1]}, {Blue, Circle[{1.15, 0}, 0.5,
{-2 Pi/3, 2   Pi/3}]}, {Dashed, Blue, Circle[{1.15, 0}, 0.5]},
Line[{{0, 0}, {0.9, 0.425}, {0.9, -0.425}, {0, 0}}], Line[{{0.9,
0.425}, {1.15, 0} , {0.9, -0.425} }] }]
```

We added legends to the output of the function using *Mathematica*'s 2D Drawing Tools: [CTRL]+D in Windows or [CTRL]+T in OS X.

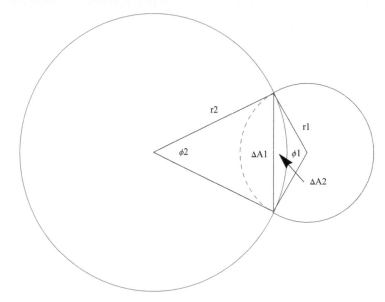

Figure 8.4   Partial eclipse diagram.

To calculate the amount of light that arrives from the eclipsed star, we have to subtract the eclipsed area from the total area. This means we have to compute $\Delta A1$ and $\Delta A2$ corresponding to the star 1 and star 2 segments, respectively, using the following formulas:

$$\Delta A1 = \frac{1}{2} R1^2 (\phi 1 - \mathrm{Sin}[\phi 1]); \quad \Delta A2 = \frac{1}{2} R2^2 (\phi 2 - \mathrm{Sin}[\phi 2])$$

The following expressions enable us to compute $\phi 1$ and $\phi 2$:

$$R2^2 = R1^2 + \rho^2 - 2 R1 \rho \, \mathrm{Cos}\left[\frac{\phi 1}{2}\right], \quad R1^2 = R2^2 + \rho^2 - 2 R2 \rho \, \mathrm{Cos}\left[\frac{\phi 2}{2}\right]$$

- The output below represents the different stages of a partial eclipse:

```
In[ ]:=  {Graphics[{Blue, Disk[{2, 0}, 1], Red, Disk[{0, 0}, 2]}],
         Graphics[{Blue, Disk[{-2, 0}, 1], Red, Disk[{0, 0}, 2]}],
         Graphics[{Red, Disk[{0, 0}, 2], Blue, Disk[{-2, 0}, 1]}],
         Graphics[{Red, Disk[{0, 0}, 2], Blue, Disk[{2.5, 0}, 1]}] }
```

Out[ ]=

The two figures on the left correspond to the scenario where star 1 (red) is the closest one, that is $z1 > z2$. In this case, the light that reaches the observer will be:

$A_1 = \pi r_1^2$ and $A_2 = \pi r_2^2 - \Delta A1 - \Delta A2$

$A_1 = \pi r_1^2$ and $A_2 = \pi r_2^2 - \Delta A1 - \Delta A2$

The rest of the figures corresponds to the case where star 1 (red) is the most distant one, that is $z1 < z2$ then:

$A_1 = \pi r_1^2 - \Delta A1 - \Delta A2$ and $A_2 = \pi r_2^2$

$A_1 = \pi r_1^2 - \Delta A1 - \Delta A2$ and $A_2 = \pi r_2^2$

Total or annular: Corresponds to the period when a star is shaded by the other one completely. In this stage: $\rho \leq r_1 - r_2$ with: $A_1 = \pi r_1^2$ and $A_2 = 0$ for $z1>z2$ and $A_1 = \pi r_1^2 - \pi r_2^2$ and $A_2 = \pi r_2^2$;

All these conditions are summarized in the table below:

□	□	z1 > z2		z1 < z2	
Stage	Condition	A_1	A_1	A_1	A_1
Without eclipse	$\rho \geq r_1 + r_2$	πr_1^2	πr_2^2	πr_1^2	πr_2^2
Partial eclipse	$r_1 + r_2 > \rho > r_1 - r_2$	πr_1^2	$\pi r_2^2 - \Delta A1 - \Delta A2$	$\pi r_1^2 - \Delta A1 - \Delta A2$	πr_2^2
Total or annular	$\rho \leq r_1 - r_2$	πr_1^2	0	$\pi r_1^2 - \pi r_2^2$	πr_2^2

Now we are going to include all the criteria described earlier in the function **area[R, i, m2/m1, r1, r2, 11, 12, θ]**, where: $m2 > m1$ and $r1 > r2$. The function returns the area that the observer will measure.

Without eclipse: When $\rho \geq r_1 + r_2$:

```
In[ ]:=  area[R_, i_, q_, r1_, r2_, 11_, 12_, θ_] :=
         Module[{A1, A2}, {A1, A2} = {Pi r1^2 , Pi r2^2 };
           11 A1 / r1^2 + 12 A2 / r2^2] /; rho[R, i, θ, q] ≥ r1 + r2
```

Partial eclipse: When $r_1 + r_2 > \rho > r_1 - r_2$:

- We first calculate $\Delta A1$ and $\Delta A2$

```
In[•]:= ΔA1[R_, i_, q_, r1_, r2_, θ_] :=
          Module[{φ1, ρ, phi, phi1}, phi1 = Last[Solve[

                  r2² ⩵ r1² + ρ² - 2 r1 ρ Cos[phi/2], phi] /. ρ → rho[R, i, θ, q]][[1, 2]];

              1/2 r1² (φ1 - Sin[φ1]) /. φ1 → phi1];


      ΔA2[R_, i_, q_, r1_, r2_, θ_] :=
          Module[{φ2, ρ, phi, phi1}, phi1 = Last[Solve[

                  r1² ⩵ r2² + ρ² - 2 r2 ρ Cos[phi/2], phi] /. ρ → rho[R, i, θ, q]][[1, 2]];

              1/2 r2² (φ2 - Sin[φ2]) /. φ2 → phi1];
```

- In this type of eclipse, for $z1 > z2$: $A1 = \pi r_1^2$ and $A2 = \pi r_2^2 - \Delta A1 - \Delta A2$ while for $z1 < z2$: $A1 = \pi r_1^2 - \Delta A1 - \Delta A2$ and $A2 = \pi r_2^2$.

```
In[•]:= area[R_, i_, q_, r1_, r2_, l1_, l2_, θ_] :=
          Module[{z1, z2, a1, a2, A1, A2}, z1 = (- R Cos[θ] Sin[i])/(1 + 1/q);

              z2 = (R Cos[θ] Sin[i])/(1 + q);

              a1 = ΔA1[R, i, q, r1, r2, θ];
              a2 = ΔA2[R, i, q, r1, r2, θ];
              {A1, A2} =
                If[z1 > z2, {Pi r1^2, Pi r2^2 - a1 - a2}, {Pi r1^2 - a1 - a2, Pi r2^2}];
              l1 A1 / r1^2 + l2 A2 / r2^2] /; r1 + r2 > rho[R, i, θ, q] > r1 - r2
```

Total or annular eclipse : When $\rho \leq r_1 - r_2$:

- When $r1 > r2$ one of the stars will be directly in front of the other one, interposing it. Then a total eclipse will happen if $z1 > z2$ with $A1 = \pi r_1^2$ and $A2 = 0$, or an annular one will occur if $z1 < z2$ with $A1' = A1 - A2$ and $A2' = A2$.

```
In[•]:= area[R_, i_, q_, r1_, r2_, l1_, l2_, θ_] :=
          Module[{z1, z2, A1, A2}, z1 = (- R Cos[θ] Sin[i])/(1 + 1/q);

              z2 = (R Cos[θ] Sin[i])/(1 + q);

              {A1, A2} = If[z1 > z2, {Pi r1^2, 0}, {Pi r1^2 - Pi r2^2, Pi r2^2}];
              l1 A1 / r1^2 + l2 A2 / r2^2] /; rho[R, i, θ, q] ≤ r1 - r2
```

Let's simulate the ideal case of an eclipsing binary. We can distinguish the repeating phases of an eclipse: one corresponds to the period without eclipse (t1 to t2), in t2 the eclipse starts and reaches its plenitude between t3 and t4, from t4 to t5 we are in the partial eclipse phase, from t6 to t9 is the transit phase, and when t9 =

t1 the cycle starts again.

- We use `Plot[area[2, 90 Degree, 0.1/0.6, 0.6, 0.3, 0.6, 0.2, 2 Pi t/8], { t, 0, 12}, AxesLabel -> {"t", "I"}, AxesOrigin -> {0, 0}]` and include the legends using the Drawing Tools option in the Graphics menu: **Graphics ▸ Drawing Tools**.

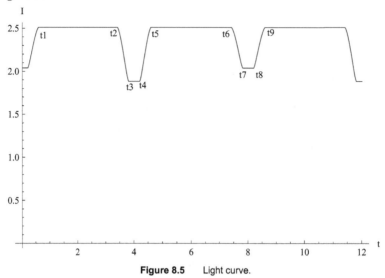

Figure 8.5   Light curve.

Example: With R = 2, i = 85°, m1 = 0.6, m2 = 0.1, R1 = 0.4, R2 = 0.3, l1 = 0.6, l2 = 0.2, we'd like to show I/Imax.

- We first calculate Imax.

In[ ]:= `max = NMaximize[area[2, 2 Pi 85 / 360, 0.1/0.6, 0.4, 0.3, 0.6, 0.2, θ], {θ, 0, 2 Pi}][[1]] // Quiet`

Out[ ]= `2.51327`

- Then we display I/Imax.

In[ ]:= `Plot[area[2, 85 Degree, 0.1/0.6, 0.4, 0.3, 0.6, 0.2, 2 Pi θ] /max, { θ, 0, 1.5}, AxesLabel → {"Phase", "I/Imax"}, AxesOrigin → {0, 0}] // Quiet`

Out[ ]=

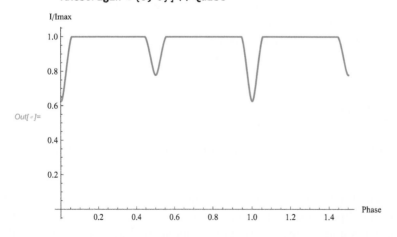

8.7.2 Application: Using Light Curves to Determine the Period-Luminosity Relation

Besides eclipsing variables or planets interposition, periodic changes in luminosity can happen for other reasons. Probably the best-known example is the change in luminosity in variable stars, such as Cepheids. The usual method for building an experimental light curve consists of measuring the luminosity at different times t and deduce the period from them.

- The following expression simulates magnitude as a function of time (in practice, since we often don't know this function we try to deduce it from the experimental data). The chosen function is very simple, and we only use it for didactic purposes.

In[ ]:= **mag[t_] = 5 + 0.14 Cos[0.3 t + 0.1];**

- Let's assume that we have already taken measurements at different times t_i.

In[ ]:= **data = Table[{t, Round[mag[t], 0.01]}, {t, 0, 200, 5}];**

- We represent graphically the previous data. If we only had the previous data points it would be very difficult to figure out the underlying pattern (remove **Joined → True**), joining the dots makes it easier to observe the cyclical nature of the data. It's normal for many intermediate points to be missing, making the analysis even more challenging.

In[ ]:= **ListPlot[data, Joined → True]**

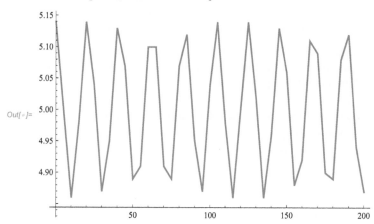

Out[ ]=

The analysis method frequently used is to represent the previous phenomenon in a phase diagram.

To calculate the phase, we need to apply this equation:

$$\phi = \text{Decimal part of}\left[\frac{t - t_0}{P}\right]$$

where:

ϕ represents the phase in cycles.

t_0 (normally called epoch) represents the origin of the time variable expressed as a Julian Date (DJ).

t is the moment when the measurement is taken (normally we will measure the apparent magnitude, m).

P is the period, the time it takes for the cycle to repeat itself.

An equivalent expression to the previous one and the one that we will be using is:

$$\phi = \text{Mod}\left[\frac{t - t_0}{P}\right]$$

■ With the function below, *Mathematica* returns the phase given the initial data points and the period.

In[ ]:= `phi[list_, p_] := Module[{data}, data = list;`
`        Transpose[{Mod[data[[All, 1]] / p, 1], data[[All, 2]]}]];`

In this example we are assuming that the period is known, but in practice that's what we are trying to find out. We can check that by using an approximate value of the period the representation is more diffuse.

■ With real observations, the period has to be determined, and that is not always easy. The next function will help us to compute it. We keep in mind that for period-magnitude curves, the order of representation in the axis OY is inverted. That is, the magnitudes are sorted in descending order not in the usual ascending one: $-3 > -2 > 0 > 1 > 2$.

In[ ]:= `curveL[data_, per_, kmax_] :=`
`        Module[{mag, a, b, k, i}, mag = Round[Transpose[data][[2]], 0.1];`
`          Manipulate[ListPlot[Join[phi[data, per + k] /. {a_, b_} → {a, -b},`
`            phi[data, per + k] /. {a_, b_} → {-(1 - a), -b}], AxesLabel →`
`              {"Period", "Magnitude"}, AxesOrigin → {-1, -Max[mag] - 0.1},`
`            Ticks → {Automatic, Table[{i, -i}, {i, -Max[mag] - 0.1,`
`                -Min[mag] + 0.1, (Max[mag] - Min[mag]) / 10}]}],`
`          {{k, 0, "Deviation from the period (in days)"}, 0, kmax}]]`

■ Let's assume a complete cycle (2π). We use the previous function to simulate the experimental measures and include a random component. We make the assumption that the period is 20 days (the duration of a complete orbit).

In[ ]:= `simulation = Table[`
`          {θ, area[2, 2 Pi 85 / 360, 0.1/0.6, 0.4, 0.3, 0.6, 0.2, 2 Pi θ / 20] +`
`            RandomReal[NormalDistribution[0, 0.005]]},`
`          {θ, 0, 60, 0.05}]; // Quiet`

■ We represent the data in a phase diagram:

In[ ]:= `simulation1 = Sort[phi[simulation, 20]];`

In[ ]:= `ListPlot[simulation1, Joined → True]`

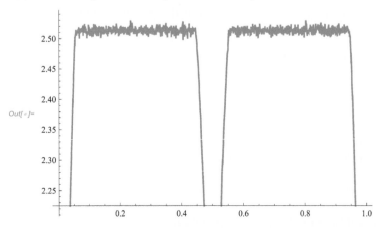

The previous luminosity data have to be multiplied by a parameter k to convert them into a magnitude measure. Remember that higher luminosity numerically implies lower magnitude. For that reason, to represent the data in a phase/luminosity diagram we assume $k = -1$.

- If you execute the following function and use the slider to choose a 5-day period, you'll see that the curve is very similar to the previous figure.

In[ ]:= `curveL[Map[Times[{1, -1}, #] &, simulation], 15, 20]`

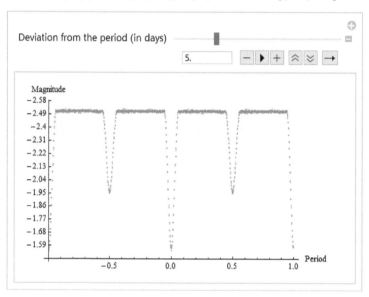

8.7.3 Application: Finding the Light Curve of a Known Eclipsing Binary

Let's apply the newly created function to the luminosity data from NSV 03199 obtained by Garcia-Melendo, E., Henden, A., 1998, IBVS, No. 4546. The data are in the file: NSV03199.xls in our data directory.

- Let's import the luminosity measures:

In[ ]:= `vsdata =`
        `Drop[Import[FileNameJoin[{NotebookDirectory[], "Data", "NSV03199.xls"}],`
            `{"XLS", "Data", 1}], 1];`

- The data have Julian dates, so we use as reference JHD=2450510.46542.

In[ ]:= `vsdata1 = vsdata /. {a_, b_} → {a - 2450510.46542, b};`

- We cannot really see the cycle pattern:

In[ ]:= `ListPlot[vsdata1]`

Out[ ]=

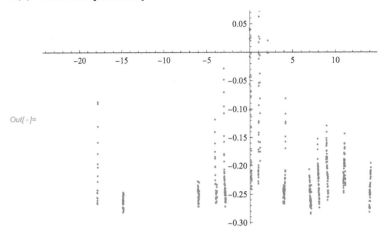

However, when using a period-magnitude curve, the periodic behavior corresponding to an eclipsing binary can be seen clearly.

- We find that the optimum period is 1.046400 days:

In[ ]:= `curveL[vsdata1, 1.046400, 0.5]`

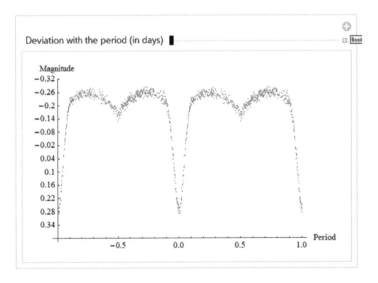

All the previous functions along with other examples in the book have been put together in a package that can be downloaded from diarium.usal.es/guillermo.

8.8. Additional Resources

Astronomy demonstrations: http://demonstrations.wolfram.com/topic.html?topic=Astronomy

Eric Weisstein's World of Astronomy: http://scienceworld.wolfram.com/astronomy

To follow objects with variable luminosity (variable stars, supernovae and many more), the best place is the website of the American Association of Variable Star Observers (AAVSO): http://www.aavso.org

Author's website astronomy section (in Spanish): http://diarium.usal.es/guillermo/astronomia

9

Nuclei and Radiations

Radioactivity is a natural phenomenon that allows us to peek into our planet's past and the history of humankind. Its effects when absorbed by human beings can also be scientifically measured. This chapter covers radioactivity and other topics related to nuclear and particle physics with examples such as the modeling of the evolution of isotopes over time or the dating of the history of humankind. The last two sections discuss the calculation of binding energies and how to design products that protect from radiation respectively. Throughout the chapter, the Mathematica functions **IsotopeData** and **ParticleData** are used extensively.

9.1 Nuclear and Particle Physics

The Entity["Isotope", name] (or IsotopeData) and ParticleData functions deal with nuclear and particle physics applications.

- Writing the isotope name inside an Inline Free-form input or IsotopeData["*isotope*", "*property*"] returns the value of the specified property for a given isotope.

In[ ]:= ▣ iodine 131 ["Properties"] // Shallow

Out[ ]//Shallow=

$\{$ atomic mass , atomic number , atomic symbol ,

binding energy per nucleon , biological half-life , biological lifetime ,

branching ratios , critical diameter , critical mass , critical organs , ≪45≫$\}$

- The next example shows some of properties of Iodine-131.

In[ ]:= IsotopeData["Iodine131", #] & /@
  {"FullSymbol", "AtomicMass", "BranchingRatios", "DecayModes",
   "DaughterNuclides", "HalfLife", "IsotopeAbundance"}

Out[ ]= $\left\{ {}^{131}_{53}I_{78} , 130.906124609\,u, \{1.00\}, \{BetaDecay\}, \{ xenon-131 \}, 6.9338 \times 10^5\,s, 0. \right\}$

The output tells us that this isotope, whose complete symbol is ${}^{131}_{53}I_{78}$, has an atomic mass of 130.09061 atomic mass units (amu), with a single type of radioactive decay, a beta decay, that disintegrates into Xenon-131 with a disintegration period of 6.9338×10^5 and that it doesn't exist in natural form (0.).

- The example below generates a dataset with information about all the properties available for Iodine-131.

$In[\ast]:=$ **DeleteMissing**$\Big[$ ⊟ iodine 131 **["Dataset"]**$\Big]$

atomic mass	130.906124609 u
atomic number	53
atomic symbol	I
binding energy per nucleon	8.422309 MeV
biological half-life	1.4×10^2 days
biological lifetime	2.0×10^2 days
branching ratios	$\{100.\%\}$
daughter nuclides	$\{$ xenon-131 $\}$
decay constant	9.9967×10^{-7} per second
Q–value	$\{0.970848 \text{ MeV}\}$
decay modes	{electron emission}
decay modes	$\{\beta^-\}$
decay products	{{{Electron, ElectronNeutrinoBar}, 1.0000}}
effective half-life	7.6 days
effective lifetime	11. days
entity classes	$\{$ ▦ boson , ▦ unstable isotopes , ▦ beta decay $\}$
Excited State Energies	$\{\cdots_{86}\}$
Excited State Half Lives	$\{\cdots_{86}\}$
excited state lifetimes	$\{\cdots_{86}\}$
excited state parities	$\{\cdots_{86}\}$

⤒ ⋀ rows 1–20 of 46 ⋁ ⤓

9.2 What Are Isotopes?

Matter consists of atoms that can be described in a simple way as a central part named the nucleus, containing neutrons and protons, surrounded by an electron cloud. Atomic elements have the same number of protons, but the number of neutrons may change. Nuclei with the same number of protons, Z, and a different one for neutrons, N, are the isotopes of that element. Normally, an isotope is written as $^A X$, where X is the symbol of the element and A is the mass number given by A = Z + N. For example: atmospheric carbon consists of 3 isotopes: Carbon-12 (or ^{12}C) with 6 protons and 6 neutrons, Carbon-13 (or ^{13}C) with 6 protons and 7 neutrons and Carbon-14 (or ^{14}C) that has 6 protons and 8 neutrons. All isotopes of the same element possess the same chemical properties but different physical ones.

In nature there are 92 elements (we can also find traces of elements with Z > 92, such as plutonium, but these are artificial elements), with an average of 2 or 3 isotopes each, although there are some elements with only one stable isotope and others with as many as 8. Isotopes can be stable or unstable. Stable nuclei, the majority on earth, last "forever", that is, they always keep the same number of neutrons and protons. Some theories predict the disintegration of protons, and therefore of any kind of atom, but they have not been proven yet. Furthermore, even if they could be validated, the disintegration speed would be so slow that for all practical purposes we could consider that stable nuclei have an almost eternal duration. Unstable nuclei (called radioisotopes) are those that over time (depending on the radioisotope the period can range from a fraction of a second to billions of years) are transmuted into other elements, called daughter isotopes. The new nucleus can in turn be stable or unstable. If it's unstable, the process continues until the daughter isotope is a stable one.

- With IsotopeData[*"element"*, *"property"*] we can obtain all the known isotopes for a given element. The element can be written as a complete word or, preferably, as a symbol using standard notation. If we type "Symbol" in *"property"*, we'll get the isotopes in $^A X$ notation. We use Short to display only one line, but the complete output includes more isotopes.

In[ ]:= **IsotopeData["C", "Symbol"] // Short**

Out[ ]//Short=

$$\{^8C, \ ^9C, \ ^{10}C, \ ^{11}C, \ ^{12}C, \ ^{13}C, \ \ll 3 \gg, \ ^{17}C, \ ^{18}C, \ ^{19}C, \ ^{20}C, \ ^{21}C, \ ^{22}C\}$$

The decay of a type of nucleus into another one usually happens due to the emission of particles α or β and radiation γ. In some radionuclides, there are also emissions of neutrons and protons and even other types of reactions such as spontaneous fission (SF), common in some heavy nuclei. Fission is the division of a nucleus into two smaller ones, each one with approximately half the mass of the original.

Using "DecayModes" or "DecayModeSymbols" as properties we can get the type of decay for a certain isotope. One isotope can display more than one decay mode although normally one of them would be the most common one. Using "BranchingRatios" and "DecayEnergies" we can see, respectively, the probability associated to each decay mode and its energy measured in keV (kiloelectron-volt).

- The command below shows the different decay modes of Uranium–238, and the symbols, probabilities and energies associated to them.

In[ ]:= **Grid[IsotopeData["Uranium238", #] & /@**
        {"DecayModes", "DecayModeSymbols", "BranchingRatios", "DecayEnergies"}]

Out[ ]=

AlphaEmission	SpontaneousFission	DoubleBetaDecay
α	SF	$2\beta^-$
1.00	5.45×10^{-7}	2.2×10^{-12}
4269.75 keV	—	1144.2 keV

Notice that the majority of the radioactive decays for U238 correspond to alpha emissions (the value 1.00 indicates that the probability is close to 100%). However, spontaneous fissions (SF), even though they have a small probability of happening, play a central role in chain reactions since they emit neutrons, essential to start a chain reaction.

In short, we can say that a radioactive isotope is like a father who has a daughter, who in turn may have a granddaughter and so on until eventually there's a descendant without offspring, the stable nucleus. The disintegration happens at a constant pace, and its characteristics depend on the particular isotope. In this process, particles (α , β, ...) with specific energies are emitted.

- You can use the free-form input to get the definition of chain reaction. Probably when you type "chain reaction definition" you will see several definitions. Just choose the most appropriate one depending on your needs.

In[ ]:=

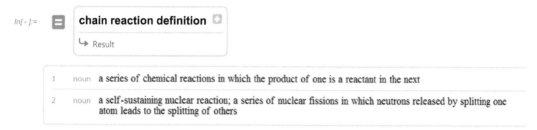

| 1 | noun | a series of chemical reactions in which the product of one is a reactant in the next |
| 2 | noun | a self-sustaining nuclear reaction; a series of nuclear fissions in which neutrons released by splitting one atom leads to the splitting of others |

- We can also get information by classes. We use Shallow to avoid displaying the complete output, which is quite large. If you want to see all the classes just remove the command from the next input.

In[ ]:= **IsotopeData["Classes"] // Shallow**

Out[ ]//Shallow=

{ alpha emission , beta decay , beta delayed alpha emission ,

beta delayed deuteron emission , beta delayed fission , beta delayed four neutron emission ,

beta delayed neutron alpha emission , beta delayed neutron emission ,

beta delayed three neutron emission , beta delayed triton emission , ≪37≫}

- Each class includes isotopes sharing a common characteristic. For example, if we want to know the isotopes with a beta emission associated to spontaneous fission (a very rare occurrence, only observed in 3 isotopes out of the more the 3,000 available), the command below gives us the answer:

In[ ]:= **IsotopeData[▦ beta delayed fission ISOTOPES]**

Out[ ]= { actinium-230 , protactinium-236 , protactinium-238 }

9.3 Decay Constants, Decay Periods, and Half-Lives

The easiest case of radioactive decay is that of a radioactive isotope A that decays into a stable nucleus B, that is, A→B, where B is not radioactive. This process can be represented by a very basic compartmental model, such as the one shown in Figure 9.1 with $x(t)$ representing the quantity of a certain radioactive isotope A present at time t in a sample and λ being the decay constant, specific to each isotope. The quantity of atoms (measured in the most appropriate way: units, grams, becquerels, etc.) that decay per unit of time is proportional to the quantity in the sample at each moment, that is: $\lambda x(t)$.

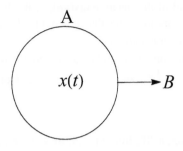

Figure 9.1   Radioactive decay of isotope A.

Let's assume that at $t = 0$ the sample contains x_0. This process can be mathematically modeled by a simple first-order differential equation:

$$\frac{dx}{dt} = -\lambda x(t), \text{ with initial condition : } x_0 \text{ at } t = 0 \tag{9.1}$$

- The solution to the previous differential equation can be obtained using DSolve. Notice that we use "=" and not ":=" Why? (normally is better to use ":=", but in this case we choose "=" since the next time we call the function it will already know the solution without having to calculate it again. In this case, the time saving is negligible).

In[ ]:= `x1[ t_, x0_, λ_] = x[t] /. DSolve[{x'[t] == -λ x[t], x[0] == x0}, x[t], t] [[1]]`

Out[ ]= $x0\,e^{\lambda(-t)}$

We type it again using the typical textbook format

$$x(t) = x_0\,e^{-\lambda t} \tag{9.2}$$

- In the next example, we use λ and the total time t as our changing parameters. We also use the previously defined function within Manipulate using the command Initialization. Alternatively, we could have used SavedDefinitions.

In[ ]:= `Manipulate[Plot[ y[t, λ, 1], {t, 0, t1}, Filling → Bottom],`
`        {{λ, 1}, 0, 5}, {{t1, 3}, 1, 10},`
`        Initialization :> (y[t_, λ_, x0_] := x0 e^(λ (-t)))]`

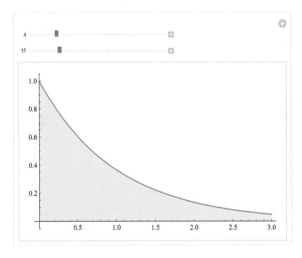

In many situations instead of the **decay constant**, λ, it's better to use the semi-decay period or **half-life**, $T_{1/2}$, defined as the time that it would take for the amount x_0 present in the sample at time $t = 0$ to become half, $x_0/2$.

- The relationship between λ and $T_{1/2}$ can be calculated as follows:

In[ ]:= `cteDes = λ /. Solve[ x1[ T, x0, λ] == x0/2 , λ, Reals][[1]]`

Out[ ]= $\dfrac{\log(2)}{T}$

Another concept is the **mean lifetime**, the life expectation (duration) of the isotopes present in a sample. It's the same as calculating the arithmetic mean of the expected duration of the individual isotopes present in a sample.

- To calculate the mean of a continuous variable t we compute $\langle t \rangle = \int_0^\infty t f(t)\, dt \big/ \int_0^\infty f(t)\, dx$, and associate the result to the symbol τ. We specify that $\lambda \in \mathbb{R}{>}0$, otherwise we'll receive a message stating that the solution is only valid if $\lambda \in \mathbb{R}{>}0$.

In[ ]:= `τ = Integrate[t x₀ E^(-t λ), {t, 0, Infinity},`
`        Assumptions → λ ∈ Reals && λ > 0] / Integrate[x₀ E^(-t λ),`
`        {t, 0, Infinity}, Assumptions → λ ∈ Reals && λ > 0]`

Out[ ]= $\dfrac{1}{\lambda}$

In summary: the radioactive decay constant λ, the half-life $T_{1/2}$, and the mean lifetime τ, are connected as follows: $T_{1/2} = \text{Log}[2]/\lambda = \tau\,\text{Log}[2]$, $\tau = 1/\lambda$.

- With IsotopeData we can find out the half-life, mean lifetime or decay constant of Iodine-131.

In[ ]:= `IsotopeData["Iodine131", #] & /@ {"HalfLife", "Lifetime", "DecayConstant" }`

Out[ ]= $\{6.9338 \times 10^5\text{ s},\ 1.0003 \times 10^6\text{ s},\ 9.9967 \times 10^{-7}\text{ per second}\}$

- By default the function returns the time in seconds, but we can convert it to other unit using UnitConvert.

```
UnitConvert[
    IsotopeData["Iodine131", #] & /@ {"HalfLife", "Lifetime" }, "days"]
```

- In the example below we can choose the isotope among several options. We use QuantityMagnitude to get only the value (in this case, the decay constant), without the unit.

In[ ]:= `Manipulate[Plot[ y[t, 3600 λ, 1 ], {t, 0, t1}, Filling → Bottom],`
`        {{t1, 3, "Hours"}, 1, 100},`
`        {λ, {QuantityMagnitude[IsotopeData["Iodine131", "DecayConstant"]] →`
`            "Iodine-131", QuantityMagnitude[`
`              IsotopeData["Iodine125", "DecayConstant"]] → "Iodine-125"}},`
`        Initialization :→ (y[t_, λ_, x0_] := x0 e^(λ (-t)))]`

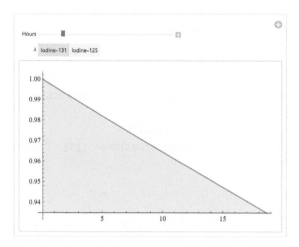

- The next example uses IsotopeData to show all the isotopes for iodine. We also include the option of selecting the unit for displaying the evolution over time for the chosen isotope. Since the decay rate for some of the isotopes is very fast we use a logarithmic scale.

```
In[ ]:= Manipulate[LogPlot[y[t, k * λ, 1], {t, 1, t1}, Filling → Bottom],
       {{t1, 3}, 1, 100}, {k, {1 → "Seconds", 3600 → "Hours", 3600 * 24 → "Days"}},
       {{λ, "", "Isotope"},
       Thread[QuantityMagnitude[IsotopeData["Iodine", "DecayConstant"]] →
       IsotopeData["Iodine"]]}, Initialization ⧴ (y[t_, λ_, x0_] := x0 ℯ^(λ (-t)))]
```

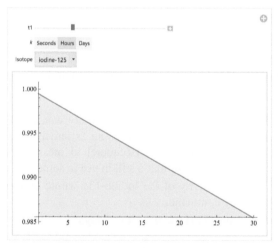

For a given radioisotope, activity A, the decay rate in 1 second (1 nuclear transformation = 1 Becquerel or Bq), can be obtained from the decay constant, $λ$ (in s^{-1}). The activity of n_{at} atoms will be $A = λ\, n_{at}$. Since what is known usually is the mass, m, we apply the conversion $n_{at} = m\, N_A / m_a$, where N_A is the Avogadro's number and m_a the atomic mass of the element expressed in the same units as m, normally in grams (g).

Iodine-131 example

Iodine-131 emissions from the Fukushima Daiichi nuclear reactor, due to the accident caused by the tsunami in March 2011, were estimated to be 1.5×10^{17} Bq as of the end of April 2011. Given this information, let's explore how much mass was emitted and how the emissions evolved over time.

Activity can be turned into mass using the formula $A = \lambda\ n_{at} = \lambda\ m\ N_A / m_a \rightarrow m = m_a A / (\lambda N_A)$.

- Iodine-131 decays into Xenon-131, a stable isotope that at room temperature is gas. We use LayeredGraphPlot to represent the decay chain.

```
In[ ]:= LayeredGraphPlot[{IsotopeData["I131", "Symbol"] →
        IsotopeData[IsotopeData["I131", "DaughterNuclides"][[1]], "Symbol"]},
    Left, VertexLabels → "Name"]
```

- We need to know the decay constant, λ131, and the atomic mass. IsotopeData will give us the latter (although we could have used 131 as an excellent approximation) and also return the mean lifetime τ (in seconds). Afterward we can calculate $\lambda = 1/\tau$.

```
In[ ]:= { mat131, λ131} = {IsotopeData["Iodine131", "AtomicMass"],
        1/IsotopeData["Iodine131", "Lifetime"] }
```

$Out[ ]= \{130.906124609\ u, 9.9967 \times 10^{-7}$ per second$\}$

- Now we use the formula $m = m_a A / (\lambda N_A)$, where N_A (Avogadro's number) is given by *Mathematica* using its natural language capabilities ("Avogadro's Number") or with the function Quantity. UnitConvert shows the SI units, in this case just the number.

```
In[ ]:= Quantity["Avogadro's Number"]
```

$Out[ ]= N_0$

```
In[ ]:= NA = UnitConvert[1 N_A]
```

$Out[ ]= 602\,214\,076\,000\,000\,000\,000\,000$

```
In[ ]:= 1.5 × 10^17 QuantityMagnitude[mat131] / (QuantityMagnitude[λ131] * NA) "grams"
```

$Out[ ]= 32.617$ grams

It may seem strange at first sight that such a big radioactive activity would only result in a very small mass. What's actually happening is that the Bq is an extremely small unit. As a matter of fact, for many years instead of using the Becquerel to measure radioactivity, people used the Curie, Ci (1 Ci = 3.7 10^{10}Bq), unit that's still in use in some places.

- To find out in how many days the radioactivity of the Iodine-131 would become just 0.1% of its initial figure, we use the following command:

```
In[ ]:= UnitConvert[t /. Solve[x1[t, 100, λ131] == 0.1, t], "days"]
```

⋯ Solve: Inverse functions are being used by Solve, so some solutions may not be found; use Reduce for complete solution information.

$Out[ ]= \{79.9774$ days$\}$

9.4 Decay Chains

```
In[ ]:= Clear["Global`*"]
```

As we've seen, a radioactive isotope is like a father who keeps on having descendants until the offspring is a stable nucleus. Sometimes, the first descendant is stable right away, as with Iodine-131 → Xenon-131, but other isotopes, like Radium-226, shown at the end of this section, have more complicated decay chains. One of Radium-226's daughter isotopes, Radon-222 (Rn222), is responsible for a large share of the natural radiation that people are exposed to. Each time we breathe, a portion of the air includes radon gas. The content of Rn222 changes substantially from one area to another. For example: granitic houses with poor ventilation usually have a high content level of radon. The reason is that granite contains Radium-226 that becomes radon gas when it disintegrates.

- An isotope can decay following different branches. For example: Radium-226 decays with a half-life period of 5.0×10^{10} s into 3 radioisotopes: ^{222}Rn, ^{212}Pb and ^{226}Th. Almost all of it decays into ^{222}Rn and an insignificant part into ^{212}Pb. Occasionally, there can also be traces of Thorium-226.

```
In[ ]:= ra226 = IsotopeData["Ra226", #] & /@
          {"DaughterNuclides", "BranchingRatios", "HalfLife"};

       TableForm[ra226, TableHeadings → {{"Isotope", "Fraction", "T (s)"}, None}]
```

Out[ ]//TableForm=

Isótopo	radon-222	lead-212	thorium-226
fracción	1.00	2.6×10^{-11}	—
T (s)	5.0×10^{10} s		

- In different contexts it is likely that when using IsotopeData you may find that the output contains the word "Missing". If you want to eliminate it, you can use DeleteMissing or DeleteCases[*list*, _*h*], which eliminate all the elements in a list whose head starts with h, as shown in the example below:

```
In[ ]:= is1 = IsotopeData["U238", "DaughterNuclides"]
```

Out[ ]= { thorium-234 , —, plutonium-238 }

```
In[ ]:= is2 = DeleteCases[is1, _Missing]
```

Out[ ]= { thorium-234 , plutonium-238 }

The set of all branches produced by the decay of a parent isotope (like Uranium-238) is known as the radioactive decay chain. However, the majority of those disintegrations happen along only one branch. IsotopeData["isotope", "BranchingRatios"] shows the results after sorting the branches from bigger to smaller in terms of their likelihood of occurrence. When an isotope decays into one or several isotopes, we can choose to keep just one: the one with the highest probability. If we apply this criterion to each daughter isotope, we will end up with a single branch named the main branch. From a practical point of view, in many decay chains it is enough to study the main branch.

- The following function lets us choose the main branch, and within it, the number of daughter nuclides *n* that we are interested in. Notice the use of *iso_String*, and *n_Integer* to limit the input type to: "*isotope name*" (between double quotation marks), and *n* integer (number of daughter nuclides that we would like to display); in any other case, the function will not be executed.

```
In[ ]:= mainbranch[iso_String, n_Integer] :=
          NestList[First[IsotopeData[#, "DaughterNuclides"]] &, iso, n]
```

- We apply it to Uranium-238 and indicate 14 daughter nuclides, the entire decay chain (one of the longest ones).

In[ ]:= `mainbranch["U238", 14]`

Out[ ]= {U238, thorium-234 , protactinium-234 , uranium-234 , thorium-230 ,

radium-226 , radon-222 , polonium-218 , lead-214 , bismuth-214 ,

polonium-214 , lead-210 , bismuth-210 , polonium-210 , lead-206 }

- This problem was posted in "comp.soft-sys.math.mathematica". The function below shows the elegant solution proposed by one of the list members (Bob Hanlon) that returns a complete branch without the need to specify *n*. However, if we are interested in limiting the number of offsprings that we wish to see, it would be better to use **mainbranch**.

In[ ]:= `mainbranch1[x_String] :=`
`    NestWhileList[IsotopeData[#, "DaughterNuclides"][[1]] &,`
`        x, UnsameQ, 2, 100, -2] // Quiet;`

- Let's apply it to Radon-226

In[ ]:= `ra226 = mainbranch1["Ra226"]`

Out[ ]= {Ra226, radon-222 , polonium-218 , lead-214 , bismuth-214 ,

polonium-214 , lead-210 , bismuth-210 , polonium-210 , lead-206 }

- The function below generates a list of isotopes showing: their disintegration period, daughter nuclides and branching ratios.

In[ ]:= `isopro[isolist_] :=`
`    Module[{vals}, vals = Table[IsotopeData[#, prop], {prop,`
`            {"Symbol", "HalfLife", "DaughterNuclides", "BranchingRatios"}}] & /@`
`        isolist;`
`    Text[Grid[Prepend[vals, {"", "Disintegration\nperiod (sec)",`
`        "Daughter nuclides", "Branching ratios"}], Frame → All,`
`        Background → {None, {{{LightBlue, White}}, {1 → LightYellow}}}]] ]`

- We use it to show the information related to the principal branch of Radon-226 obtained in the previous calculation.

In[ ]:= `isopro[ra226]`

Out[ ]=

	Disintegration period (sec)	Daughter nuclides	Branching ratios
^{226}Ra	5.0×10^{10} s	{ radon-222 , lead-212 , thorium-226 }	$\{1.00, 2.6 \times 10^{-11}, —\}$
^{222}Rn	330 350. s	{ polonium-218 }	{1.00}
^{218}Po	185.9 s	{ lead-214 , astatine-218 }	{1.00, 0.00020}
^{214}Pb	1610. s	{ bismuth-214 }	{1.00}
^{214}Bi	1.19×10^3 s	{ polonium-214 , thallium-210 , lead-210 }	{1.0, 0.00021, 0.000030}
^{214}Po	0.0001643 s	{ lead-210 }	{1.00}
^{210}Pb	7.01×10^8 s	{ bismuth-210 , mercury-206 }	$\{1.00, 1.9 \times 10^{-8}\}$
^{210}Bi	4.33×10^5 s	{ polonium-210 , thallium-206 }	$\{1.00, 1.32 \times 10^{-6}\}$
^{210}Po	1.19557×10^7 s	{ lead-206 }	{1.00}
^{206}Pb	∞	{}	{}

The functions below are part of the documentation related to LayeredGraphPlot. Sometimes we may get lucky and find examples in the documentation that perfectly match what we want to do. This is one of those occasions.

- The following functions (from ref/LayeredGraphPlot) let us input the parent isotope (the first in the chain) and get as a result all the daughter nuclides from all the branches except those that have an insignificant contribution ("Missing").

```
In[ ]:= DaughterNuclides[s_List] :=
    DeleteCases[Union[Apply[Join, Map[IsotopeData[#, "DaughterNuclides"] &,
        DeleteCases[s, _Missing]]]], _Missing]
```

```
In[ ]:= ReachableNuclides[s_List] :=
    FixedPoint[Union[Join[#, DaughterNuclides[#]]] &, s]
```

```
In[ ]:= DecayNetwork[iso_List] :=
    Apply[Join,
        Map[Thread[# → DaughterNuclides[{#}]] &, ReachableNuclides[iso]]]
```

```
In[ ]:= DecayNetworkPlot[s_] :=
    LayeredGraphPlot[Map[IsotopeData[#, "Symbol"] &, DecayNetwork[{s}], {2}],
        PerformanceGoal → "Quality", VertexShapeFunction →
        (Text[Framed[Style[#2, 8, Black], Background → LightYellow], #1] &)]
```

- We use the last one to show the decay chain of Rn222, the immediate decay product of Ra226. You can also apply the same function to see the complete chain of the latter. You may have to adjust the output image size to see it properly. To do that just place the cursor in one of the corners of the image and resize it accordingly.

In[ ]:= **DecayNetworkPlot["Rn222"]**

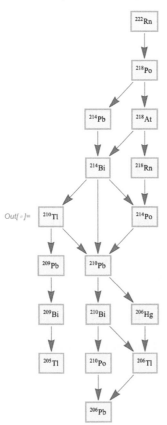

Out[ ]=

- The following expression returns all the isotopes that have the same number of neutrons:

In[ ]:= **isoneu[neut_] :=**
    Select[IsotopeData[], IsotopeData[#, "NeutronNumber"] == neut &]

In[ ]:= **isoneu[14]**

Out[ ]= { boron-19 , carbon-20 , nitrogen-21 , oxygen-22 , fluorine-23 , neon-24 ,
        sodium-25 , magnesium-26 , aluminum-27 , silicon-28 , phosphorus-29 ,
        sulfur-30 , chlorine-31 , argon-32 , potassium-33 , calcium-34 }

- Using the previous function we display the decay chains of all the isotopes with 4 neutrons.

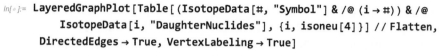

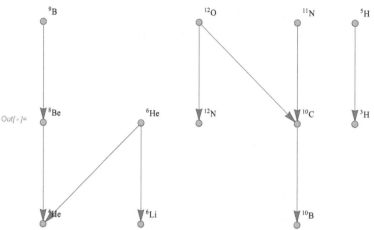

9.5 Application: Modeling the Evolution of a Chain of Isotopes over Time

A radioactive decay chain can be considered as a system of compartments in which the flow is one-directional, as shown in the figure below. This is an example of what is known in compartment modeling as a catenary model.

```
Graphics[{Circle[{0, 0}, 0.5] , Text["1", {0, 0}, {0, 0}],
    Text["λ₁", {1.75, .4}, {0, 0}], Arrow[{{0.5, 0}, {3, 0}}],
            Circle[{3.5, 0}, 0.5] , Text["2", {3.5, 0}, {0, 0}],
    Circle[{7, 0}, 0.5] , Text["3", {7, 0}, {0, 0}],
    Arrow[{{4, 0}, {6.5, 0}}], Text["λ₂", {5.25, .4}, {0, 0}],
    {Dashed, Arrow[{{7.5, 0}, {10.5, 0}}]}, Text["λᵢ", {9, .4}, {0, 0}]},
    AspectRatio → Automatic, Axes → None]
```

The first circle represents the first isotope in the chain, with a decay constant λ_1, the second circle refers to the next isotope, with a decay constant λ_2, and so on.

Let's consider the case of a chain of 3 isotopes, with $q_1(t)$, $q_2(t)$, and $q_3(t)$ the quantities present of each one at time t, satisfying the following set of differential equations:

$$q_1'(t) = -\lambda_1 q_1(t)$$
$$q_2'(t) = \lambda_1 q_1(t) - \lambda_2 q_2(t)$$
$$q_3'(t) = \lambda_2 q_2(t) - \lambda_3 q_3(t)$$

- As an initial condition let's assume that at $t = 0$ there's a quantity (atoms, gm, Bq) of isotope 1 in compartment 1 and nothing in the rest. This means: q1[0] = b1, q2[0] = q3[0] = 0 then:

```
In[*]:= eq = DSolve[{q1'[t] == -λ1 q1[t],
         q2'[t] == λ1 q1[t] - λ2 q2[t], q3'[t] == λ2 q2[t] - λ3 q3[t],
         q1[0] == b1, q2[0] == 0, q3[0] == 0}, {q1[t], q2[t], q3[t]}, t];
```

- The retention in compartments 1, 2, and 3 is given by:

```
In[*]:= {q1[t_], q2[t_], q3[t_]} = {q1[t], q2[t], q3[t]} /. eq[[1]] // Simplify
```

$$Out[*]= \left\{ b1\, e^{\lambda 1\,(-t)},\; -\frac{b1\,\lambda 1\,\left(e^{\lambda 1\,(-t)} - e^{\lambda 2\,(-t)}\right)}{\lambda 1 - \lambda 2}, \right.$$

$$\left. \frac{b1\,\lambda 1\,\lambda 2\,e^{-t(\lambda 1+\lambda 2+\lambda 3)}\left((\lambda 1 - \lambda 2)\,e^{t(\lambda 1+\lambda 2)} + (\lambda 3 - \lambda 1)\,e^{t(\lambda 1+\lambda 3)} + (\lambda 2 - \lambda 3)\,e^{t(\lambda 2+\lambda 3)}\right)}{(\lambda 1 - \lambda 2)\,(\lambda 1 - \lambda 3)\,(\lambda 2 - \lambda 3)} \right\}$$

- Grouping identical exponential terms we get:

$$q3\,(t) = \frac{b1\,e^{-t\lambda 1}\,\lambda 1\,\lambda 2}{(\lambda 1 - \lambda 2)\,(\lambda 1 - \lambda 3)} + \frac{b1\,e^{-t\lambda 3}\,\lambda 1\,\lambda 2}{(\lambda 1 - \lambda 3)\,(\lambda 2 - \lambda 3)} + \frac{b1\,e^{-t\lambda 2}\,\lambda 1\,\lambda 2}{(-\lambda 1 + \lambda 2)\,(\lambda 2 - \lambda 3)} =$$

$$b1\lambda 1\,\lambda 2\,\frac{e^{-t\lambda 1}}{(\lambda 1 - \lambda 2)\,(\lambda 1 - \lambda 3)} + \frac{e^{-t\lambda 3}}{(\lambda 1 - \lambda 3)\,(\lambda 2 - \lambda 3)} + \frac{e^{-t\lambda 2}}{(-\lambda 1 + \lambda 2)\,(\lambda 2 - \lambda 3)} =$$

$$= b_1 \prod_{j=1}^{3-1}\lambda_j \sum_{j=1}^{3} \frac{e^{-\lambda_j t}}{\prod_{p=1,p\neq j}^{n}(\lambda_p - \lambda_j)}$$

Following the same approach, the previous equation can be extended for *n* daughter isotopes:

$$q_n(t) = b_1 \prod_{j=1}^{n-1}\lambda_j \sum_{j=1}^{n} \frac{e^{-\lambda_j t}}{\prod_{p=1,p\neq j}^{n}(\lambda_p - \lambda_j)},\; \text{with initial conditions}: q_1(t) = b_1,\, q_2(0) =\ldots = q_n(0) = 0 \qquad (9.3)$$

This equation is known as Bateman's equation, in honor of the mathematician who first published it.

- We can type the equation as follows:

```
In[*]:= q[n_] := b
```
$$\left(\sum_{j=1}^{n} \frac{e^{-k_j t}}{\left(\prod_{p=1}^{j-1}(k_p - k_j)\right)\prod_{p=j+1}^{n}(k_p - k_j)} \right) \prod_{j=1}^{n-1} k_j$$

- Here we apply it to the parent and the first 3 daughter isotopes (of the main branch) of Uranium-238.

```
In[*]:= u3 = mainbranch["U238", 3]
```

$$Out[*]= \{\text{U238},\; \text{thorium-234},\; \text{protactinium-234},\; \text{uranium-234}\}$$

- We need the decay constants λ. Remember that $\tau = 1/\lambda$, with τ in seconds. However, since we prefer to measure *t* in days we multiply it by 24 × 3600.

```
In[*]:= λs = Thread[{k₁, k₂, k₃, k₄} → Evaluate[
         24×3600/QuantityMagnitude[IsotopeData[#, "Lifetime"]] & /@ u3]]
```

$$Out[*]= \{k_1 \to 4.248\times10^{-13},\, k_2 \to 0.029,\, k_3 \to 2.5,\, k_4 \to 7.735\times10^{-9}\}$$

- Next we're going to calculate the evolution of the radioactivity, A_i, in the previous isotopes. We assume that at time $t = 0$ there's 1 gm of U238 in the sample, which is expressed as the number of its atoms, N_A (Avogadro's number)/238. Now we can compute A_i, using the general formula $A = \lambda N_A$.

In[ ]:= `{A1[t_], A2[t_], A3[t_], A4[t_]} =`
   `{k₁ q[1], k₂ q[2], k₃ q[3], k₄ q[4]} /. λs /.`
      `b → UnitConvert[1 Nₐ] / 238 // ExpandAll`

Out[ ]= $\left\{ 1.075 \times 10^9 \, e^{-4.248 \times 10^{-13} \, t}, \; 1.1 \times 10^9 \, e^{-4.248 \times 10^{-13} \, t} - 1.1 \times 10^9 \, e^{-0.029 \, t}, \right.$

$1.3 \times 10^7 \, e^{-2.5 \, t} - 1.1 \times 10^9 \, e^{-0.029 \, t} + 1.1 \times 10^9 \, e^{-4.248 \times 10^{-13} \, t},$

$\left. -0.04 \, e^{-2.5 \, t} + 3. \times 10^2 \, e^{-0.029 \, t} - 1.1 \times 10^9 \, e^{-7.735 \times 10^{-9} \, t} + 1.1 \times 10^9 \, e^{-4.248 \times 10^{-13} \, t} \right\}$

- We subtract 10^8 Bq from the activity of Pa234 with the only purpose of differentiating it from Th234. Otherwise both graphs would be superimposed since they are in equilibrium as we will see shortly.

In[ ]:= `Plot[{A1[t], A2[t] - 10^8, A3[t] - 10}, {t, 0, 180},`
   `PlotRange → All, AxesLabel → {"t (days)", "A(t) in Bq"},`
   `PlotLegends → Placed[{"U238", "Th234", "Pa234"}, Center]]`

Out[ ]=
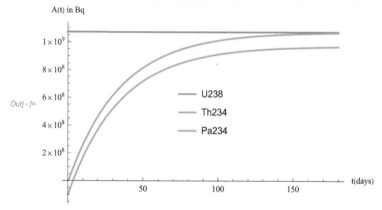

In the graph above we can see that after 180 days, $A1 \approx A2 \approx A3$. This is known as secular equilibrium. In this example, it's enough to calculate the radioactivity of the isotope parent to know the daughter isotopes' radioactivity. This happens when $\lambda_1 < \lambda_2$.

- The next example shows that given enough time, in this case thousands of years, U238 and U234 will reach secular equilibrium. Normally we consider that the secular equilibrium has been reached after a period of time longer than 8 half-life periods of the daughter isotope.

```
In[ ]:= Plot[{A1[t], A4[t]}, {t, 0, 10^9},
    PlotRange → All, AxesLabel → {"t (days)", "A(t) in Bq"},
    PlotLegends → Placed[{"U238", "U234"}, Center]]
```

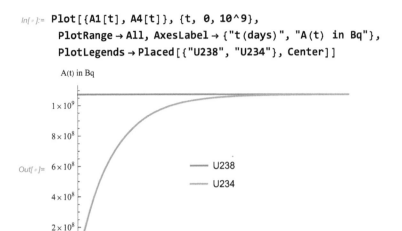

9.6 Application: Dating the History of Humankind

The fact that radioactive isotopes decay following a well-known physics law even under extreme conditions, makes them the ideal candidates for dating different type of samples. The actual isotopes used for that purpose will depend, among other factors, on the estimated age of the object that we want to date. We are going to discuss two methods: Carbon-14 (^{14}C) dating is suitable for organic samples less than 50,000 years old and the lead-lead method is used when dating rocks thousands or even millions of years old.

Radiocarbon Dating

The ^{14}C dating method was discovered in 1947 by Willard Libby. It can be applied to samples that are no more than 50,000 years old. This method has become the most frequently used one for measuring the age of organic objects, especially archeological ones.

- The table that follows displays the isotopic abundance (proportion of isotopes) of the different carbon isotopes and their half-lives (If "HalfLife"< 1 s is not shown). An isotopic abundance of 0 indicates that the isotope doesn't exist in nature or is present only as traces. For example: the proportion of ^{14}C is very small although, as we will see, it plays a crucial role in dating.

```
In[ ]:= vals = DeleteCases[Table[IsotopeData[#, prop],
        {prop, {"Symbol", "HalfLife", "IsotopeAbundance"}}] & /@
        IsotopeData["C"], {a_, b_ /; b < 1 s, c_}];
```

```
Text[
  Grid[Prepend[vals, {"", "Half-lives", "Isotope abundance"}], Frame → All,
    Background → {None, {{{LightBlue, White}}}, {1 → LightYellow}}}]]
```

Out[∘]=

	Half-lives	Isotope abundance
^{10}C	19.29 s	0.
^{11}C	1220.0 s	0.
^{12}C	∞	0.9889
^{13}C	∞	0.0111
^{14}C	1.8×10^{11} s	0.
^{15}C	2.449 s	0.

This method is based on the following principle: 98.9% of the carbon in the Earth's atmosphere is Carbon-12, 1.1% Carbon-13, and the rest, 1.18 10^{-14} % is Carbon-14. This last isotope is the key to the method.

- The command below returns the decay scheme of ^{14}C. It decays into ^{14}N with a beta emission (1 electron) and has a half-life period of 1.8×10^{11} s (approximately 5,700 years)

In[∘]:= `c14 = IsotopeData["C14", #] & /@`
          `{"DaughterNuclides", "BranchingRatios", "DecayModes", "HalfLife"}`

Out[∘]= $\{\{$ nitrogen-14 $\}, \{1.00\}, \{BetaDecay\}, 1.8 \times 10^{11}$ s$\}$

The nitrogen present in the atmosphere is subject to a continuous neutron bombardment from cosmic radiation. Some of these neutrons strike the nitrogen atoms with enough energy to transform them into ^{14}C that will eventually decay into ^{14}N.

If we assume that the cosmic radiation has not changed significantly during the last 50,000 years, the production rate of ^{14}C atoms will be constant. Since its disintegration period is relatively short (approximately 5,700 years), the quantity of atoms that are produced will be similar to the ones that disintegrate, so the percentage of this isotope out of the total amount of carbon in the atmosphere will not have changed over the last millennia. Living organisms continuously absorb carbon when breathing or performing photosynthesis but when they die, this absorption stops. Two (Carbon-12 and Carbon-13) out of the three carbon isotopes are stable and will remain so; the other one, Carbon-14, will decay due to its radioactivity and its percentage of the total amount of carbon will get smaller. Measuring the content of this latter isotope in an organic sample and comparing it with the total amount of carbon in the entire sample will allow us to know its age. In reality, the method is not that simple since the proportion of ^{14}C in the atmosphere is not constant, but fluctuates with solar activity, introducing the need to make adjustments. Furthermore, we have to take into account ^{14}C originated from non-organic sources, such as nuclear explosions, mainly during the fifties and beginning of the sixties in the 20th century.

The evolution of ^{14}C can be inferred by measuring its content in the trunks of very old trees. Tree trunks consist of layers or rings, with each one corresponding to one year in the life of the tree. By counting the number of rings starting from the bark we can know the age of a tree. This dating technique is known as dendrochronology. Using this method, scientists have been able to "calibrate" the ^{14}C method for samples up to 9,000 years old. This calibration has recently been extended to 40,000 years. This has been possible thanks to the comparison between this method and the measurement of U234 and Th230 in corals. When corals are born, their U234 content is known. From that moment on Th230 starts to accumulate. Comparing the content of C14 with Th230, we can determine the age of the coral. Since corals can be dated using the ^{14}C method, it's possible to compare both techniques.

Example: There are 2 grams of carbon in a piece of wood found in an archeological site and we discover that it has an activity of 10 nuclear transformations per minute per gm. How old is the piece of wood? For simplification purposes we will assume that the C14 content in the wood at the time it was cut is the same as in wood recently cut: 15 nuclear transformations per minute per gm. However, in real cases, as explained before, we may have to make some adjustments regarding this assumption.

- This problem is a case of a decay chain with only one daughter isotope: $A = A_0\, e^{-t\, k_1}$, with k_1 being the decay constant of C14. Therefore, the age of the sample would be:

In[ ]:= `Solve[{ A == A0 e`$^{-\lambda t}$ `}, t, Reals] /.`

     `{λ → 1/UnitConvert[IsotopeData["C14", "Lifetime"], "year"],`

     `A → 10, A0 → 15}`

Out[ ]= $\{\{t \to 3.3 \times 10^3 \text{ yr}\}\}$

The Age of the Earth

To estimate the age of the Earth, we need to know its origins. The mainstream theory of the formation of the solar system and probably, of other planetary systems similar to ours, is as follows: there are certain areas in galaxies where clouds of dust and gas accumulate, and once these clouds reach certain mass and density they may condense as a result of gravitational attraction forces. This phenomenon can benefit from, or depends on?, the explosion of a star (supernova) in its proximity. During the condensation, the clouds may break into smaller units that become protostars. These protostars continue contracting rapidly until their centers reach very high temperatures and densities causing the fusion (union) of the lightest atomic nuclei. This fusion emits enough energy to stop the gravitational collapse. At this moment, stars are born. Surrounding them there are still many fragments. Some of these fragments start clustering during successive collisions until they reach a certain size and become planets. This is probably the origin of the planets closest to the Sun, such as the Earth. The fragments that didn't become part of a bigger mass were left wandering around the solar system, originating some of the currently existing meteorites.

Over long periods of time, orogenic and sedimentary processes have destroyed the remains of the rocks that originally formed the Earth (primordial), making it impossible for us to date the Earth directly from them. Fortunately, we've been able to calculate its age using visitors from outer space: meteorites. Their analyses supports the idea that after the formation of the solar system, with some exceptions, the distribution of isotopes was homogeneous.

Among meteorites, the carbonaceous chondrite type is particularly interesting. Chondrites are made of chondrules, molten droplets created during the collisions that gave birth to planets, that have the isotopic composition of the planet-forming period. Furthermore, they didn't experience any further heating intense enough to melt them again. It's been possible to determine their age using dating techniques based on radioactive isotopes with long half-lives. The most commonly used methods are: the lead-lead method (Pb-Pb) and the rubidium-strontium one (Rb-Sr). Next, we're going to discuss the Pb-Pb method.

- The decay chains of Uranium-238 and Uranium-235 are of particular interest among natural isotopes. Their final nuclides, are respectively: ^{206}Pb and ^{207}Pb.

In[ ]:= `Row[IsotopeData[#, "Symbol"] & /@ mainbranch1["U238"], "->"]`

Out[ ]= ^{238}U → ^{234}Th → ^{234}Pa → ^{234}U →
^{230}Th → ^{226}Ra → ^{222}Rn → ^{218}Po → ^{214}Pb → ^{214}Bi → ^{214}Po → ^{210}Pb → ^{210}Bi → ^{210}Po → ^{206}Pb

In[ ]:= `Row[IsotopeData[#, "Symbol"] & /@ mainbranch1["U235"], "->"]`

Out[ ]= ^{235}U → ^{231}Th → ^{231}Pa → ^{227}Ac → ^{227}Th → ^{223}Ra → ^{219}Rn → ^{215}Po → ^{211}Pb → ^{211}Bi → ^{207}Tl → ^{207}Pb

- In nature, apart from ^{206}Pb and ^{207}Pb, we also have other stable lead isotopes in the following proportions:

In[ ]:= `pbestable = DeleteCases[ Transpose[{IsotopeData["Pb"],`
`        IsotopeData[#, "Stable"] & /@ IsotopeData["Pb"] }], {_, False}];`

In[ ]:= `vals =`
`      Table[IsotopeData[#, prop], {prop, {"Symbol", "IsotopeAbundance"}}] & /@`
`      First[Transpose[pbestable]];`

In[ ]:= `Text[Grid[Prepend[vals, {"Isotope", "Isotope Abundance"}], Frame → All,`
`        Background → {None, {{{LightBlue, White}}, {1 → LightYellow}}}]]`

Out[ ]=

Isotope	Isotope Abundance
^{204}Pb	0.014
^{206}Pb	0.241
^{207}Pb	0.221
^{208}Pb	0.524

These proportions may vary widely depending on the origin of the sample.

- ^{208}Pb is a stable daughter isotope of the decay chain of Th232, and doesn't play any role in the dating method we are about to describe. ^{204}Pb, however, is a primordial isotope. This means that the amount of this nuclide present on Earth has not changed since the formation of our planet. Therefore it's useful for estimating the fraction of the other lead isotopes in a given sample that are also primordial since their relative fractions are constant everywhere.

In[ ]:= `Row[IsotopeData[#, "Symbol"] & /@ mainbranch1["Th232"], "->"]`

Out[ ]= ^{232}Th → ^{228}Ra → ^{228}Ac → ^{228}Th → ^{224}Ra → ^{220}Rn → ^{216}Po → ^{212}Pb → ^{212}Bi → ^{212}Po → ^{208}Pb

At the moment of the formation of the solar system, U238 and U235, after going through an homogenization process, started to decay, respectively, into Pb-206 and Pb-207 thus adding to the existing amount in the rocks.

The quantity from the parent isotope that decays into a daughter isotope is: $N_h(t) = 1 - N_p(t)$, that is:

$$N_h = N_p \, e^{\lambda t}$$

As components of rocks or meteorites, the total number of atoms in the systems U238 + Pb206 and U235 + Pb207 remains constant.

$$\left(^{207}\text{Pb}\right)_P = \left(^{207}\text{Pb}\right)_I + \left(^{235}\text{U}\right)\left(e^{\lambda 235\, t} - 1\right)$$

$$\left(^{206}\text{Pb}\right)_P = \left(^{206}\text{Pb}\right)_I + \left(^{238}\text{U}\right)\left(e^{\lambda 238\, t} - 1\right)$$

with the P and I subscripts indicating, respectively, the present and initial quantities, $\lambda 235$ and $\lambda 238$ the decay constants of U235 and U238, and t the elapsed time.

There will also be a certain amount of Pb204 that will not change. Under these hypotheses we obtain the following relationships:

$$\left(\frac{^{207}\text{Pb}}{^{204}\text{Pb}}\right)_P = \left(\frac{^{207}\text{Pb}}{^{204}\text{Pb}}\right)_I + \left(\frac{^{235}\text{U}}{^{204}\text{Pb}}\right)\left(e^{\lambda 235\, t} - 1\right)$$

$$\left(\frac{^{206}\text{Pb}}{^{204}\text{Pb}}\right)_P = \left(\frac{^{206}\text{Pb}}{^{204}\text{Pb}}\right)_I + \left(\frac{^{238}\text{U}}{^{204}\text{Pb}}\right)\left(e^{\lambda 238\, t} - 1\right)$$

Rearranging the previous equations we get:

$$\left(\frac{^{207}\text{Pb}}{^{204}\text{Pb}}\right)_P - \left(\frac{^{207}\text{Pb}}{^{204}\text{Pb}}\right)_I = \left(\frac{^{235}\text{U}}{^{204}\text{Pb}}\right)\left(e^{\lambda 235\, t} - 1\right)$$

$$\left(\frac{^{206}\text{Pb}}{^{204}\text{Pb}}\right)_P - \left(\frac{^{206}\text{Pb}}{^{204}\text{Pb}}\right)_I = \left(\frac{^{238}\text{U}}{^{204}\text{Pb}}\right)\left(e^{\lambda 238\, t} - 1\right)$$

After dividing the first one by the second one:

$$\left[\left(\frac{^{207}\text{Pb}}{^{204}\text{Pb}}\right)_P - \left(\frac{^{207}\text{Pb}}{^{204}\text{Pb}}\right)_I\right] / \left[\left(\frac{^{206}\text{Pb}}{^{204}\text{Pb}}\right)_P - \left(\frac{^{206}\text{Pb}}{^{204}\text{Pb}}\right)_I\right] = \frac{1}{^{238}\text{U}/^{235}\text{U}} \frac{\left(e^{\lambda 235\, t} - 1\right)}{\left(e^{\lambda 238\, t} - 1\right)} = K$$

- The ratio $^{238}\text{U}/\,^{235}\text{U}$ is:

```
In[ ]:= IsotopeData["U238", "IsotopeAbundance"] /
        IsotopeData["U235", "IsotopeAbundance"]
```

```
Out[ ]= 137.80
```

However, in the scientific literature: $^{238}\text{U}/\,^{235}\text{U} = 137.88$, and this is the value we are going to use.

Since $\dfrac{1}{137.88} \dfrac{\left(e^{\lambda 235\, t} - 1\right)}{\left(e^{\lambda 238\, t} - 1\right)}$ is constant for samples from the same period, the relation $\frac{^{207}\text{Pb}}{^{204}\text{Pb}} / \frac{^{206}\text{Pb}}{^{204}\text{Pb}}$ is a straight line, as Figure 9.2 shows:

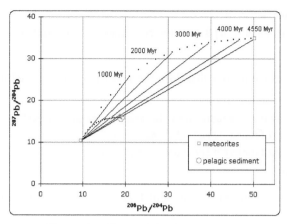

Figure 9.2   Paterson's lead isotope isochron.

Source: http://en.wikipedia.org/wiki/File:Paterson_isochron _animation.gif

- The different lines obtained depend on the original U/Pb ratio. For samples coming from the oldest meteorites (see the graph), K = 0.61, meaning that *t*, in eons (billions of years) is:

$$In[\circ]:= \quad K = \frac{1}{137.88} \frac{\left(e^{\lambda 235\,t} - 1\right)}{\left(e^{\lambda 238\,t} - 1\right)} \,/.$$

```
{λ235 → 1/QuantityMagnitude[UnitConvert[IsotopeData["U235",
        "Lifetime"], "eons"]], λ238 → 1/QuantityMagnitude[
   UnitConvert[IsotopeData["U238", "Lifetime"], "eons"]]};
```

$In[\circ]:=$ **FindRoot[K == 0.61 , {t, 0.5}]**

$Out[\circ]=$ $\{t \to 4.54709\}$

- We can compare this result with the one given by *Mathematica* using the free-form input and see that is quite similar.

$In[\circ]:=$ ▤ **Age of Earth** »   ⊞

      PlanetData[Entity["Planet", "Earth"], "Age"]

$Out[\circ]=$ 4.54×10^9 yr

Orogenic phenomena have left us without remains of rocks from the time of the formation of our planet. The oldest ones have been found in zircon crystals from Mount Narryer, in Western Australia. They are estimated to be more than 4,000 years old, proving that at that time the Earth already had a continental crust.

9.7 Application: Calculating Binding Energies

The mass of an atom is always smaller than the sum of its component particles. This difference in mass is known as binding energy and is responsible for keeping the particles together to form the element. To obtain some property for a subatomic particle we use IsotopeData.

- For example, the iron isotope Fe56 consists of N neutrons, and Z protons, whose actual numbers can be calculated as follows:

In[ ]:= `{Nfe56, Zfe56} =`
    `IsotopeData["Fe56", #] & /@ {"NeutronNumber", "AtomicNumber"}`

Out[ ]= `{30, 26}`

- The sum of the particle masses that form the iron isotope (expressed in MeV/c^2): $m_A = N\, m_n + Z\,(m_p + m_e)$:

In[ ]:= `fe56particlesmass = (Nfe56 ParticleData["Neutron", "Mass"] +`
    `Zfe56 ParticleData["Proton", "Mass"] +`
    `Zfe56 ParticleData["Electron", "Mass"])`

Out[ ]= $52\,595.320\ \text{MeV}/c^2$

- We calculate below the conversion factor of 1 amu (atomic mass unit) to 1 MeV:

In[ ]:=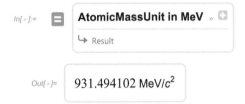

Out[ ]= $931.494102\ \text{MeV}/c^2$

- We find the mass of the iron isotope Fe56 and express it in MeV:

In[ ]:= `fe56mass =`
    `UnitConvert[IsotopeData["Fe56", "AtomicMass"], ` 931.494 MeV/c^2 ✓ `]`

Out[ ]= $52\,103.0644\ \text{MeV}/c^2$

- Next we calculate the binding energy per nucleon B, in MeV as well:

In[ ]:= `(fe56particlesmass - fe56mass) / (Nfe56 + Zfe56)`

Out[ ]= $8.79027\ \text{MeV}/c^2$

- We can get approximately the same value directly with (sometimes in MeV/c^2 we omit c^2 or in natural units we consider $c = 1$):

In[ ]:= `IsotopeData["Fe56", "BindingEnergy"]`

Out[ ]= $8.790323\ \text{MeV}$

- Another commonly used concept is "excess of mass" calculated as:

In[ ]:= `(QuantityMagnitude[IsotopeData["Fe56", "AtomicMass"]] -`
    `IsotopeData["Fe56", "MassNumber"]) 931.494`

Out[ ]= `-60.6054`

- We can also calculate it directly:

In[ ]:= `IsotopeData["Fe56", "MassExcess"]`

Out[ ]= $-60.605352\ \text{MeV}$

- The function below displays the binding energies as a function of the mass number. With Tooltip we can see the information related to a point when placing the cursor on it:

In[ ]:= `bindingenergy = Tooltip[{IsotopeData[#, "MassNumber"],`
    `IsotopeData[#, "BindingEnergy"]}] & /@ IsotopeData[];`

```
In[ ]:= ListPlot[bindingenergy, Frame → True,
        FrameLabel → {"Mass number A", "Binding energy per nucleon (B), in MeV"},
        Axes → False, ImageSize → {300, 250}]
```

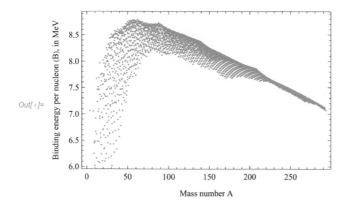

- The following command shows the binding energy per nucleon, as a function of the atomic number and the number of neutrons until $Z = 92$ (uranium):

```
In[ ]:= bindingenergyZNB = Flatten[
          Table[{ z, a - z, IsotopeData[{z, a}, "BindingEnergy"]}, {z, 1, 92},
            {a, IsotopeData[#, "MassNumber"] & /@ IsotopeData[z]}], 1];
```

- Here we visualize it in 3D:

```
In[ ]:= ListPlot3D[bindingenergyZNB, AxesLabel → {"A", "N", "B"}]
```

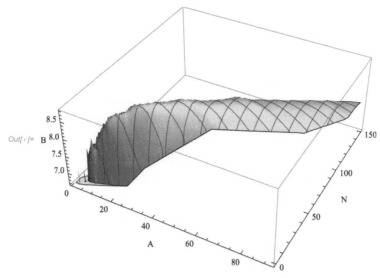

- Next we show an example with 224 nuclides with less than 15 protons each ($Z < 15$) and plot it using the function ColorData["Atoms"] that associates each element to one color. We can find the value of Z and N by placing the cursor inside the graph, right clicking with the mouse button and selecting "Get Coordinates".

```
In[ ]:= lessthan15 = Take[bindingenergyZNB, 224];
```

In[ ]:= `ListDensityPlot[lessthan15, FrameLabel → {"N", "Z"},`
`ColorFunction → Map[ColorData["Atoms", ElementData[#, "Abbreviation"]] &,`
`Transpose[lessthan15]⟦1⟧]]`

Out[ ]=

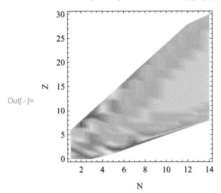

- In this case, we present the same information but using a color code that associates redder colors to higher binding energies:

In[ ]:= `ListDensityPlot[lessthan15, FrameLabel → {"N", "Z"},`
`InterpolationOrder → 1, ColorFunction → "Rainbow"]`

Out[ ]=

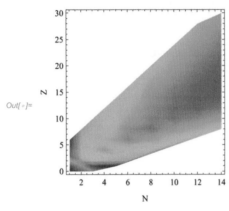

Decays and nuclear transformations are related to binding energies. There are two types of nuclear transformations that are particularly useful: fusion and fission.

Fusion is the process of joining nuclei. This process is the most predominant one in elements with low atomic numbers, specifically, until iron. The lower the atomic number the lower the binding energy associated to it. Fusion is the normal method by which stars generate energy.

Fission is the opposite process of fusion. It consists of splitting apart nuclei to turn them into lighter ones.

Some isotopes of uranium and plutonium represent a special case where the process accelerates when interacting with neutrons.

- Some very heavy nuclei may experience spontaneous fission. The next function displays the first five such elements:

In[•]:= `Take[Transpose[Cases[Table[IsotopeData[#, prop],`
`        {prop, {"Symbol", "SpontaneousFission"}}]] & /@`
`      IsotopeData[], Except[{{_, False}}]]][[1]], 5]`

Out[•]= $\{^{230}\text{Th}, \, ^{232}\text{Th}, \, ^{231}\text{Pa}, \, ^{234}\text{Pa}, \, ^{230}\text{U}\}$

- Since IsotopeData includes "Spontaneous Fission" as a property, we can also get those elements as follows (we don't show the complete output):

In[•]:= `IsotopeData[` ⬛ spontaneous fission ISOTOPES `] // Shallow`

Out[•]//Shallow=

{ thorium-230 , thorium-232 , protactinium-231 , protactinium-234 , uranium-230 ,

uranium-232 , uranium-233 , uranium-234 , uranium-235 , uranium-236 , ≪193≫}

- Many of them are transuranic elements obtained by artificial means, although there are also some isotopes in nature that exhibit this property. They are largely responsible for the presence of free neutrons that have lifetimes of just a few minutes:

In[•]:= `ParticleData["Neutron", "Lifetime"]`

Out[•]= 885.6 s

- This can be explained by the fact that both, neutrons and protons, are not really elementary particles but are made of two types of quarks:

In[•]:= `ParticleData[#, "QuarkContent"] & /@ {"Proton", "Neutron"}`

Out[•]= $\begin{pmatrix} \{ \text{down quark} , \ \text{up quark} , \ \text{up quark} \} \\ \{ \text{down quark} , \ \text{down quark} , \ \text{up quark} \} \end{pmatrix}$

- Let's see some of these quarks' characteristics:

In[•]:= `values =`
`    Table[ParticleData[#, prop], {prop, {"Symbol", "Charge", "Mass"}}] & /@`
`      {"DownQuark", "UpQuark"}`

Out[•]= $\begin{pmatrix} d & -\frac{1}{3} & 5.0\,\text{MeV}/c^2 \\ u & \frac{2}{3} & 2.2\,\text{MeV}/c^2 \end{pmatrix}$

In[•]:= `Text[Grid[Prepend[values, {"", "Charge", "Mass (MeV/c²)"}], Frame → All,`
`      Background → {None, {{{LightBlue, White}}, {1 → LightYellow}}}]]`

Out[•]=

	Charge	Mass(MeV/c²)
d	$-\frac{1}{3}$	$5.0\,\text{MeV/c}^2$
u	$\frac{2}{3}$	$2.2\,\text{MeV/c}^2$

- Notice how small the mass of the quarks is compared to that of a neutron (two quarks down and one quark up) or a proton (two quarks up and one quark down):

In[•]:= `{qd, qu} = ParticleData[#, "Mass"] & /@ {"DownQuark", "UpQuark"}`

Out[•]= $\{5.0\,\text{MeV}/c^2, \, 2.2\,\text{MeV}/c^2\}$

```
In[ ]:= {quarksproton, quarkneutron} =
        100 { (2 qu + 2 × 1 qd) / ParticleData["Proton", "Mass"],
              (2 qu + 1 qd) / ParticleData["Neutron", "Mass"]} "%"
```

```
Out[ ]= {1.5 %, 1.0 %}
```

This means that the quarks only represent between 1% and 1.5% of the atomic nuclei mass with the rest coming from their interactions. This is the reason why quarks are not present as free particles in nature.

The mass of quarks and other particles such as electrons is attributed to the Higgs field (more accurately to the Brout–Engler–Higgs mechanism), whose existence was confirmed by the detection of the Higgs boson, officially announce by CERN on July 4, 2012. In that case, what about our mass? Does it come as well from the Higgs field? Only a small portion of it.

As we've seen, quarks only represent between 1% and 1.5% of the nuclei mass and that mass is explained by the Higgs field. Therefore, for a person weighing 80 kg, the Higgs field explains approximately 1 kg. What about the rest? It comes from gluons, g, that keep the photons and quarks joined. In this case, we should refer to mass equivalence, $m = E/c^2$. These particles don't interact with the Higgs field or BEH, but they interact with the gravitational field. However, without the Higgs field, electrons would move at the speed of light, atoms would not have formed, and therefore we wouldn't exist.

The *Standard Model of Particle Physics*, apart from quark d and quark q, also includes 10 additional types of quarks and antiquarks, the leptons, the gauge bosons, and the Higgs boson. We will use the EntityClass function, which we saw in Chapter 5, to represent some properties of these particles.

```
In[ ]:= TextSentences[WikipediaData["Standard_Model"]] [[ ;; 2]]
```

```
Out[ ]= {The Standard Model of particle physics is the theory describing three of the four
         known fundamental forces (electromagnetic, weak and strong interactions,
         omitting gravity) in the universe and classifying all known elementary particles.,
       It was developed in stages throughout the latter half of the 20th century, through the
         work of many scientists worldwide, with the current formulation being finalized
         in the mid–1970s upon experimental confirmation of the existence of quarks.}
```

- A few properties of the gauge bosons, quarks, and leptons are summed up in the next tables.

```
In[ ]:= Dataset[EntityValue[EntityClass["Particle", "GaugeBoson"],
         {EntityProperty["Particle", "Lifetime"], EntityProperty["Particle",
          "Width"]}, "EntityPropertyAssociation"]]
```

gluon	mean lifetime	—
	width	—
graviton	mean lifetime	∞ s
	width	—
photon	mean lifetime	∞ s
	width	0. MeV
Z boson	mean lifetime	2.6379×10^{-25} s
	width	2.4952×10^{9} eV
W− boson	mean lifetime	3.076×10^{-25} s
	width	2.140×10^{9} eV
W+ boson	mean lifetime	3.076×10^{-25} s
	width	2.140×10^{9} eV

In[ ]:= **Dataset[EntityValue[EntityClass["Particle", "Lepton"],**
  {"Symbol", "Charge", "Mass"}, "EntityPropertyAssociation"]]

	Symbol	Charge	Mass
electron	e	−1 e	510.9989461 keV/c^2
electron neutrino	"ν"$_{e}$"	0 e	—
electron antineutrino	"$\overline{ν}$"$_{e}$"	0 e	—
muon	μ	−1 e	105.6583745 MeV/c^2
anti-muon	"$\overline{μ}$"	e	105.6583745 MeV/c^2
muon neutrino	"ν"$_{μ}$"	0 e	—
muon antineutrino	"$\overline{ν}$"$_{μ}$"	0 e	—
positron	"$\overline{e}$"	e	510.9989461 keV/c^2
tau	τ	−1 e	1.77686 GeV/c^2
anti-tau	"$\overline{τ}$"	e	1.77686 GeV/c^2
tau neutrino	"ν"$_{τ}$"	0 e	—
tau antineutrino	"$\overline{ν}$"$_{τ}$"	0 e	—

In[ ]:= **Dataset[EntityValue[EntityClass["Particle", "Quark"],**
  {"Symbol", "Charge", "Mass"}, "EntityPropertyAssociation"]]

	Symbol	Charge	Mass
bottom quark	b	−1/3 *e*	4.18 GeV/*c*²
bottom antiquark	"b̄"	1/3 *e*	4.18 GeV/*c*²
charm quark	c	2/3 *e*	1.275 GeV/*c*²
charm antiquark	"c̄"	−2/3 *e*	1.275 GeV/*c*²
down quark	d	−1/3 *e*	4.7 MeV/*c*²
down antiquark	"d̄"	1/3 *e*	4.7 MeV/*c*²
strange quark	s	−1/3 *e*	95. MeV/*c*²
strange antiquark	"s̄"	1/3 *e*	95. MeV/*c*²
top quark	t	2/3 *e*	173.0 GeV/*c*²
top antiquark	"t̄"	−2/3 *e*	173.0 GeV/*c*²
up quark	u	2/3 *e*	2.2 MeV/*c*²
up antiquark	"ū"	−2/3 *e*	2.2 MeV/*c*²

9.8 Radiation Attenuation

Other function with applications in nuclear physics and radiological protection is: StoppingPowerData. Let's take a look at a few examples based on the documentation pages.

- One way to reduce radiation exposure is through the use of shielding. For example, the thickness of a water layer needed to reduce photon radiation by a tenth is:

In[ ]:= **StoppingPowerData["Water", {"Particle" → "Photon",**
    "Energy" → Quantity[0.1, "Megaelectronvolts"]}, "HalfValueLayer"]

Out[ ]= 4.06062 cm

- In the design of radiation protection products it is important to take into account the mass attenuation coefficients of the materials used. This attenuation depends on the incident particles (protons, alpha, protons, electrons, neutrons) and their energy. Here we compare the mass attenuation coefficient for various substances and energies:

```
In[ ]:= materials = {"Water", lead ELEMENT, iron ELEMENT, aluminum ELEMENT};
    ListLogLogPlot[
      Table[{Quantity[x, "Megaelectronvolts"], StoppingPowerData[#,
          {"Particle" → "Photon", "Energy" → Quantity[x, "Megaelectronvolts"]},
          "MassAttenuationCoefficient"]}, {x, 0.01, 100, .5}] & /@ materials,
      Joined → True, PlotLegends → materials, AxesLabel → Automatic]
```

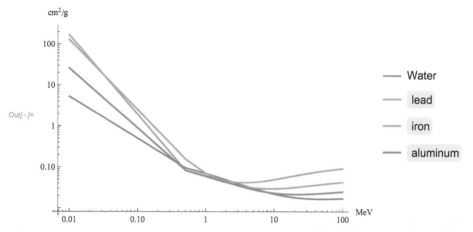

- To accurately estimate absorbed doses, we measure the amount of the energy deposited in human tissues. The example below shows the stopping power changes, dE/dx, for an alpha particle emitted from U234 as it travels through biological tissue:

```
In[ ]:= initialKE = QuantityMagnitude @
      IsotopeData[Entity["Isotope", "Uranium234"], "BindingEnergy"]
```

```
Out[ ]= 7.600708
```

```
In[ ]:= dEByDxVal[k_Real ? Positive] :=
      QuantityMagnitude @ StoppingPowerData["A150TissueEquivalentPlastic",
        {"Particle" → Entity["Particle", "AlphaParticle"], "Energy" →
          Quantity[k, "Megaelectronvolts"]}, "LinearStoppingPower"] / 10 000;
    dEByDxVal[k_Real] := 0
```

```
In[ ]:= sol = NDSolveValue[{KiM'[x] == -dEByDxVal[KiM[x]], KiM[0] == initialKE,
      WhenEvent[KiM[x] ≤ 0, "StopIntegration"]}, KiM, {x, 0, 60},
      PrecisionGoal → 3];
```

In[ ]:= `Plot[dEByDxVal[sol[x]], {x, 0, 60},`
    `Frame → True, GridLines → Automatic, FrameLabel →`
      `{Row[{"Distance Travelled", Quantity[None, "Micrometers"]}, "  "], Row[`
        `{"dE/dx", Quantity[None, "Megaelectronvolts"/"Micrometers"]}, "  "]}]`

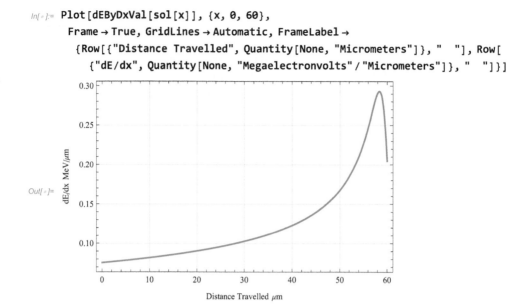

9.9 Additional Resources

On the Wolfram Demonstrations website you can find quite sophisticated examples about nuclear physics and isotopes:

Nuclear physics demonstrations:
http://demonstrations.wolfram.com/search.html?query=Nuclear
Isotopes demonstrations: http://demonstrations.wolfram.com/search.html?query=isotopes

Individual demonstrations worth mentioning:

Table of Nuclides: http://demonstrations.wolfram.com/TableOfNuclides by Enrique Zeleny
Isotope Browser: http://demonstrations.wolfram.com/IsotopeBrowser by Theodore Gray and Yifan Hu
Binding Energies of Isotopes:
http://demonstrations.wolfram.com/BindingEnergiesOfIsotopes by Stephen Wolfram and Jamie Williams

10

Modeling: Applications in Biokinetics, Epidemiology, and more

This chapter shows how to use Mathematica for biokinetic, pharmacokinetic, physiological and epidemiological modeling, mainly in the context of building multi-compartment and nonlinear models using differential equations (DEs). These models also have many applications in a wide variety of fields such as chemical kinetics, clinical and nuclear medicine, ecology, and internal dosimetry, among others. There is also a section covering the statistical analysis of experimental data with a special emphasis on how to estimate parameters and perform optimal experimental designs. A short section about partial differential equations (PDEs) is also included. The chapter concludes with a very brief introduction to System Modeler, Wolfram's program for modeling and simulating physical systems. The reader may find the contents of this chapter useful for learning about modeling and how to fit nonlinear data in systems of differential equations.

10.1 Compartmental Modeling

10.1.1 An Introduction to Compartmental Analysis

Compartmental analysis has applications in chemical kinetics, clinical medicine, ecology, internal dosimetry, nuclear medicine, and pharmacokinetics. It can be described as the analysis of a system when it is separated into a finite number of component parts, which are called compartments. Compartments interact through the exchange of materials or energies. These materials or energies can be chemical substances, hormones, individuals in a population and so on. A compartmental system is usually represented by a flow or a block diagram.

■ The following code generates a diagram representing a basic compartmental system:

```
In[ ]:= Show[Graphics[{Arrow[{{
                              0, 1.5}, {0, 0.5}}]],
         Text["Input", {0.4, 1}, {0, -1}], Circle[{0, 0}, 0.5],
         Text["Compartment 1", {0, 0}, {0, 0}], Arrow[{{0.5, 0.}, {2.5, 0.}}]],
         Text["Flow from 1 to 2", {1.5, .4}, {0, 0}],
         Circle[{3, 0}, 0.5], Text["Compartment 2", {3, 0}, {0, 0}],
         Arrow[{{3, -0.5}, {3, -1.5}}], Text["Output", {2.4, -1}, {0, -1}]},
       AspectRatio → Automatic, Axes → None]]
```

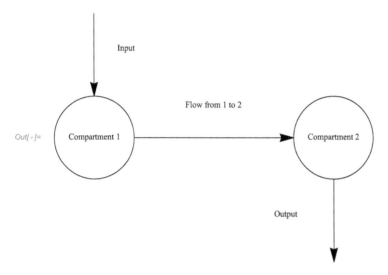

We adopt the convention of representing compartments with circles or rectangles. The flow in or out of the compartments is represented by arrows. The ith compartment of a system of n compartments is labeled i and the size (amount or content) of the component in compartment i as $x_i(t)$. The exchange between compartments, or between a compartment and the environment is labeled k_{ij}, where ij represents the flow from i to j. If there is no ambiguity, k_i will be used in place of k_{ij}. Additionally, if $i < 10$ and $j < 10$, k_{ij} will be written in place of k_{ij}. The environment refers to the processes that are outside the system and is usually represented by zero, so k_{i0} is the fractional excretion coefficient from the ith compartment to the environment. If we assume that the substance introduced into the system is a radioactive isotope, we must consider the radioactive decay, which is given by a constant rate represented by λ (this constant is specific for each isotope). The decay constant can be interpreted as an equal flow going out of the system in each compartment. The input from the environment into the jth compartment is called $b_j(t)$. With regards to the environment, we only need to know the flow, $b_j(t)$, into the system from the outside. The k_{ij}s are called fractional transfer rate coefficients. They are usually assumed to be constants, but in some cases they can be functions of time (that is $k_{ij}(t)$).

Based on the definitions in the previous paragraph, we can describe the two-compartment model in Figure 10.1 as follows: $b_1(t)$ is the input from the environment to compartment 1, k_{12} the transfer rate coefficient from compartment 1 to compartment 2 and k_{21} from compartment 2 to 1, k_{20} the transfer rate coefficient from compartment 1 to the environment (output), and $x_1(t)$ and $x_2(t)$ represent the quantities in compartments 1 and 2 at time t:

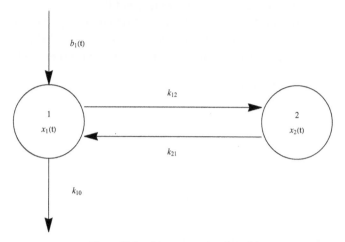

Figure 10.1 A two-compartment model.

A Two-compartment Model with Input and Output from Compartment 1

The variables $x_1(t)$ and $x_2(t)$ are called the state variables of the system, and their evolution over time is described by a system of ordinary differential equations (SODE). In this case:

$$\dot{x}_1(t) = -k_{10}\,x_1(t) - k_{12}\,x_1(t) + k_{21}\,x_2(t) + b_1(t)$$
$$\dot{x}_2(t) = \quad k_{12}\,x_1(t) - k_{21}\,x_2(t)$$

(10.1)

or in matrix notation:

$$\begin{pmatrix} \dot{x}_1(t) \\ \dot{x}_2(t) \end{pmatrix} = \begin{pmatrix} -k_{10} - k_{12} & k_{21} \\ k_{12} & -k_{21} \end{pmatrix} + \begin{pmatrix} b_1(t) \\ 0 \end{pmatrix}$$

We also need to know the initial conditions. In this case, the contents of $x_1(t)$ and $x_2(t)$ at $t = 0$.

$$x_1(0) = x_{10}; \; x_2(0) = x_{20}$$

This SODE can be solved using DSolve or DSolveValue.

■ For instance, in day^{-1}, for $k_{10} = 0.1$, $k_{12} = 0.2$, $k_{21} = 0.3$, $b_1(t) = 0.2\,t\,\text{Exp}[-2.1\,t]$, with $x_1(0) = x_2(0) = 0$. The solution is:

```
In[ ]:=  {x1[t_], x2[t_]} =
           DSolveValue[{x₁'[t] == -0.1 x₁[t] - 0.2 x₁[t] + 0.3 x₂[t] + 0.2 t Exp[-2.1 t],
               x₂'[t] == 0.2 x₁[t] - 0.3 x₂[t], x₁[0] == 0, x₂[0] == 0},
               {x₁[t], x₂[t]}, t] // ExpandAll // Chop
```

Out[ ]= $\{-0.113208\,e^{-2.1\,t}\,t - 0.0652664\,e^{-2.1\,t} + 0.0413534\,e^{-0.544949\,t} + 0.0239131\,e^{-0.055051\,t},$
$0.0125786\,e^{-2.1\,t}\,t + 0.0142399\,e^{-2.1\,t} - 0.0337649\,e^{-0.544949\,t} + 0.0195249\,e^{-0.055051\,t}\}$

We have used "**ExpandAll // Chop**" to eliminate insignificant terms.

■ The evolution of $x_1(t)$ and $x_2(t)$ over time can be visualized as follows (we show two graphs using different ways to include the legends). In the graphic on the right we use Callout:

In[ ]:= `GraphicsRow[`
    `{Plot[{x1[t], x2[t]}, {t, 0, 30}, AxesOrigin → {0, 0}, AxesLabel →`
        `{"Days", "x"}, PlotLegends → Placed[{"x₁(t)", "x₂(t)"}, Center],`
      `PlotLabel → Style["Using PlotLegends", Bold]],`
     `Plot[{Callout[x1[t], "x₁(t)", Above], Callout[x2[t], "x₂(t)", 4]},`
       `{t, 0, 30}, AxesOrigin → {0, 0}, AxesLabel → {"Days", "x"},`
       `PlotLabel → Style["Using Callout", Bold]]}, ImageSize → 400 ]`

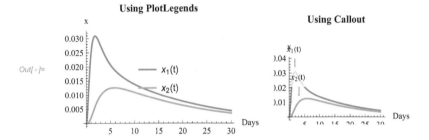

Out[ ]=

10.1.2 The Iodine ICRP 78 Biokinetic Model

In[ ]:= `ClearAll["Global`*"]`

The flow diagram in Figure 10.2 represents the iodine model (obtained from ICRP 78) where compartment 1 is the blood, compartment 2 is the thyroid, compartment 3 is the rest of the body, and compartment 4 is the bladder. Additionally, $3 \to 0$, i.e., a transfer from compartment 3 to the environment, represents the output to the gastrointestinal tract (GIT) and $4 \to 0$ represents the output, via urine excretion, to the environment.

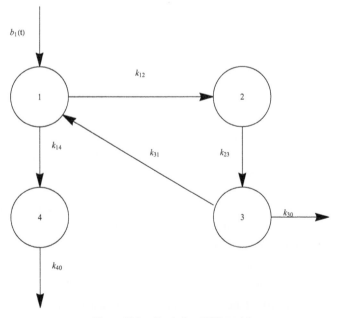

Figure 10.2   The Iodine ICRP model.

- We build the compartmental matrix **A** for iodine given by:

In[ ]:= `A = {{-k₁₂ - k₁₄, 0, k₃₁, 0},`
        `{k₁₂, -k₂₃, 0, 0}, {0, k₂₃, -k₃₀ - k₃₁, 0}, {k₁₄, 0, 0, -k₄₀}};`

In[ ]:= `A`

Out[ ]=
$$
\begin{pmatrix}
-k_{12} - k_{14} & 0 & k_{31} & 0 \\
k_{12} & -k_{23} & 0 & 0 \\
0 & k_{23} & -k_{30} - k_{31} & 0 \\
k_{14} & 0 & 0 & -k_{40}
\end{pmatrix}
$$

- We need to know the transfer constants k_{ij} values. According to the ICRP 78, the transfer rates for iodine, in days^{-1}, are $k_{12} = 0.3 \, \text{Log}(2)/0.25$, $k_{14} = 0.7 \, \text{Log}(2)/0.25$, $k_{23} = \text{Log}(2)/80$, $k_{30} = 0.2 \, \text{Log}(2)/12$, $k_{31} = 0.8 \, \text{Log}(2)/12$ and $k_{40} = 12$;

In[ ]:= `kidine = {k₁₂ →` $\dfrac{\text{0.3 Log[2]}}{\text{0.25}}$ `, k₁₄ →` $\dfrac{\text{0.7 Log[2]}}{\text{0.25}}$ `, k₂₃ →` $\dfrac{\text{Log[2]}}{\text{80}}$ `,`

`k₃₀ →` $\dfrac{1}{12}$ `×0.2 Log[2], k₃₁ →` $\dfrac{1}{12}$ `×0.8 Log[2], k₄₀ → 12};`

In[ ]:= `Aiodine = A /. kidine`

Out[ ]=
$$
\begin{pmatrix}
-2.77259 & 0 & 0.0462098 & 0 \\
0.831777 & -\frac{\log(2)}{80} & 0 & 0 \\
0 & \frac{\log(2)}{80} & -0.0577623 & 0 \\
1.94081 & 0 & 0 & -12
\end{pmatrix}
$$

In[ ]:= `X[t_] = {x1[t], x2[t], x3[t], x4[t]};`

- We assumed a single input $b_1 = 1$, in t= 0. Then the initial conditions are: x1(t) = 1, x2(t)=x3(t), x4(t)=0

In[ ]:= `system = X'[t] == Aiodine.X[t];`

In[ ]:= `ic = Thread[X[0] == {1, 0, 0, 0}];`

In[ ]:= `sol =`
        `DSolveValue[{system, ic}, {x1[t], x2[t], x3[t], x4[t]}, t] // ExpandAll`

Out[ ]= $\{1.00003 \, e^{-2.77254 t} - 0.000840995 \, e^{-0.0601471 t} + 0.000808587 \, e^{-0.00632391 t},$
$-0.300955 \, e^{-2.77254 t} + 0.0135875 \, e^{-0.0601471 t} + 0.287368 \, e^{-0.00632391 t},$
$0.00096051 \, e^{-2.77254 t} - 0.0493651 \, e^{-0.0601471 t} + 0.0484045 \, e^{-0.00632391 t},$
$-0.210331 \, e^{-12. t} + 0.210337 \, e^{-2.77254 t} - 0.000136703 \, e^{-0.0601471 t} + 0.000130845 \, e^{-0.00632391 t}\}$

- The evolution of the iodine retention in the blood (compartment 1) and in the thyroid (compartment 2) is plotted as follows:

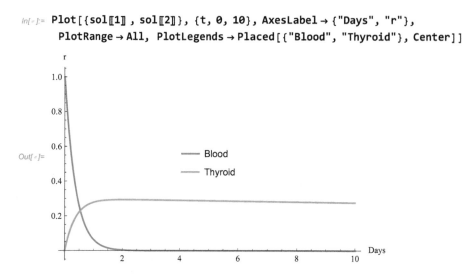

In[ ]:= `Plot[{sol〚1〛 , sol〚2〛}, {t, 0, 10}, AxesLabel → {"Days", "r"},`
`    PlotRange → All, PlotLegends → Placed[{"Blood", "Thyroid"}, Center]]`

10.1.3 Developing Functions for Solving Compartmental Systems

Instead of using DSolve or DSolveValue directly, we are going to develop some functions for solving multi-compartment models using a friendly notation.

In[ ]:= `ClearAll["Global`*"]`

Figure 10.3 shows a more complex compartmental model: The general multi-input multi-output (MIMO) three-compartment model. We have assumed a radioactive decay with a disintegration constant λ (it can be considered as an output from each compartment):

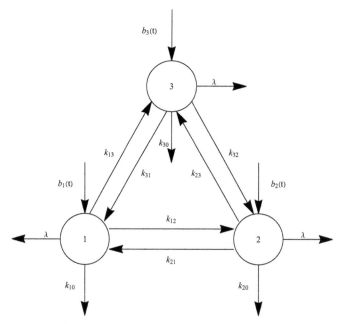

Figure 10.3   The generic three-compartment model.

$$\dot{x}_1(t) = -(\lambda + k_{10} + k_{12} + k_{13}) x_1(t) + k_{21} x_2(t) + k_{31} x_3(t) + b_1(t)$$
$$\dot{x}_2(t) = k_{12} x_1(t) - (\lambda + k_{20} + k_{21} + k_{23}) x_2(t) + k_{32} x_3(t) + b_2(t) \qquad (10.2)$$
$$\dot{x}_3(t) = k_{13} x_1(t) + k_{23} x_2(t) - (\lambda + k_{30} + k_{31} + k_{32}) x_3(t) + b_3(t)$$

The model can be expressed in matrix notation as follows:

$$\dot{\mathbf{x}}(t) = \mathbf{A}\,\mathbf{x} + \mathbf{b}(t), \ t \geq 0$$

where $\mathbf{x}(t)$ is a column vector representing the contents of the compartments:

$$\mathbf{x}(t) = \{x_1(t), \ x_2(t), \ x_3(t)\}^T$$

$b(t)$ is a column vector representing the inputs to compartments 1, 2, and 3:

$$\mathbf{b}(t) = \{b_1(t), \ b_2(t), \ b_3(t)\}^T$$

This pattern can be expanded to systems of n compartments or n state variables (in the case of physiological models). The equation for any compartment i is given by:

$$\dot{\mathbf{x}}(t) = \mathbf{A}\,\mathbf{x} + \mathbf{b}(t), \ t \geq 0$$
$$\mathbf{x}(0) = \mathbf{x}_0 \qquad (10.3)$$

where :

$\mathbf{x}(t) = \{x_1(t), \ x_2(t), \ ..., x_n(t)\}^T$ is a column vector and $x_i(t)$ denotes the amount or content of materials or energies in compartment i at time t.

$\mathbf{A}$ is a $n \times n$ is usually known as the compartmental matrix or system matrix.

$\mathbf{b}(t) = \{b_1(t), \ b_2(t), \ ..., b_n(t)\}^T$ is a column vector where $\{b_i(t)\}$ is the input rate into compartment i from an outside system.

$\mathbf{x}(0) = \{x_1(0), \ x_2(0), \ ..., x_n(0)\}^T$ are the initial conditions, so $x_i(0)$ represents the amount or content of materials or energies in compartment i at time $t = 0$.

These models are known as Systems of Ordinary Differential Equations with Constant Coefficients (SODECC).

In the rest of this section, we're going to develop several functions that will enable users, even those with minimal knowledge of differential equations, to solve SODEs related to compartmental and physiological systems. If you're interested in programming, you may want to pay close attention to how the functions have been created. Otherwise, you just need to follow the examples to see how to use the commands properly.

We develop the function **CompartMatrix1**$[n, \textit{matrixA}, \lambda]$ that returns the matrix of coefficients of a compartmental system, where n is the number of compartments of the system, $\textit{matrixA}$ represents the transfer rates, and λ is the radioactive decay constant (by default $\lambda = 0$, which means that it is not a radioactive isotope.).The transfer rates $\{k_{ij}\}$ from compartment i to compartment j are written: $\{\{1, 2, k_{12}\}, ..., \{i, j, k_{ij}\}, ...\}$, by default $k_{ij} = 0$.

```
In[ ]:= CompartMatrix1[(n_)?IntegerQ, matrixA_?MatrixQ, lambda_ : 0] :=
        Module[{k, A},
          Apply[Set, ({k @@ Take[#1, 2], Last[#1]} &) /@ matrixA, {1}];
          k[i_, j_] := 0;
          A = DiagonalMatrix[Table[-Sum[k[i, j], {j, 0, n}], {i, 1, n}]] +
            Table[k[j, i], {i, 1, n}, {j, 1, n}] - lambda * IdentityMatrix[n]];
```

Notice the use of: ?IntegerQ, matrixA_?MatrixQ to check that the input contains an integer and a matrix as the first two arguments. Otherwise, the function will not be executed. By using the structure: lambda_:0, we give lambda (the disintegration constant) a default value of 0 in case the user doesn't enter any value for it.

■ Using this function we can construct the Iodine model (Figure 10.2).

```
In[ ]:= iodine131matrix = CompartMatrix1[4, {{1, 2, k12}, {1, 4, k14},
          {4, 0, k40}, {2, 3, k23} , {3, 0, k30} , {3, 1, k31}}, λ];
```

```
In[ ]:= % // MatrixForm
```

Out[ ]//MatrixForm=

$$\begin{pmatrix} -\lambda - k_{12} - k_{14} & 0 & k_{31} & 0 \\ k_{12} & -\lambda - k_{23} & 0 & 0 \\ 0 & k_{23} & -\lambda - k_{30} - k_{31} & 0 \\ k_{14} & 0 & 0 & -\lambda - k_{40} \end{pmatrix}$$

■ We also create the function **SystemDSolve1** to solve the model where $matrixA$ is the coefficients matrix; *initcond* are the initial conditions: $\{c_1, ..., c_n\}$; input are the inputs $\{b_1(t), ..., b_n(t)\}$; x is the symbol that represents the retention variables. The function returns the analytical solution $x_i(t)$. We can use $t1 = t$ to get the solution as a function of t instead of as a numerical value. We have used Chop and ExpandAll to eliminate the terms with values very close to 0.

```
In[ ]:= SystemDSolve1[matrixA_?MatrixQ,
          incond_List, input_List, t_, t1_, x_, opts___?OptionQ] :=
        Module[{A, n, var, i, system, ic, X, X0, sol }, A = matrixA;
          n = Dimensions[A]〚1〛;
          X = Table[x_i[t], {i, 1, n}]; X0 = Table[x_i[0], {i, 1, n}];
          system = D[X, t] == A.X + input;
          ic = Thread[X0 == incond];
          sol = DSolve[{system, ic}, X, t, opts];
          Chop[ExpandAll[sol]] /. t → t1];
```

■ To apply the previous function, to the iodine model (Fig 10.2), we also assume as the initial condition: $\{1, 0, 0, 0\}$ and as the input function: $\{1 + 0.5 \cos(0.3\,t), 0, 0, 0\}$. Also, in the example we refer to Iodine–131, then we need the decay constant λ of the I-131 (in days^{-1} because k_{ij} are in days^{-1}). It can be obtained as follow:

```
In[ ]:= λ131 = QuantityMagnitude[

          UnitConvert[ iodine-131 ISOTOPE [ decay constant ], 1 / "Days"]]
```

Out[ ]= 0.086371

$In[\circ]:=$ **iodine131matrix1 =**

iodine131matrix $/.$ $\left\{ k_{12} \rightarrow \dfrac{0.3\,\mathsf{Log[2]}}{0.25}, \; k_{14} \rightarrow \dfrac{0.7\,\mathsf{Log[2]}}{0.25}, \; k_{23} \rightarrow \dfrac{\mathsf{Log[2]}}{80}, \right.$

$\left. k_{30} \rightarrow \dfrac{1}{12} \times 0.2\,\mathsf{Log[2]}, \; k_{31} \rightarrow \dfrac{1}{12} \times 0.8\,\mathsf{Log[2]}, \; k_{40} \rightarrow 12, \; \lambda \rightarrow \lambda 131 \right\}$

$Out[\circ]=$ $\left\{ \{-2.85896, 0, 0.0462098, 0\}, \; \{0.831777, -0.095036, 0, 0\}, \right.$

$\left\{ 0, \; \dfrac{\mathsf{Log[2]}}{80}, \; -0.144134, 0 \right\}, \; \{1.94081, 0, 0, -12.086371\} \right\}$

$In[\circ]:=$ **{x1[t1_], x2[t1_], x3[t1_], x4[t1_]} =**
 {x_1[t1], x_2[t1], x_3[t1], x_4[t1]} $/.$ **SystemDSolve1[iodine131matrix1,**
  {1, 0, 0, 0}, {1 + 0.5 Cos[0.3 t], 0, 0, 0}, t, t1, x][[1]]

$Out[\circ]=$ $\left\{ 0.352778 + 0.477246\,e^{-2.85892\,t1} + 0.00545159\,e^{-0.146518\,t1} - \right.$
 $0.00829459\,e^{-0.0926952\,t1} + 0.17282\,\mathsf{Cos[0.3\,t1]} + 0.0182514\,\mathsf{Sin[0.3\,t1]},$
 $3.0876 - 0.143625\,e^{-2.85892\,t1} - 0.0880781\,e^{-0.146518\,t1} -$
 $2.94785\,e^{-0.0926952\,t1} + 0.0919583\,\mathsf{Cos[0.3\,t1]} + 0.450027\,\mathsf{Sin[0.3\,t1]},$
 $0.185606 + 0.000458385\,e^{-2.85892\,t1} + 0.319999\,e^{-0.146518\,t1} -$
 $0.49654\,e^{-0.0926952\,t1} - 0.00952309\,\mathsf{Cos[0.3\,t1]} + 0.00723119\,\mathsf{Sin[0.3\,t1]},$
 $0.0566485 - 0.184237\,e^{-12.0864\,t1} + 0.100379\,e^{-2.85892\,t1} +$
 $0.00088615\,e^{-0.146518\,t1} - 0.00134223\,e^{-0.0926952\,t1} +$
 $0.0276571\,\mathsf{Cos[0.3\,t1]} + 3.03897 \times 10^{-9}\,e^{-11.9937\,t1}\,\mathsf{Cos[0.3\,t1]} -$
 $5.25381 \times 10^{-9}\,e^{-11.9399\,t1}\,\mathsf{Cos[0.3\,t1]} + 8.6242 \times 10^{-6}\,e^{-9.22746\,t1}\,\mathsf{Cos[0.3\,t1]} +$
 $\left. 0.00361739\,\mathsf{Sin[0.3\,t1]} - 4.92571 \times 10^{-8}\,e^{-9.22746\,t1}\,\mathsf{Sin[0.3\,t1]} \right\}$

- The evolution of the iodine retention in the blood (compartment 1) and in the thyroid (compartment 2) is plotted below:

$In[\circ]:=$ **Plot[{x1[t], x2[t]}, {t, 1, 200}, PlotStyle →**
  {{RGBColor[1, 0, 0]}, {AbsoluteDashing[{4, 4}], RGBColor[0, 0, 1]}},
  AxesLabel → {"Days", "r"}, PlotLegends →
  Placed[{"Blood", "Thyroid"}, Center, (Framed[#, RoundingRadius → 5] &)],
  PlotLabel → Style[Framed["b1(t)= 1 + 0.5 Cos[0.3 t]"],
  12, Blue, Background → Lighter[Yellow]]]

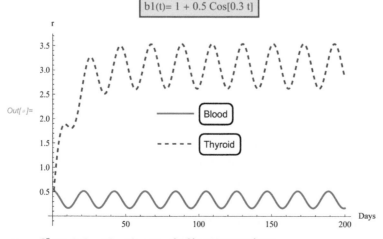

$In[\circ]:=$ **Clear[x1, x2, x3, x4, iodine131matrix1]**

- The following function computes numerically the SODE using NDSolve and returns the solution in the interval $\{t, t_{min}, t_{max}\}$:

```
In[ ]:= SystemNDSolve1[matrixA_?MatrixQ, incond_List,
        input_List, {t_, tmin_, tmax_}, t1_, x_, opts___?OptionQ] :=
    Module[{A, n, var, i, system, ic, X, X0, sol }, A = matrixA;
    n = Dimensions[A][[1]];
    X = Table[xᵢ[t], {i, 1, n}]; X0 = Table[xᵢ[0], {i, 1, n}];
    system = D[X, t] == A.X + input;
    ic = Thread[X0 == incond];
    sol = NDSolve[{system, ic}, X, {t, tmin, tmax}, opts];
    Chop[ExpandAll[sol]] /. t → t1];
```

- This function can be used when **SystemDSolve1** does not find a solution, for instance, if some transfer coefficients are variables:

```
In[ ]:= iodine131matrix2 = iodine131matrix /. {k₁₂ → (0.3 Log[2])/0.25 (1 + Cos[0.2 t]),

        k₁₄ → (0.7 Log[2])/0.25, k₂₃ → Log[2]/80 (1 + Cos[0.3 t]),

        k₃₀ → 1/12 × 0.2 Log[2], k₃₁ → 1/12 × 0.8 Log[2], k₄₀ → 12, λ → λ131};
```

```
In[ ]:= {q1[t1_], q2[t1_], q3[t1_], q4[t1_]} =
    {q₁[t1], q₂[t1], q₃[t1], q₄[t1]} /. SystemNDSolve1[ iodine131matrix2,
        {1, 0, 0, 0}, {1 + 0.5 Cos[0.3 t], 0, 0, 0}, {t, 0, 100}, t1, q][[1]];
```

```
In[ ]:= Plot[{Callout[q1[t], "Blood", Above], Callout[q2[t], "Thyroid", Above]},
    {t, 1, 100}, AxesOrigin → {0, 0}, AxesLabel → {"Days", "r"}]
```

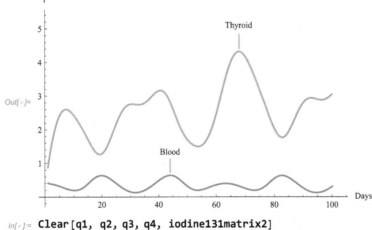

```
In[ ]:= Clear[q1, q2, q3, q4, iodine131matrix2]
```

10.1.4 Laplace Transforms and Identifiability Analysis

The transfer rates k_{ij} are usually estimated using experimental data as shown in the next section. The problem that often arises is that there is no unique value of k_{ij} that satisfies the model but a finite number of values. This issue is addressed by a group of mathematical methods named identifiability analysis. The Laplace transforms are very useful in this kind of analysis.

The solution of a SODE with constant coefficients (eqn. 10(3)) can be found as follows:

$$x(t) = x_0\, e^{At} + \int_0^t e^{A(t-\tau)}\, b(\tau)\, d\tau$$

Applying Laplace transforms we get:

$$X(s) = (sI - A)^{-1}\, x_0 + (sI - A)^{-1}\, B(s)$$

where $X(s)$ and $B(s)$ are the Laplace transforms of $x(t)$ and $b(t)$ respectively.

- This equation can be written in *Mathematica* as:

```
In[ ]:= SystemLTSolve1[matrixA_List, initcond_List, input_List, t_, s_, X_] :=
        Module[{B, A1, R, R1, P, r, n}, A1 = matrixA; n = Dimensions[A1][[1]];
          B = s*IdentityMatrix[n] - A1; R = Inverse[B]; P = Apart[R];
          r = LaplaceTransform[input, t, s];
          R1 = Dot[P, initcond] + Dot[P, r];
          Thread[Subscript[X, #1][s] & /@ Range[n] -> R1]];
```

Example: In Figure 10.4 (Godfrey, 1983, Example 6.5) the response of the central compartment 1 to a single input of 1 unit at $t = 0$ has been fitted (no noise is assumed) to the function $x_{1\,exp}(t) = 0.7\exp(-5\,t) + 0.2\exp(-t) + 0.1\exp(-0.1\,t)$. The problem consists of determining the unknown transfer rates, $\{k_{10}, k_{12}, k_{13}, k_{21}, k_{31}\}$, assumed to be constants.

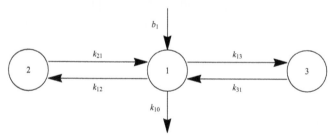

Figure 10.4   Catenary three-compartment model.

- The Laplace transform $X(s)$ of an impulsive input of 1 (we can use any mass unit, so if 1 is in mg the solution will be in mg) in compartment 1 at $t = 0$ can be modeled using **SystemLTSolve1** with the initial conditions $\{1, 0, 0\}$ and $b(t) = \{0, 0, 0\}$. To solve the problem, we will also need to define the compartmental matrix of the system.

```
In[ ]:= tricompartment = CompartMatrix1[3,
          {{1, 2, k12}, {2, 1, k21}, {1, 3, k13}, {3, 1, k31}, {1, 0, k10}}];
```

```
In[ ]:= {X1[s_], X2[s_], X3[s_]} = {X1[s], X2[s], X3[s]} /.
          SystemLTSolve1[tricompartment, {1, 0, 0}, {0, 0, 0}, t, s, X] // Simplify
```

$$Out[ ]= \left\{ \frac{(k_{21} + s)(k_{31} + s)}{k_{12}\, s\, (k_{31} + s) + k_{10}\, (k_{21} + s)(k_{31} + s) + s\,(k_{21} + s)(k_{13} + k_{31} + s)}, \right.$$

$$\frac{k_{12}\,(k_{31} + s)}{k_{12}\, s\, (k_{31} + s) + k_{10}\, (k_{21} + s)(k_{31} + s) + s\,(k_{21} + s)(k_{13} + k_{31} + s)},$$

$$\left. \frac{k_{13}\,(k_{21} + s)}{k_{12}\, s\, (k_{31} + s) + k_{10}\, (k_{21} + s)(k_{31} + s) + s\,(k_{21} + s)(k_{13} + k_{31} + s)} \right\}$$

- The Laplace transform $X_{1\,\text{exp}}(s)$ of $x_{1\,\text{exp}}(t) = 0.7\exp(-5\,t) + 0.2\exp(-t) + 0.1\exp(-0.1\,t)$ is:

In[ ]:= **X1exp = LaplaceTransform[0.7 Exp[-5 t] + 0.2 Exp[-t] + 0.1 Exp[-0.1 t], t, s]**

Out[ ]= $\dfrac{0.2}{s + 1} + \dfrac{0.7}{s + 5} + \dfrac{0.1}{s + 0.1}$

- Now, the transfer rate $\{k_{10}, k_{12}, k_{13}, k_{21}, k_{31}\}$ can be obtained solving $X_{1\,\text{exp}}(s) = X_1(s)$. The following procedure is applied:

In[ ]:= **SolveAlways[X1exp == X1[s], s]**

Out[ ]= $\{\{k_{10} \to 0.746269, k_{12} \to 1.25488, k_{13} \to 1.70885, k_{21} \to 0.324354, k_{31} \to 2.06565\},$
$\{k_{10} \to 0.746269, k_{12} \to 1.70885, k_{13} \to 1.25488, k_{21} \to 2.06565, k_{31} \to 0.324354\}\}$

In this case, there are two sets of possible solutions. Notice that neither set contains any negative rate constants, which would permit the rejection of such set.

All the functions described so far are part of *Sysmodel,* a package included in BIOKMOD that we will use in the next section.

In[ ]:= **Clear[X1, X2, X3, X1exp]**

10.2 Epidemiological Models

In[ ]:= **ClearAll["Global`*"]**

Compartmental models can also be used to model pandemics. The SIR model is the most basic one. It consists of three compartments representing three situations: (S) Susceptible, (I) Infectious, and (R) Recovered. The model can be visualized as follows:

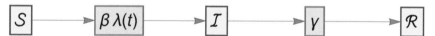

A very simple version of this model consists of replacing S, *I*, and *R* with the variables s, *i*, and *r,* each one representing fractions of the entire population (values between 0 and 1).

Initially, we assume that almost all of the population is susceptible of getting infected (i.e., s(0) ≅ 1). Over time, that fraction will go down as the infection spreads. We also assume that the infectious population at t = 0 will be almost 0 (i.e., i(0) ≅ 0) and that at the beginning, there will be no people yet who have recovered (i.e., r(0) = 0). At each point in time, the following equation must hold: s + i + r = 1. Each infectious individual is also assumed to transmit the disease to *b* new people daily so the higher the value of *b*, the faster the epidemic will spread. Finally, *k* represents the fraction of the infected population that recovers every day. For example, if the disease lasts 7 days, k is 1/7.

- The equation below describes how the percentage of the population susceptible of getting infected changes over time. We can assume that the rate of change is going to be negative since we start with a value close to 1 that goes down over time. Additionally, the rate of change is proportional to *b*, the number of contacts, (the higher the number of contacts, the higher the number of people becoming infectious and therefore the faster s goes down) and to s itself, the fraction of the population that is susceptible of becoming infectious at any time *t*.

```
In[ ]:= equationS =
        s'[t] == -b s[t] × i[t];
```

- The second equation, **equationI**, describes how *i* changes over time On the right side of the equation there are two terms. The first one refers to right side of **equationS** but has the opposite sign, representing the increase in the infectious population, (a flow from S to I) and the second term corresponds to the decrease in infectious cases (recovered people):

```
In[ ]:= equationI =
        i'[t] == b s[t] × i[t] - k i[t];
```

- The last equation models people (as a fraction of the total population) that have recovered. Expressed in compartmental terms, it indicates a flow from the I compartment to the R compartment.

```
In[ ]:= equationR =
        r'[t] == k i[t];
```

- Let's solve now this system of equations in terms of *b* and *k* by assuming some initial conditions regarding the variables *s*, *i*, and *r* at t = 0.

```
In[ ]:= sirmodel2 = { equationS, equationI, equationR};
```

```
In[ ]:= {sols2, solr2, soli2} = {s, r, i} /. ParametricNDSolve[
        Join[sirmodel2, {s[0] == 0.9999, i[0] == 0.0001, r[0] == 0.0000}],
        {s, i, r}, {t, 0, 100}, { b, k}]
```

Out[ ]= {ParametricFunction[⊞ M Expression: s
                                  Parameters: {b, k}],

         ParametricFunction[⊞ M Expression: r
                                  Parameters: {b, k}],

         ParametricFunction[⊞ M Expression: i
                                  Parameters: {b, k}]}

Imagine that we got some data *{t,i}* with *t* representing the number of days since the beginning of the epidemic and *i* the fraction of infectious people, and we would like to calculate the values for *b* and *k*.

- The data has been generated assuming b = 0.5 and k = 1/14 and adding a random component to both parameters: (SeedRandom[31415]; Evaluate[Table[{t, soli2[0.5 + RandomReal[{-0.1, 0.1}], 1/14 + RandomReal[{-0.01, 0.01}]][t]}, {t, 10, 100, 5}]] /. {a_, b_} -> {a, Round[b, 0.001]}]

```
In[ ]:= sample = {{10, 0.007}, {15, 0.108}, {20, 0.188}, {25, 0.471},
        {30, 0.516}, {35, 0.456}, {40, 0.395}, {45, 0.153}, {50, 0.168},
        {55, 0.158}, {60, 0.071}, {65, 0.048}, {70, 0.036}, {75, 0.015},
        {80, 0.011}, {85, 0.012}, {90, 0.015}, {95, 0.006}, {100, 0.002}};
```

- We use FindFit to estimate *b* and *k*.

```
In[ ]:= fit = FindFit[sample, soli2[ b, k][t],
        {{b, 0.5}, {k, 0.01}}, t, PrecisionGoal → 10 ]
```

Out[ ]= {b → 0.460394, k → 0.071466}

According to the solution, each infectious person transmits the disease to another person every two days (b = 0.46 ≈1/2),   and an infected person will recover in 14 days (k =

0.071466 ≈ 1/14).

- Let's plot the fitted model against the actual data:

```
In[ ]:= Plot[Evaluate[{sols2[b, k][t], soli2[b, k][t], solr2[b, k][t]} /. fit],
        {t, 0, 100}, PlotRange → {0, 1.01}, PlotStyle → {Blue, Red, Green},
        PlotLegends → {"Susceptible", "Infectious", "Recovered"},
        Epilog → {Point@sample}]
```

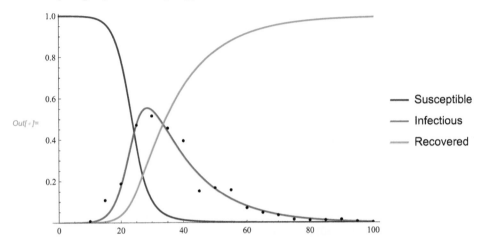

We can now make the plot interactive to see how the model changes when we change *b* and *k*. Their default values are the ones from the initial model. This method is quite useful since we can visually adjust the model to improve the fit.

```
In[ ]:= Manipulate[Plot[{sols2[b, k][t], solr2[b, k][t], soli2[b, k][t]},
        {t, 0, 100}, PlotStyle → {Blue, Red, Green},
        PlotLegends → {"Susceptible", "Infectious", "Recovered"},
        Epilog → {Point@sample}], {{b, 0.46}, 0.1, 1}, {{k, 1/14}, 0.01, 1}]
```

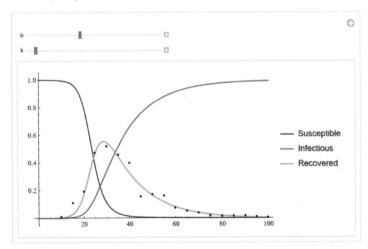

- We can also analyze the sensitivity of the parameters by evaluating the derivative with respect to them. The higher the sensitivity, the higher the impact on the model when the parameter values change. For visualization purposes we add and subtract the derivative of the model multiplied by 0.1 before plotting it.

```
In[ ]:= Manipulate[
    GraphicsColumn[{Plot[{sols2[b, k][t], solr2[b, k][t], soli2[b, k][t]},
       {t, 0, 100}, PlotStyle → {Blue, Red, Green}, PlotLegends →
        Placed[{"Susceptible", "Infectious", "Recovered"}, Above]], Plot[
       Evaluate[(soli2[b1, k1][t] + {0, .1, -.1} D[soli2[b1, k1][t], b1]) /.
         {b1 → b, k1 → k}], {t, 0, 100}, PlotRange → All,
       Filling → {2 → {3}}, PlotLegends → {"Infectious, b"}], Plot[
       Evaluate[(soli2[b1, k1][t] + {0, .1, -.1} D[soli2[b1, k1][t], k1]) /.
         {b1 → b, k1 → k}], {t, 0, 100}, PlotRange → All,
       Filling → {2 → {3}}, PlotLegends → {"Infectious, k"}]],
     {{b, 0.5}, 0.1, 1}, {{k, 1/14}, 0.01, 1}]
```

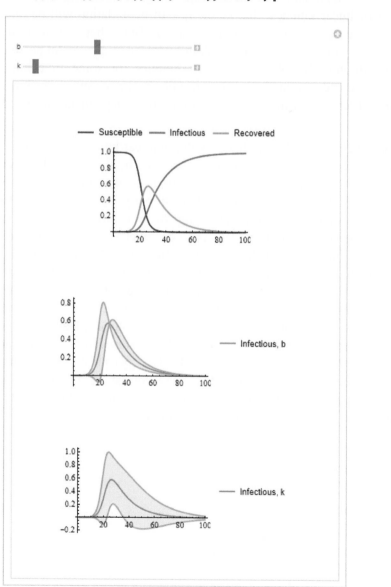

10.3 Physiological Modeling

10.3.1 Michaelis–Menten (M–M) Models

Most drugs are metabolized by enzymes. The enzyme levels will change under different physiological cases (e.g., when many drugs are administered at the same time, they might have affinity for the same enzymes and thus compete for the same enzyme sites) or pathological states (e.g., in case of disease). This change in enzyme levels produces fluctuations in the concentration levels of drugs in the blood. For example, when given low-concentration/tolerable doses of Phenytoin, an anti-epileptic drug, its degradation by the corresponding metabolizing enzymes will follow first order or linear kinetics, that is, the rate of degradation will be directly proportional to the drug or substrate concentration. But when given a higher concentration of Phenytoin, as in the case of drug overdose or accidental poisoning, the metabolism will initially follow first-order kinetics for some time until all enzyme sites get occupied/saturated, and then it will change to zero-order kinetics, that is, the rate of change will become constant or independent of the drug concentration.

This mixed-order reaction (combining linear with nonlinear kinetics when there are no more sites available for drug binding) can be depicted by the Michaelis–Menten (M–M) equation. A example of the M–M model is shown in Figure 10.5. It describes the disposition in the animal or human body of a drug distributed according to a mammillary model, consisting of a central component with peripheral components connecting to it. This is a convenient model for many drugs for which the equilibrium between drug concentrations in different tissues is not achieved rapidly. It tries to describe mathematically the concentration data after the administration of a drug by depicting the body as a two-compartment open model where the drug is both introduced into, and exits from, the central compartment, named compartment 1, which can be sampled through the blood (plasma or serum). It may consist of organs or tissues which, being highly perfused with blood, reach a rapid equilibrium distribution with the blood. The peripheral compartment, compartment 2, usually cannot be sampled. It may consist of organs or tissues which, being poorly perfused with blood, reach equilibrium distribution with the blood only slowly. The concentration in compartment i is denoted x_i.

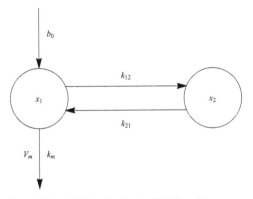

Figure 10.5   A Michaelis–Menten (M–M) model.

In this case, the drug movement between compartments will be considered as a linear kinetic process described by the transfer coefficients k_{12} and k_{21}. However, the elimination process will be nonlinear, as it happens for instance in the hepatic metabolism, and the elimination rate of the drug can be mathematically expressed by the M–M equation with parameters V_m = maximum transformation speed and k_m = M–M constant. The drug administration will be assumed to be an impulsive input (bolus) b_0. Thus:

$$\dot{x}_1(t) = -k_{21} x_1(t) + k_{12} x_2(t) - \frac{V_m x_1(t)}{k_m + x_1(t)}$$

$$\dot{x}_2(t) = k_{21} x_1(t) - k_{12} x_2(t)$$

$$x_1(0) = b_0, \; x_2(0) = 0$$

(10.4)

In our example: $b_0 = 1$, Vm = 0.1, km = 0.3, and k_{12} and k_{21} are the unknown parameters. They can be estimated using experimental data.

- We define a model $x_1(t, k_{12}, k_{21})$ and then use the `ParametricNDSolve` function to obtain a numerical solution as a function of k12 and k21:

```
In[ ]:= ClearAll["Global`*"]

    Vm = 0.1; km = 0.3;

    eq1 = x1'[t] == -k₁₂ x1[t] + k₂₁ x2[t] - (Vm x1[t])/(km + x1[t]);

    eq2 = x2'[t] == k₁₂ x1[t] - k₂₁ x2[t];

In[ ]:= sol = ParametricNDSolve[
        {eq1, eq2, x1[0] == 1, x2[0] == 0}, {x1, x2}, {t, 0, 100}, {k₁₂, k₂₁}]
```

Out[ ]= {x1 → ParametricFunction[▦ 〽 Expression: x1 Parameters: {k₁₂, k₂₁}],

x2 → ParametricFunction[▦ 〽 Expression: x2 Parameters: {k₁₂, k₂₁}]}

- $x_1(t, k_{12}, k_{21})$ and $x_2(t, k_{12}, k_{21})$ can be represented as functions of k_{12} and k_{21}.

```
In[ ]:= Manipulate[Plot[Evaluate[{x1[k12, k21][t], x2[k12, k21][t]} /. sol],
        {t, 0, 30}, PlotRange → All, PlotLegends →
         Placed[{"x1", "x2"}, Center, (Framed[#, RoundingRadius → 5] &)]],
        {{k12, 0.03, "k12"}, 0.01, 0.1}, {{k21, 0.02, "k21"}, 0.01, 0.1}]
```

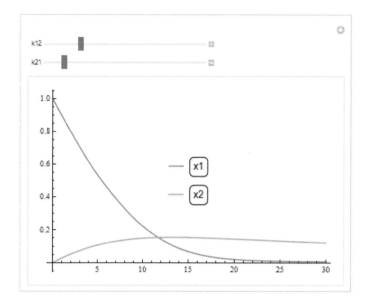

10.3.2 A Nonlinear Physiological Model

In this section, we present an example of a physiological nonlinear model (M. S. Roberts, M. Rowland, A dispersion model of hepatic elimination, *Journal of Pharmacokim. Biopharm 14*: pp 227-260, 1986.):

An hepatic drug dispersion model in the liver can be represented as a cylindrical vessel (Figure 10.6). Let's consider an element of differential thickness dz at distance z along its length L. Inside this element there is blood, with a volume $V_{B,z}$ flowing at rate Q through a cross-sectional area A and a velocity v = Q/A, containing a drug concentration $C_B(z, t)$, part of which is unbounded at a concentration Cu in the plasma.

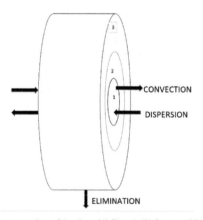

Figure 10.6   Schematic cross section of the liver:(1) Blood, (2) Space of Disse, (3) The hepatocyte.

The model is represented by a partial differential equation, where the axial dispersion flux is given by term (1) and the convective flux by term (2). $fu_B = C_u / C_B$, CL_{int} is a clearance constant, ρ is a permeability constant, and V_H is the apparent volume of the distribution of the drug.

$$\overbrace{V_B D \frac{\partial C_B(z,\,t)}{\partial z^2}}^{(1)} - \overbrace{V_B v \frac{\partial C_B(z,\,t)}{\partial z}}^{(2)} - \text{fu}_b\,\text{CL}_{\text{int}}\,\rho\,C_B(z,\,t) = V_H \frac{\partial C_B(z,\,t)}{\partial t}$$

To scale the equations and make the variables dimensionless we make the following replacement:

$$Z = \frac{z}{L};\ d_n = \frac{D}{vL} = \frac{DA}{QL};\ r_n = \frac{\text{fu}_b\,\text{CL}_{\text{int}}\,\rho}{Q};\ c(z,\,t) = C_B(z,\,t)$$

After which we obtain:

$$\frac{\partial^2 c}{\partial z^2} - \frac{1}{d_n}\frac{\partial c}{\partial z} - \frac{r_n}{d_n}c = \frac{1}{d_n}\frac{\partial c}{\partial t} \tag{10.5}$$

with the following the initial conditions (IC):

$$t = 0,\ c(z,\,0) = 0,\ 0 \le z \le 1$$

■ In *Mathematica* notation, the above result can be written as:

In[ ]:= **roweq = D[c[z, t], {z, 2}] - $\dfrac{1}{d}$ D[c[z, t], z] - $\dfrac{r}{d}$ c[z, t] == $\dfrac{1}{d}$ D[c[z, t], t]**

Out[ ]:= $-\dfrac{r\,c[z,\,t]}{d} - \dfrac{c^{(1,0)}[z,\,t]}{d} + c^{(2,0)}[z,\,t] == \dfrac{c^{(0,1)}[z,\,t]}{d}$

■ We can solve it using the Laplace transform method turning the partial differential equation into an ordinary differential equation:

In[ ]:= **rowlandeqLT = LaplaceTransform[roweq, t, s]**

Out[ ]:= $-\dfrac{r\,\text{LaplaceTransform}[c[z,\,t],\,t,\,s]}{d} -$

$\dfrac{\text{LaplaceTransform}\!\left[c^{(1,0)}[z,\,t],\,t,\,s\right]}{d} + \text{LaplaceTransform}\!\left[c^{(2,0)}[z,\,t],\,t,\,s\right] ==$

$\dfrac{-c[z,\,0] + s\,\text{LaplaceTransform}[c[z,\,t],\,t,\,s]}{d}$

■ In the above output, the replacement **LaplaceTransform**[f(z,t), t,s] → LT[z], takes also into account the initial condition (c[z,0]=0):

In[ ]:= **rowlandeqLT1 = rowlandeqLT /. {LaplaceTransform$\left[c^{(1,0)}[z, t], t, s\right]$→cT'[z],**
LaplaceTransform$\left[c^{(2,0)}[z, t], t, s\right]$→cT''[z],
LaplaceTransform[c[z, t], t, s]→cT[z], c[z, 0]→0}

Out[ ]:= $-\dfrac{r\,cT[z]}{d} - \dfrac{cT'[z]}{d} + cT''[z] == \dfrac{s\,cT[z]}{d}$

■ The above equation is an ordinary differential equation that can be solved using **DSolve**:

In[ ]:= **sol1 = cT[z] /. DSolve[rowlandeqLT1 , cT[z], z]**

Out[ ]:= $\left\{ e^{\frac{1}{2}\left(\frac{1}{d} - \frac{\sqrt{1+4\,d\,r+4\,d\,s}}{d}\right)z}\, c_1 + e^{\frac{1}{2}\left(\frac{1}{d} + \frac{\sqrt{1+4\,d\,r+4\,d\,s}}{d}\right)z}\, c_2 \right\}$

The coefficients C_1 and C_2 can be determined using the boundary conditions, which in this case are: at $z = 0$, $c = \delta(z)$, and at $z = \infty$, $c = 0$, that is, $c(0,\,t) = \delta(z)$ and $c(\infty,\,t) = 0$.

- For convenience we replace C_1 by c1 and C_2 by c2.

In[ ]:= **sol2 = sol1⟦1⟧ /. {C[1] → c1, C[2] → c2}**

Out[ ]= $c1\, e^{\frac{1}{2}\left(\frac{1}{d}-\frac{\sqrt{1+4dr+4ds}}{d}\right)z} + c2\, e^{\frac{1}{2}\left(\frac{1}{d}+\frac{\sqrt{1+4dr+4ds}}{d}\right)z}$

- c1 can be computed using the boundary condition $c(0, t) = \delta(z)$:

In[ ]:= **solc1 = Solve[**
   Evaluate[sol2 /. z → 0] == LaplaceTransform[DiracDelta[z], z, s], c1]⟦1⟧

Out[ ]= $\{c1 \to 1 - c2\}$

- c2 can be computed using the boundary condition $c(\infty, t) = 0$:

In[ ]:= **solc2 = Solve[Evaluate[sol2 /. solc1] == 0, c2]⟦1⟧ // Simplify**

Out[ ]= $\left\{c2 \to -\dfrac{1}{-1 + e^{\frac{\sqrt{1+4d\,(r+s)}\,z}{d}}}\right\}$

In[ ]:= **Limit[solc2⟦1, 2⟧, z → Infinity, Assumptions → s > 0 && r > 0 && d > 0]**

Out[ ]= 0

- Hence:

In[ ]:= **sol3 = Evaluate[sol2 /. c1 → 1 - c2] /. c2 → 0**

Out[ ]= $e^{\frac{1}{2}\left(\frac{1}{d}-\frac{\sqrt{1+4dr+4ds}}{d}\right)z}$

In[ ]:= **InverseLaplaceTransform[sol3, s, t] // Simplify**

Out[ ]= $\dfrac{e^{-\frac{t^2+4drt^2-2tz+z^2}{4dt}}\,z}{2\,d^2\,\sqrt{\pi}\,\sqrt{\dfrac{t^3}{d^3}}}$ if $d\,z > 0$

In[ ]:= **csol[z_, t_] =** $\dfrac{z\, e^{-\frac{4drt^2+t^2-2tz+z^2}{4dt}}}{2\,\sqrt{\pi}\,d^2\,\sqrt{\dfrac{t^3}{d^3}}}$ **;**

- The plot below displays $c(z, t)$ for different values of d and $r = 0$:

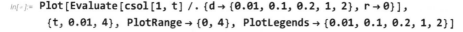

In[ ]:= `Plot[Evaluate[csol[1, t] /. {d → {0.01, 0.1, 0.2, 1, 2}, r → 0}],`
   `{t, 0.01, 4}, PlotRange → {0, 4}, PlotLegends → {0.01, 0.1, 0.2, 1, 2}]`

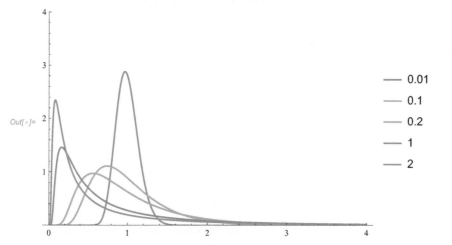

In[ ]:= `Manipulate[ContourPlot[Evaluate[csol[z, t] /. {d → d1, r → r1}],`
   `{z, 0, 30}, {t, 0, 30}, FrameLabel → {"z", "t"},`
   `PlotLabel → Style[c(z,t), Blue, 20], PlotLegends → Automatic],`
   `{{d1, 0.2, "d"}, 0.01, 1}, {{r1, 0.2, "r"}, 0, 0.5}]`

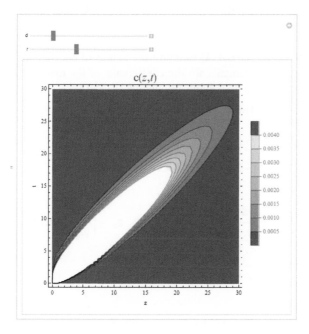

10.4 Fitting a Model

In this section, with the help of examples, we describe how to fit biokinetic parameters using experimental data. For more details about the model see A. Sánchez-Navarro, C. Casquero, and M. Weiss, "Distribution of Ciprofloxacin and Ofloxacin in the isolated hindlimb of the rat", *Pharmaceutical Research*, 16: 587-591, 1999. We will use the functions developed in Section 10.1.

$In[•]:=$ **ClearAll["Global`*"]**

Let's suppose that we have a physiological model represented by the following SODE:

$$\dot{x}_1(t) = -\frac{Q + PS}{V_p}\, x_1(t) \;+\; \frac{PS}{V_p}\, x_2(t) \;+\; \frac{Q}{V_p}\, b_1(t)$$

$$\dot{x}_2(t) = \frac{PS}{V_{Tu}}\, x_1(t) \;-\; \left(\frac{PS}{V_{Tu}} + k_{on}\right) x_2(t) \;+\; k_{off}\, \frac{V_{Tb}}{V_{Tu}}\, x_3(t) \tag{10.6}$$

$$\dot{x}_3(t) = k_{on}\, \frac{V_{Tu}}{V_{Tb}}\, x_2(t) \;-\; k_{off}\, x_3(t)$$

where the drug is transported from V_p (vascular volume), by perfusate flow Q, to V_{Tu} (tissue water space) across a permeability barrier (permeability-surface product *PS*), and its binding is described by binding/unbinding constants k_{on} and k_{off}

When the coefficients a_{ij} of a SODE are associated with physiologically meaningful values corresponding to measured physiological parameters or a function of them, we should use a physiological model instead of a compartmental one. In those cases, we will directly enter the values for a_{ij} to the **A** matrix, instead of the k_{ij} values.

- We build the function **CoeffMatrix1**[n, *matrixA*, λ] to return the matrix of coefficients for n retention variables, where the individual elements are the coefficients of the matrix: $\{\{1, 2, a_{12}\}, ..., \{i, j, a_{ij}\}, ...\}$ for the ith row and the jth column. By default, $a_{ij} = 0$.

$In[•]:=$ **CoeffMatrix1[(n_) ?IntegerQ, (matrixA_) ?MatrixQ] :=**
        Module[{i, j, k},
        Normal[SparseArray[matrixA /. {i_, j_, k_} → {i, j} → k, {n, n}]]];

- The coefficients matrix using the function **CoeffMatrix1** is:

$In[•]:=$ **physiomodel1 = CoeffMatrix1$\Big[$3,**

$$\left\{\left\{1,\, 1,\, -\left(\frac{Q}{V_p} + \frac{PS}{V_p}\right)\right\},\, \left\{1,\, 2,\, \frac{PS}{V_p}\right\},\, \left\{2,\, 1,\, \frac{PS}{V_{Tu}}\right\},\, \left\{2,\, 2,\, -\left(\frac{PS}{V_{Tu}} + kon\right)\right\},\right.$$

$$\left.\left\{2,\, 3,\, koff\, \frac{VTb}{V_{Tu}}\right\},\, \left\{3,\, 2,\, kon\, \frac{V_{Tu}}{VTb}\right\},\, \{3,\, 3,\, -koff\}\right\}\Big];$$

$In[•]:=$ **physiomodel1**

$Out[•]=$ $\left\{\left\{-\dfrac{PS}{V_p} - \dfrac{Q}{V_p},\, \dfrac{PS}{V_p},\, 0\right\},\, \left\{\dfrac{PS}{V_{Tu}},\, -kon - \dfrac{PS}{V_{Tu}},\, \dfrac{koff\, VTb}{V_{Tu}}\right\},\, \left\{0,\, \dfrac{kon\, V_{Tu}}{VTb},\, -koff\right\}\right\}$

We have a system of ODE, where some parameters *par*: $\{k_{on}, k_{off}\}$ are constants whose values are unknown:

$$X(t,\, par) = A(t,\, par)\, X(t) + B(t) \tag{10.7}$$

- Here is developed a function to solve (10.7) using ParametricNDSolve, where the inputs are given in an easy way

```
In[ ]:= ParametricSystemNDSolve1[matrixA_, incond_, input_, {t_, tmin_, tmax_},
          x_, par_, opts___] := Module[{B, A1, R, P, n, i}, A1 = matrixA;
        n = Dimensions[A1][[1]];
        R = Join[Thread[Table[xᵢ'[t], {i, 1, n}] ==
               Table[xᵢ[t], {i, 1, n}].Transpose[matrixA] + input],
           Thread[Table[xᵢ[0], {i, 1, n}] == incond]];
        B = Table[xᵢ, {i, 1, n}];
        Flatten[ParametricNDSolve[R, B, {t, tmin, tmax}, par, opts]]];
```

where:

matrixA is coeffsmatrix **A**; *incond* (the initial conditions): $\{x_1(0),.., x_n(0)\}$; *input*: B(t) = $\{b_1(t),..., b_n(t)\}$ x is the symbol that represents x_i; *par*: $\{p_1, ...,p_n\}$ are parameters, options are the options of **ParametricNDSolve**.

- The following parameters values of the model are known:

```
In[ ]:= Q = 3; VTb = 1; V_Tu = 6.411; V_p = 0.973; PS = 2.714;
```

- The mathematical expression for $b_1(t)$ was unknown initially, but an experiment was made where $b_1(t)$ was given by the best fit of the input function to experimental data $\{\{t_1, b_1\}, ..., \{t_n, b_n\}\}$ obtained via sampling from an arterial catheter. Here are the data:

```
In[ ]:= dataCateter =
        {{0., 0.}, {0.05, 402.7}, {0.1, 430.3}, {0.15, 375.4}, {0.2, 292.4},
         {0.25, 202.2}, {0.3, 148.4}, {0.35, 96.4}, {0.4, 64.9}, {0.45, 41.7},
         {0.5, 25.3}, {0.55, 17.8}, {0.6, 8.8}, {0.65, 6.6}, {0.7, 3.2},
         {0.75, 2.5}, {0.8, 1.4}, {0.85, 0.9}, {0.9, 0.5}, {1., 0.2}, {1.1, 0.07},
         {1.2, 0.03}, {1.3, 0.01}, {1.4, 0.003}, {1.45, 0.001}, {1.5, 0.001}};
```

```
In[ ]:= ListPlot[dataCateter, Joined → True, PlotRange → All, ImageSize → Small]
```

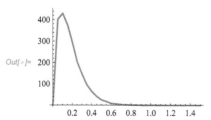

- The shape of the graph suggests that the function below could provide a good fit:

```
In[ ]:= b[t_] = a t Exp[c t] /. FindFit[dataCateter, a t Exp[c t], {a, c}, t]
```

$$Out[ ]= 13\,610.1\, e^{-11.216\,t}\, t$$

- Here are the experimental data and the fitted function:

```
In[ ]:= Plot[b[t], {t, 0, 1}, Epilog →
           {Hue[0], PointSize[0.02`], Point /@ dataCateter}, ImageSize → Small]
```

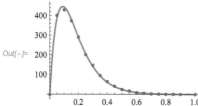

- We also measured $x_1(t)$ by sampling from the following experimental data $\{\{t_1, x_1(t_1)\}, ..., \{t_n, x_1(t_n)\}\}$.

```
In[ ]:= X1sample = {{0.03, 27.5}, {0.08, 48.61}, {0.28, 133.39}, {0.33, 106.18},
           {0.38, 92.11}, {0.48, 60.47}, {0.55, 56.54}, {0.65, 27.73},
           {0.75, 23.25}, {0.85, 15.83}, {0.95, 13.82}, {1.05, 10.58}, {1.15, 7.83},
           {1.25, 7.35}, {1.35, 6.08}, {1.45, 5.21}, {1.6, 4.18}, {1.8, 3.48},
           {2., 2.86}, {2.2, 2.42}, {2.4, 2.11}, {2.6, 2.01}, {2.8, 1.7}, {3., 1.58},
           {3.4, 1.58}, {3.75, 1.44}, {4.25, 1.35}, {4.75, 1.3}, {5.25, 1.29},
           {6.75, 0.81}, {7.25, 0.78}, {7.75, 0.73}, {8.25, 0.75}, {11., 0.68}}};
```

We build the model:

- The input function and the vector of the initial conditions are {0, 0, 0}:

```
In[ ]:= inputcatheter1 = {Q/Vp b[t] , 0, 0};
```

```
In[ ]:= initialcondition = {0, 0, 0};
```

- Then, the solution for x_1 where the measurements are taken is:

```
In[ ]:= X1a = x1 /. ParametricSystemNDSolve1[physiomodel1,
           {0, 0, 0}, inputcatheter1, {t, 0.01, 12}, x, { koff, kon}]
```

```
Out[ ]= ParametricFunction[           Expression: x1
                                        Parameters: {koff, kon}           ]
```

- Now we want obtained k_{off} and k_{on} for fitting $x_1(t)$ samples:

```
In[ ]:= FindFit[X1sample , X1a[koff, kon][t], {{koff, 0.1, 1}, {kon, 0.1, 1}}, t]
```

```
Out[ ]= {koff → 0.142112, kon → 0.832626}
```

- Also we can used NonlinearModelFit

```
In[ ]:= f[koff_ ?NumericQ, kon_ ?NumericQ, t_] = X1a[koff, kon][t];
```

```
In[ ]:= nlm = NonlinearModelFit[X1sample ,
           f[koff, kon, t], {{koff, 0.1, 1}, {kon, 0.1, 1}}, t];
```

```
In[ ]:= nlm["ParameterTable"]
```

	Estimate	Standard Error	t-Statistic	P-Value
koff	0.142112	0.291949	0.486772	0.629738
kon	0.832626	0.471213	1.76698	0.0867685

- Once again, we quit the session before proceeding to the next section.

```
In[ ]:= Quit[]
```

10.5 Optimal Experimental Designs (OED)

10.5.1 OED Introduction

Many scientific processes, specially in pharmacokinetics (PK) and pharmacodynamics (PD) studies, as we have seen before, are defined by systems of ordinary differential equations (ODEs). If there are unknown parameters that need to be estimated, the optimal experimental design (OED) approach offers a way to obtain optimum bias-free and minimum-variance estimators for those parameters based on specific criteria.

The unknown parameters are usually estimated by fitting experimental data. A typical model has a solution given by a function $r(t, \beta)$, where $\beta = \{\beta_1, ..., \beta_p\}$ are the unknown parameters that need to be fitted. The objective is to choose the best moments $\{t_0, ..., t_i, ..., t_n\}$ based on the data, and this can be done using optimal experimental designs (OED). Here, "best" depends on the objectives of the practitioner, which are usually linked to a particular optimality criterion, such as obtaining accurate estimators of the parameters, minimizing the variance of the predicted response, or optimizing a different characteristic of the model. Many optimality criteria are based on the Fisher information matrix (FIM). The FIM has a simple expression for linear models and observations that come from exponential-family distributions. For more complex models, some formulas can be obtained from the general information matrix. The usual procedure for a nonlinear model is to linearize it and apply the well-known toolbox for linear models. The most widely-used criterion is D-optimality, which focuses on the determinant of the FIM.

In the following pages, we're going to describe one example using the D-optimality criterion to obtain the $\{t_0, ..., t_i, ..., t_n\}$ values. For further information we refer the interested reader to: J. Lopez-Fidalgo, J. Rodriguez-Diaz, JM. Sanchez and G. Santos-Martin, "Optimal designs for compartmental models with correlated observations", *Journal of Applied Statistics*: 32(10), pp.1075-1088, 2005. A D-optimal design will be a design that maximizes the determinant of the FIM.

- We first define $r(t, \beta)$. In our example: $r(t, \beta_1, \beta_2) = \beta_1 \left(e^{-2.0\,t - 0.09\,\beta_1} + e^{-0.001\,t - 0.2\,\beta_2} \right)$ being $\beta = \{\beta_1, \beta_2\}$ the unknown parameters.

$In[\circ]:=$ `r[β1_, β2_] = β1 (`$e^{-2.0\,t-0.09\,β2}$` + `$e^{-0.001\,t-0.2\,β2}$`); β = {β1, β2};`

- Then, we compute $\nabla(r(t, \beta_1, \beta_2)) = \{ \frac{\partial r(t)}{\partial \beta_1}, \frac{\partial r(t)}{\partial \beta_2} \}$:

$In[\circ]:=$ `g[t_] =`∇_β`r[β1, β2]`

$Out[\circ]=$ $\left\{ e^{-0.001\,t-0.2\,β2} + e^{-2.\,t-0.09\,β2}, \left(-0.2\,e^{-0.001\,t-0.2\,β2} - 0.09\,e^{-2.\,t-0.09\,β2} \right) β1 \right\}$

- Next, we specify the number of points to be used in the optimal design. We assume a 3-point design with the first one, t_0, defined by the user.

$In[\circ]:=$ `n = 3;`

- For computational purposes, we prefer to use the distance $d_i = t_i - t_{i-1}$, instead of t_i, then $t_i = \sum_i d_i$ with $d_0 = t_0$ defined by the user. That is, the points to be estimated are:

$In[\circ]:=$ `dd = Table[d`$_i$`, {i, n}]`

$Out[\circ]=$ `{d`$_1$`, d`$_2$`, d`$_3$`}`

$In[\circ]:=$ `tm = Accumulate[Flatten[{d`$_0$`, dd}]]`

$Out[\circ]=$ `{d`$_0$`, d`$_0$` + d`$_1$`, d`$_0$` + d`$_1$` + d`$_2$`, d`$_0$` + d`$_1$` + d`$_2$` + d`$_3$`}`

- We evaluate $\nabla(r(t), \beta)$ at points $t:\{t_0, ..., t_n\}$, obtaining $X = \{X_1, ..., X_p\}$ with $X_1 = \{ \frac{dr(t_0)}{d\beta_1}, ..., \frac{dr(t_n)}{d\beta_1} \}, ..., X_p = \{ \frac{dr(t_0)}{d\beta_p}, ..., \frac{dr(t_n)}{d\beta_p} \}$:

$In[\circ]:=$ `X = g[tm];`

- A typical choice for the covariance matrix is $\Gamma = \{l_{ij}\}$ with $l_{ij} = \exp\{\rho|t_j - t_j|\}$ (meaning the relationship between samples decays exponentially as the time-distance between them increases), that is:

$In[\circ]:=$ `ff[i_, j_] := Which`$\left[i == j, 1, i < j, e^{-\rho \sum_{k=i}^{j-1} d_k}, i > j, e^{-\rho \sum_{k=j}^{i-1} d_k} \right]$`;`

$In[\ ]:=$ **Γ = Array[ff, {n + 1, n + 1}]**

$Out[\ ]:=$ $\left\{\left\{1,\ e^{-\rho\,d_1},\ e^{-\rho\,(d_1+d_2)},\ e^{-\rho\,(d_1+d_2+d_3)}\right\},\ \left\{e^{-\rho\,d_1},\ 1,\ e^{-\rho\,d_2},\ e^{-\rho\,(d_2+d_3)}\right\},\right.$
$\left.\left\{e^{-\rho\,(d_1+d_2)},\ e^{-\rho\,d_2},\ 1,\ e^{-\rho\,d_3}\right\},\ \left\{e^{-\rho\,(d_1+d_2+d_3)},\ e^{-\rho\,(d_2+d_3)},\ e^{-\rho\,d_3},\ 1\right\}\right\}$

- We assume that all the measures have approximately the same uncertainty level, that is $\sigma^2 \simeq \sigma_i^2$, then covariance matrix is $\Sigma = \sigma^2\ \Gamma$. In our example, we assume $\rho = 1$ and $\sigma = 2$.

$In[\ ]:=$ **σ = 2; ρ = 1; Σ = σ² * Γ;**

- We also need to give the initial values to β_1 and β_2. The ultimate purpose is to find their real values empirically. However, the method requires an initial guess, (we can use previous experiments, information derived from similar experiments, etc.) and to find out how sensible our choice of values has been we can perform a sensitivity analysis. We assume the following values:

$In[\ ]:=$ **β1 = 100; β2 = 5;**

- The moment $d_0 = t_0$ for taking the first measurement is defined by the user. In this case, we take the measurement at $t_0 = 0.5$.

$In[\ ]:=$ **d₀ = 0.5;**

- Then we can obtain the information matrix (FIM): M = $X^T\ \Sigma^{-1}\ X$.

$In[\ ]:=$ **m = X . Inverse[Σ] . Transpose[X];**

- A D-optimal design will be a design that maximizes det|M|, the determinant of the information matrix. This can be done using **Maximize** as follows (in the next example, we show that usually it is better to use **FindMaximum** or **NMaximize**):

$In[\ ]:=$ **deter1[x_] := Det[m /. Thread[dd → x]];**

$In[\ ]:=$ **sol = Maximize[{deter1[dd], Thread[dd > 1]}, dd]**

$Out[\ ]:=$ $\{0.16647,\ \{d_1 \to 1.48704,\ d_2 \to 6.18228,\ d_3 \to 6.91368\}\}$

- This indicates that the samples should be taken in $\{t_i\}$. The same solution can be directly obtained using Biokmod (Section 10.4.4):

$In[\ ]:=$ **optimumt = tm /. sol⟦2⟧**

$Out[\ ]:=$ $\{0.5,\ 1.98704,\ 8.16931,\ 15.083\}$

- In practice, since it's very difficult to take samples at exactly the precise time, we need to measure the robustness of our calculations: how sensitive they are to small changes. For that purpose, we calculate their efficiency using the following formula:

$\sqrt[N]{\dfrac{Det[m[d]]}{Det[m[max]\ ]}}$ (N is the total number of estimated points).

For example, to measure the efficiency of d_3:

$In[\ ]:=$ **d3 = Table[i, {i, -3 + d₃ /. sol⟦2, 3⟧, 3 + d₃ /. sol⟦2, 3⟧, 0.1 }];**

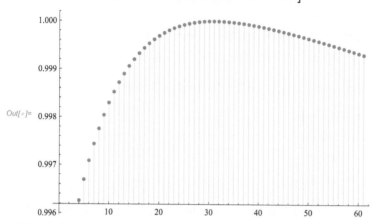

```
In[ ]:= ListPlot[n-1√((Det[m /. Thread[{sol[[2, 1]], sol[[2, 2]], d3 → #}]]) / (Det[m /. sol[[2]]])) & /@ d3,

            Filling → Axis, ColorFunctionScaling → True]
```

- The optimum point is the one with efficiency 1. In our case, this corresponds to **Det**[m[max]], obtained when t:

```
In[ ]:= optimumt[[4]]
```

```
Out[ ]= 15.083
```

- As usual, to avoid any problems in the next section with previously defined variables we quit the session.

10.5.2 Michaelis–Menten Models (M–M) and OED

We will now apply OED to the model described in Section 10.1.4 to choose the best moments $\{t_1, ..., t_i\}\}$ to take the samples. In this case, the problem is that we do not have an exact solution of the model so we cannot obtain an analytical expression for $\{x_1(t, k_{12}, k_{21}), x_2(t, k_{12}, k_{21})\}$. Because of that, we apply the D-optimal method described in the previous section. We want to design an experiment to take the samples in compartment 1.

- We repeat the process discussed in 10.1.4 to obtain $x_1(t, k_{12}, k_{21})$:

```
In[ ]:= Vm = 0.1; km = 0.3;

        eq1 = x1'[t] == -k12 x1[t] + k21 x2[t] - (Vm x1[t]) / (km + x1[t]);

        eq2 = x2'[t] == k12 x1[t] - k21 x2[t];
```

```
In[ ]:= sol = ParametricNDSolve[
            {eq1, eq2, x1[0] == 1, x2[0] == 0}, {x1, x2}, {t, 0, 100}, {k12, k21}] ;
```

- We then proceed to compute numerically $\nabla x_1(t, k_{12}, k_{21}) = \{\frac{\partial x_1(t)}{\partial k_{12}}, \frac{\partial x_1(t)}{\partial k_{21}}\}$. For convenience, we call $k_{12} = a$ and $k_{21} = b$:

```
In[ ]:= fa[a1_?NumberQ, b_?NumberQ, t_?NumberQ] := D[x1[a, b], a][t] /. a → a1 /. sol
```

```
In[ ]:= fb[a_?NumberQ, b1_?NumberQ, t_?NumberQ] := D[x1[a, b], b][t] /. b → b1 /. sol
```

```
In[ ]:= X1[a_, b_, ti_] := {fa[a, b, ti], fb[a, b, ti]}
```

- Next, we define the number of points to be used in the optimal design. We assume a 5-point design:

In[ ]:= `n = 5 - 1;`

- After that, we compute $\Gamma = \{l_{ij}\}$ with $l_{ij} = \exp\{\rho|t_j - t_j|\}$ (remember that $d_i = t_i - t_{i-1}$ is used instead of t_i:

In[ ]:= `dd = Table[d_i, {i, n}]; tt = FoldList[Plus, Subscript[d, 0], dd]`

Out[ ]= $\{d_0, d_0 + d_1, d_0 + d_1 + d_2, d_0 + d_1 + d_2 + d_3, d_0 + d_1 + d_2 + d_3 + d_4\}$

In[ ]:= `ff[i_, j_] := Which[i == j, 1, i < j, e^{-\rho \sum_{k=i}^{j-1} d_k}, i > j, e^{-\rho \sum_{k=j}^{i-1} d_k}];`

In[ ]:= `Γ = Array[ff, {n + 1, n + 1}];`

- Once we know Γ, we compute the covariance matrix $\Sigma = \sigma^2\ \Gamma$. We assume $\rho = 1, \sigma = 1$:

In[ ]:= `ρ = 1; σ = 1; Σ = σ^2 * Γ;`

- The next to last step is the calculation of the information matrix $M = X^T \Sigma^{-1} X$. Initially, we assign arbitrary values to k_{12} and k_{21}. As mentioned in the previous section, their real values are unknown, but we have to make an initial estimation based on the information that we have. In this case we assume $k_{12} = 0.03$ and $k_{21} = 0.02$:

In[ ]:= `m1[ti_] := Transpose[Map[X1[0.03, 0.02, #] &, ti]].`
`            Inverse[Σ].Map[X1[0.03, 0.02, #] &, ti]`

- `Maximize` tries to find a global maximum using analytical methods that may take too long to compute. For this reason, it may be more convenient to use `NMaximize` or `FindMaximum`, *Mathematica* functions that take a vector of numbers as their inputs. In our example, with $n = 4$.

In[ ]:= `obj[d0_?NumericQ, d1_?NumericQ,`
`        d2_?NumericQ, d3_?NumericQ, d4_?NumericQ] :=`
`    Det[m1[{d0, d1 + d0, d0 + d1 + d2, d0 + d1 + d2 + d3, d0 + d1 + d2 + d3 + d4}]]`

- Finally, we find the solution. Remember that a D-optimal design is a design that maximizes det|M|, the determinant of the information matrix. We can show the intermediate calculations with `StepMonitor`:

In[ ]:= `sol1 = FindMaximum[{obj[d_0, d_1, d_2, d_3, d_4],`
`        0 < d_0 < 10, 0 < d_1 < 10, 0 < d_2 < 10, 0 < d_3 < 10, 0 < d_4 < 10},`
`        {{d_0, 3}, {d_1, 3}, {d_2, 3}, {d_3, 3}, {d_4, 3}}, StepMonitor :>`
`        Print[{"d0:", d_0, "d1:", d_1, "d2:", d_2, "d3:", d_3, "d4:", d_4}]]`

  {d0:, 3., d1:, 3., d2:, 3., d3:, 3., d4:, 3.}

  {d0:, 3.96374, d1:, 3.00914, d2:, 3.56722, d3:, 3.59301, d4:, 3.19179}

  {d0:, 4.30936, d1:, 2.84337, d2:, 3.98828, d3:, 3.71813, d4:, 3.38737}

  {d0:, 4.34705, d1:, 2.80134, d2:, 4.16208, d3:, 3.66308, d4:, 3.45106}

  {d0:, 4.34836, d1:, 2.79967, d2:, 4.1764, d3:, 3.65599, d4:, 3.45459}

  {d0:, 4.34836, d1:, 2.79963, d2:, 4.17644, d3:, 3.65596, d4:, 3.45455}

  {d0:, 4.34836, d1:, 2.79963, d2:, 4.17644, d3:, 3.65596, d4:, 3.45455}

Out[ ]= $\{12.8081, \{d_0 \to 4.34836, d_1 \to 2.79963, d_2 \to 4.17644, d_3 \to 3.65596, d_4 \to 3.45455\}\}$

- This indicates that the samples should be taken at $\{t_i\}$:

In[ ]:= **bestsampling = tt /. sol1〚2〛**

Out[ ]= {4.34836, 7.14799, 11.3244, 14.9804, 18.4349}

Sensitivity Analysis of k_{12}, k_{21}

As we have explained before, to solve D-optimal designs we need to give the initial values to the starting points k_{12} and k_{21}. To learn more about how different values for those points would impact the design, we need to perform a sensitivity analysis. One way to proceed would be to calculate the efficiency, as we did in the example of Section 10.3.1, but in this example we will do it using derivatives.

- To analyze the sensibility of x_1 with respect to k_{12}, we keep k_{21} constant at 0.02 and then calculate $\frac{\partial x1(k_{12}, 0.02)}{\partial k_{12}}$, that we call $s(t, k_{12})$, to see how it changes over time. The closer the derivative is to zero, the less sensitive x_1 will be to changes in initial values of k_{12}. We can do the same for k_{21}, calculating in this case $s(t, k_{21}) = \frac{\partial x1(0.03, k_{21})}{\partial k_{21}}$ (Notice that *Mathematica* actually calculates the numerical derivatives of x1[k12, 0.02][t] and x1[0.03, k21][t], since both are numerical expressions).

In[ ]:= **Plot[{ D[x1[k12, 0.02] [t], k12] /. k12 → 0.03 /. sol,**

    D[x1[0.03, k21] [t], k21] /. k21 → 0.02 /. sol},

    {t, 0, 20}, AxesLabel → {"t", "s(t)"},

    PlotLegends → { " $\frac{\partial x1(k_{12}, 0.02)}{\partial k_{12}}$ ", " $\frac{\partial x1(0.03, k_{21})}{\partial k_{21}}$ "}]

- The derivative gives an idea of the absolute sensibility, but we are more interested in the relative sensitivity (rs). This sensibility can be analyzed using the expressions $rs(t, k_{12}) = x_1(0.03, 0.02)(t) \pm \Delta\, s(t, k_{12})$ and $rs(t, k_{12}) = x_1(0.03, 0.02)(t) \pm \Delta\, s(t, k_{21})$. A good choice for Δ is 0.1.

```
In[ ]:= GraphicsRow[{ Plot[
          Evaluate[(x1[0.03, 0.02][t] + {0, .1, -.1} D[x1[k12, 0.02], k12][t] /.
            k12 → 0.03) /. sol], {t, 0, 20},
          Filling → {2 → {3}}, AxesLabel → {"t", "rs(t, k₁₂)"}], Plot[
          Evaluate[(x1[0.03, 0.02][t] + {0, .1, -.1} D[x1[k21, 0.02], k21][t] /.
            k21 → 0.02) /. sol], {t, 0, 20}, Filling → {2 → {3}},
          AxesLabel → {"t", "rs(t, k₂₁)"}]}, ImageSize → {500, 200}]
```

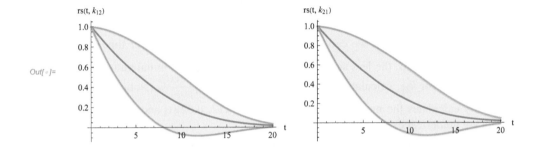

The left graph indicates that the sensitivity of the solution with respect to k_{12} increases with t until $t = 10$. We can see the pattern by looking at the difference in the measurement of rs(t, k_{12}) between the 3 potential values of k_{12} around t = 10. After that, it decreases. Basically, this means that measurements taken at the beginning or at the end of the interval {0, 20} would not change significantly for different values of k_{12}. However, measurements taken around $t = 10$ may show large discrepancies if our initial estimate for the constant was not the correct one. For k_{21}, we obtain a similar conclusion (right graph).

10.6 BIOKMOD: The New Iodine OIR Model (ICRP 137)

```
In[ ]:= ClearAll["Global`*"]
```

In this section, we are going to use the *Mathematica* package SysModel2 included in the application Biokmod.

Biokmod can be downloaded at: http://diarium.usal.es/guillermo/biokmod/

To install it, extract the file "biokmodXX.zip" and copy the folder named Biokmod to the AddOns\Applications directory in the *Mathematica* installation folder ($InstallationDirectory).

- Load the package SysModel2 as usual:

```
In[ ]:= Needs["Biokmod`SysModel2`"]
```

    SysModel, version 2.0.b7   2020–11–30

- Here are the package functions:

$In[\circ]:=$ **? "Biokmod`SysModel2`*"**

⌄ Biokmod`SysModel2`

AcuteInput	IsotopeChainData	Modelfit2	MultiExpInput	qMultiple1
Assayfit	IsotopeChainPlot	Modelfit3	ParametricSystemNDSolve	qMultipleSingle
Catenary	IsotopeDecayModes	ModelfitLogLSM	qConstant	qMultipleUnitCte
CoefMatrix	matrixExp	ModelfitLSM	qContinuous	qMultipleUnitCte1
CompartMatrix	Modelfit	ModelfitX2	qMultiple	qRandom

One of the most important applications of biokinetic models is related to internal dosimetry. The most commonly used dosimetry calculation methods are the MIRD (Medical Internal Radiation Dose) and the ICRP (International Commission on Radiological Protection). Although both are similar, they use different terminology and symbols for fundamental quantities, such as the absorbed fraction, specific absorbed fraction, and various dose coefficients. The MIRD is devoted to the application of internal dosimetry in diagnostic and therapeutic nuclear medicine. The International Commission on Radiological Protection (ICRP) is an organization that also supplies dosimetric models and technical data for use in providing recommendations regarding limits on ionizing radiation exposure to workers and members of the general public. MIRD Pamphlet No. 21 describes a integrated method for MIRD and ICRP.

MIRD Pamphlet No. 21: A Generalized Schema for Radiopharmaceutical Dosimetry—Standardization of Nomenclature. Wesley E. Bolch, Keith F. Eckerman, George Sgouros and Stephen R. Thomas. *Journal of Nuclear Medicine* March 2009, 50 (3) 477-484;

DOI: https://doi.org/10.2967/jnumed.108.056036

Both schema usually require knowledge of the disintegration in a region (or compartment). The isotope contents $x_i(t)$ in a compartment i can be calculated following the methods that we have described previously. Therefore, the number of transformations, in Bqs, at time τ in a source region S following an acute intake is given by the following integral:

$$U_s(\tau) = \int_0^\tau x_i(t)\, dt \tag{10.8}$$

The ICRP has updated the biokinetic models ICRP 130, ICRP 134, ICRP 137, ICRP 141:

https://journals.sagepub.com/doi/suppl/10.1177/ANIB_48_2-3/suppl_file/OIR_Data_Viewer_for_P134-P137-P141.zip

The flow diagram for both, ingestion and injection, is presented in the figure below:

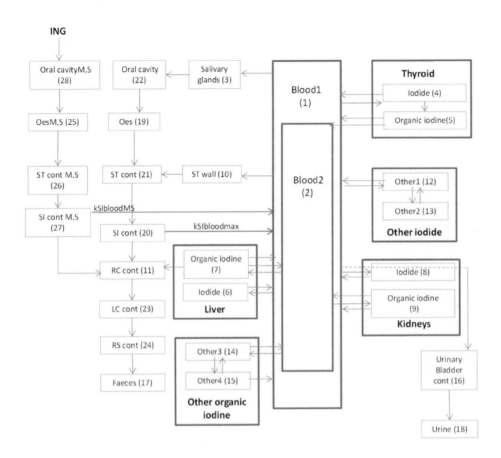

Figure 10.7   New iodine model ICRP 137.

- The transfer coefficients are:

```
In[ ]:= transfercoeffI = {{1, 4, 7.26}, {1, 16, 11.84}, {1, 3, 5.16},
         {1, 10, 8.6}, {1, 12, 600},   {1, 8, 25}, {1, 6, 15},
         {3, 22, 50}, {10, 21, 50}, {4, 5, 95}, {4, 1, 36},
        {5, 2, 0.0077}, {5, 1, 0}, {12, 1, 330},
         {12, 13, 35}, {13, 12, 56}, {8, 1, 100},   {6, 1, 100},
         {2, 14, 15}, {14, 2, 21}, {14, 15, 1.2}, {15, 14, 0.62},
        {15, 1, 0.14}, {2, 9, 3.6}, {9, 2, 21}, {9, 1, 0.14},
         {2, 7, 21}, {7, 2, 21},   {7, 1, 0.14}, {7, 11, 0.08},
         {22, 19, 7200}, {19, 21, 2160}, {21, 20, 20.57},
        {20, 11, 6}, {20, 1, 594}, {16, 18, 12}, {11, 23, 2},
         {23, 24, 2.01}, {24, 17, 2.02}, {25, 26, 2160}, {26, 27, 20.57},
         {27, 1, (6*fA) / (1 - fA)}, {27, 11, 6}, {28, 25, 7200}};
```

where:

f_A is the absorption factor from the alimentary tract. Its value depends on the chemical form of the iodine intake.

λ_I refers to the decay constant. Its value depends on the isotope.

- Then the compartmental matrix as function of f_A and λ_I is:

```
In[ ]:= iodinematrix[fA_, λI_] = CompartMatrix[28, transfercoeffI , λI ];
```

- The package function **AcuteInput** computes the content in each compartment as a function of the time for a single intake at $t = 0$. Alternatively, **SystemDSolve** can also be used.
  AcuteInput[A, IC,t1, x] gives the solution X(t) = {x1(t), ..., xn(t) at time t1 of an ODE system X'(t) = A X(t) , with initial conditions X(0) = IC

In[ ]:= `qIodine[fa_, lambda_, initialcondition_, t_] :=`
       `AcuteInput[iodinematrix[fa, lambda], initialcondition, t, x];`

- The response function for each compartment in case of ingestion at $t = 0$ ($b_i = 0$ with $i \neq 28$ and $b_{28} = 1$) is created as follows:

In[ ]:= `initcondition = PadLeft[{1}, 28]`

Out[ ]= {0, 0, 0, 0, 0, 0, 0, 0, 0, 0, 0, 0, 0, 0, 0, 0, 0, 0, 0, 0, 0, 0, 0, 0, 0, 0, 0, 1}

- If we were dealing with injection instead, the b_1 coefficient would have been 1 and the rest 0.
- Let's consider the case of I131 ingestion with $f_A = 0.99$ (the value recommended by default for ICRP 134)

In[ ]:= `lambdaIodine131 =`

       `QuantityMagnitude[` iodine-131 ISOTOPE ⚬⚬⚬ ✓ `["DecayConstant"], "Per day"]`

Out[ ]= 0.086371

In[ ]:= `iodinecontent =`
       `qIodine[ 0.99, lambdaIodine131, initcondition, t] // ExpandAll // Chop;`

- The $U_s(\tau)$ (number of transformations, in Bqs, at time τ) for 1 Bq in t $= 0$ ($b_{28}=1$) in each compartment is given by the expression below:

In[ ]:= `UsI131[t1_] = 86400 * (qContinuous[1, {#1}, t, t1] & ) /@ Evaluate[`
       `Evaluate[Table[x_i[t], {i, 1, 28}]] /. iodinecontent]] // ExpandAll;`

- We can use the same approach to compute the accumulated disintegrations Us (rounded to the nearest integer) in all compartments for different values of τ (in days).

In[ ]:= `tm = {0.1, 0.5, 1.0, 5, 10, 100, 1000};`

In[ ]:= `nround[x_] := MapAll[If[NumberQ[#], Round[#], #] &, x];`

In[ ]:= `DisintegrationLabel[dis_, tafter_] :=`
       `TableForm[Map[Flatten[#] &, nround[dis]],`
         `TableHeadings → {None, Flatten[{"Compartment", tafter "day"}]}];`

In[ ]:= `dis = Transpose[Join[{Range[28]}, UsI131[#] & /@ tm]] // Quiet;`

In[ ]:= **DisintegrationLabel[dis, tm]**

Out[ ]//TableForm=

Compartment	0.1 day	0.5 day	1. day	5 day	10 day	100 day	1000 day
1	931	3695	4579	4816	4827	4861	4861
2	0	3	16	404	1081	2939	2940
3	73	373	470	496	497	501	501
4	47	203	253	267	267	269	269
5	173	5409	16050	95151	160639	271834	271862
6	123	548	685	722	723	728	728
7	0	2	14	392	1060	2897	2897
8	204	914	1142	1203	1206	1214	1214
9	0	0	2	67	182	498	499
10	122	622	784	827	829	835	835
11	46	335	563	741	767	842	842
12	1592	6682	8315	8752	8773	8834	8834
13	770	4094	5173	5462	5474	5513	5513
14	0	1	10	275	749	2059	2060
15	0	0	3	260	904	2920	2920
16	340	3242	4406	4718	4729	4762	4762
17	0	8	98	3889	8121	17918	17928
18	127	9269	32207	213297	368634	661425	661562
19	2	9	11	11	12	12	12
20	7	77	103	110	110	111	111
21	220	2265	3005	3202	3210	3232	3232
22	1	3	3	3	3	3	3
23	3	117	340	704	729	803	803
24	0	31	165	668	694	767	767
25	40	40	40	40	40	40	40
26	3646	4182	4183	4183	4183	4183	4183
27	124	143	143	143	143	143	143
28	12	12	12	12	12	12	12

- The graph that follows displays the disintegration in the thyroids (5). Note that most of the disintegrations happen during the first few days.

In[ ]:= **LogLogPlot[Evaluate[UsI131[t1] [[5]]], {t1, 0, 100}]**

Out[ ]=

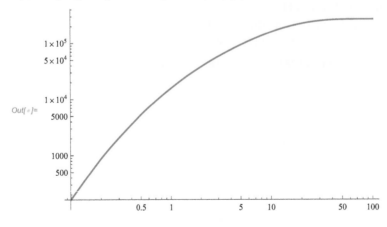

10.7 Additional Modeling Examples

Mathematica includes many more capabilities for modeling using differential equations. In this section, we show some of those capabilities using basic examples.

```
In[ ]:= ClearAll["Global`*"]
```

10.7.1 DEs Using WhenEvent

We can use WhenEvent to solve DEs with variable conditions:

https://reference.wolfram.com/language/ref/WhenEvent.html

- The instructions below simulate and visualize the first 20 seconds of a bouncing ball dropped from a height of 10 m that retains 90% of its velocity in each bounce.

```
In[ ]:= sol1 = NDSolve[{y''[t] == -9.81, y[0] == 10, y'[0] == 0,
          WhenEvent[y[t] == 0, y'[t] → -0.9 y'[t]]}, y, {t, 0, 20}];
```

```
In[ ]:= Plot[y[t] /. sol1, {t, 0, 20}]
```

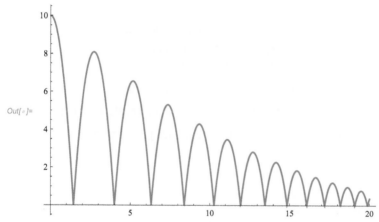

- This next example models a ball falling down a 1-meter-wide staircase from a height of 11 m and an initial velocity of 3 m/s. The first stair is located at h = 10 m and the ball retains 80% of its velocity in each bounce.

```
In[ ]:= sol2 =
      Block[{c = .8}, NDSolve[{y''[t] == -9.81, y[0] == 11, y'[0] == 3, a[0] == 10,
          WhenEvent[y[t] - a[t] == 0, y'[t] → -c y'[t]], WhenEvent[Mod[t, 1],
            a[t] → a[t] - 1]}, {y, a}, {t, 0, 8}, DiscreteVariables → {a}] ];
      Plot[Evaluate[{y[t], a[t]} /. sol2], {t, 0, 8}, Filling → {2 → 0}]
```

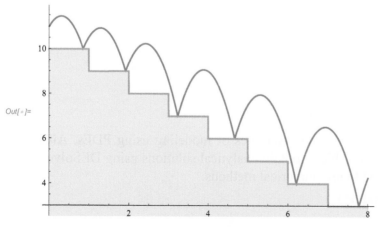

10.7.2 Delay Differential Equations (DDEs)

We can also model systems in *Mathematica* using delay differential equations (DDEs), a type of differential equation in which the derivative of the unknown function at a certain time is given in terms of the values of the function at previous times. DDEs are also called time-delay systems, systems with after-effect or dead-time, hereditary systems, equations with deviating argument, or differential-difference equations.

https://reference.wolfram.com/language/tutorial/NDSolveDelayDifferentialEquations.html

A typical example of these type of models are the Lotka–Volterra systems. These systems model the growth and interaction of animal species assuming that the effect of one species on another is delayed by introducing time lags in the interaction terms.

$Y_1'(t) = Y_1(t)(Y_2(t - \tau_2) - 1)$, $Y_2'(t) = Y_2(t)(2 - Y_1(t - \tau_1))$.

The code below solves and visualizes the system with and without delay:

```
In[*]:= lvsystem[τ1_, τ2_] := {Y₁'[t] == Y₁[t] (Y₂[t - τ1] - 1),
         Y₁[0] == 1, Y₂'[t] == Y₂[t] (2 - Y₁[t - τ2]), Y₂[0] == 1};
```

```
In[*]:= lv = First[NDSolve[lvsystem[0, 0], {Y₁, Y₂}, {t, 0, 25}]];
```

```
In[*]:= lvd = Quiet[First[NDSolve[lvsystem[.01, 0], {Y₁, Y₂}, {t, 0, 25}]]];
```

```
In[*]:= ParametricPlot[Evaluate[{{Y₁[t], Y₂[t]} /. {lv, lvd}], {t, 0, 25}]
```

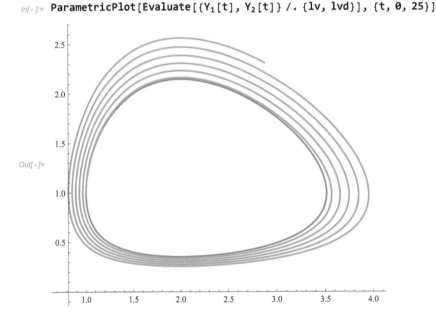

10.8 Modeling Using PDEs

In this section, we explore some examples of modeling using PDEs. Although in certain cases, we may be able to obtain analytical solutions using DESolve, in most situations we will need to use numerical methods.

```
In[*]:= ClearAll["Global`*"]
```

10.8.1 The Transport Equation

.

The transport equation describes the transportation of a scalar quantity in a space. In general, it applies to the transportation of a scalar field (e.g., temperature, material properties or chemical concentration) inside an incompressible fluid. In mathematics, this equation is also known as the convection-diffusion equation, a first-order PDE. This is the equation used in the most common transportation models.

$\frac{\partial u}{\partial x} + c\,\frac{\partial u}{\partial y} = 0$ with c constant

- Here is an example of a first-order linear partial differential equation modeling a wave:

In[ ]:= `u1[x_, t_] = DSolveValue[{D[u[x, t], {t}] + D[u[x, t], {x}] == 0,`
`u[x, 0] == Exp[-x^2] }, u[x, t], {x, t}]`

Out[ ]= $e^{-(x-t)^2}$

- The command below simulates its movement:

In[ ]:= `Animate[Plot[u1[x, t], {x, 0, 10},`
`PlotRange → All, Ticks → False, Filling → Axis],`
`{t, 0, 10}, DefaultDuration → 15, AnimationRepetitions → 1]`

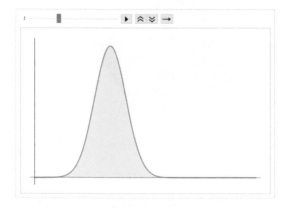

10.8.2 Tsunami Modeling

We can also model tsunamis with the following PDE:

$$\frac{\partial^2 u(t, x)}{\partial t^2} = \frac{\partial^2 u(t, x)}{\partial x^2} + a\,u(t, x)^3 + b\,u(t, x)^2 + c\,u(t, x) + d,$$

$$\text{con } u(0, x) = e^{-x^2}, \quad \frac{\partial u(t, x)}{\partial t} = 0 \; u(t, -x_0) = u(t, x_0)$$

- The example below solves and visualizes the previous equation for a given set of parameters:

```
In[•]:= Manipulate[
        Plot3D[Evaluate[u[t, x] /. Quiet[NDSolve[Evaluate[{D[u[t, x], t, t] ==
                D[u[t, x], x, x] + a u[t, x]^3 + b u[t, x]^2 + c u[t, x] + d,
                u[0, x] == E^-x^2, D[u[t, x], t] == 0 /. t → 0,
                u[t, -x0] == u[t, x0]}], u, {t, 0, x0}, {x, -x0, x0}]]],
         {x, -x0, x0}, {t, 0, x0}, MeshFunctions → {#3 &},
         Mesh → None,
         ColorFunction → "Rainbow",
         PlotPoints → 20],
        {{a, -0.3}, -4, 0},
        {{b, -0.6}, -4, 0},
        {{c, 0.8}, -4, 1},
        {{d, 0}, -1, 1},
        {{x0, 16, "Solution Range"}, 5, 20},
        ControlPlacement → Left]
```

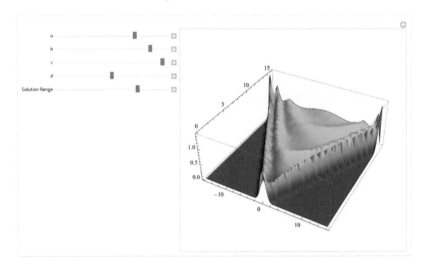

10.8.3 Burgers' Equation

Burgers' equation is a PDE commonly used in applied mathematics when solving problems in areas such as fluid mechanics, nonlinear acoustics, or gas dynamics.

```
In[•]:= ClearAll["Global`*"]
```

$$\frac{\partial u}{\partial t} - \nu \frac{\partial^2 u}{\partial x^2} - u \frac{\partial u}{\partial x} = 0$$

■ Here's an example of the equation:

```
In[•]:= ufun = NDSolveValue[
        {D[u[t, x], t] == 1/100 D[u[t, x], x, x] + u[t, x] × D[u[t, x], x],
         u[t, 0] == 1, u[t, 1] == 1, u[0, x] == Cos[2 Pi x]}, u, {t, 0, 2},
         {x, 0, 1}, Method → {"MethodOfLines", "SpatialDiscretization" →
             {"FiniteElement", "MeshOptions" → {"MaxCellMeasure" → 0.01}}}];
```

```
In[•]:= list = Table[Plot[ufun[t, x], {x, 0, 1},
         PlotRange → {-1.1, 1.1}, ImageSize → Large], {t, 0, 2, 1/20}];
        ListAnimate[list, SaveDefinitions → True]
```

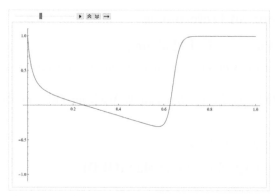

Notice the use of the finite element method (FEM) to solve the equation. To learn more about it, the "Finite Element User Guide" provides an overview of how to solve PDEs using this method:
https://reference.wolfram.com/language/FEMDocumentation/tutorial/FiniteElementOvervie w.html

10.9 *System Modeler*

For more advanced modeling needs, we can use *System Modeler*, a program for modeling and simulating physical systems: https://www.wolfram.com/system-modeler/.

It is a full graphical modeling and simulation environment where by using drag and drop, you can build realistic, multi-domain models in areas as diverse as aerospace, industrial manufacturing, and life sciences.

Fortunately, we can explore some of the software's capabilities from within *Mathematica* using functions such as SystemModelExamples and SystemModelPlot:

- We first select the desired example and then plot its behavior:

In[ ]:= **SystemModelExamples[]**

In[ ]:= **SystemModelPlot["IntroductoryExamples.Systems.InvertedPendulum"]**

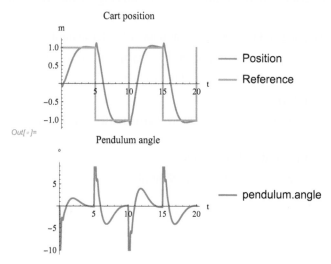

10.10 Additional Resources and references

About Compartmental and Physiological Modeling

J.A. Jacquez, *Compartmental Analysis in Biology and Medicine*, The University of Michigan Press, 1985.

K. Godfrey, *Compartmental Models and Their Application,* Academic Press, London, 1983.

About Epidemic Modeling

https://community.wolfram.com/groups/-/m/t/1872608

About Biokmod and Optimal Experimental Design (OED)

Visit the author's website:

https://diarium.usal.es/guillermo/publicaciones/especializadas/

Biokmod can be downloaded from: http://diarium.usal.es/guillermo/biokmod. Extensive documentation is included.

Some of its features can be run directly over the Internet with the help of *webMathematica*: http://oed.usal.es/webMathematica/Biokmod/index.html.

J. Lopez-Fidalgo, J.M. Rodriguez-Diaz, J. M. Sanchez, G. Santos-Martin, MT, Optimal designs for compartmental models with correlated observations, *Journal of Applied Statistics*: 32(10) pp. 1075-1088, 2005.

J. M. Sanchez, J. M. Rodríguez-Díaz, Optimal design and mathematical model applied to establish bioassay programs, *Radiation Protection Dosimetry,* doi:10.1093/rpd/ncl499, 2007.

J. M. Rodríguez-Díaz, G. Sánchez-León, Design optimality for models defined by a system of ordinary differential equations, *Biometrical Journal* 56 (5), pp. 886-900, 2014.

About Differential Equations

The general guide for DifferentialEquations:

https://reference.wolfram.com/language/guide/DifferentialEquations.html

Working with DSolve: A User's Guide:

https://reference.wolframcloud.com/language/tutorial/DSolveWorkingWithDSolve.html

Symbolic Differential Equation Solving:

https://reference.wolframcloud.com/language/tutorial/DSolveOverview.html

Introduction to Advanced Numerical Differential Equation Solving in the Wolfram Language:

https://reference.wolfram.com/language/tutorial/NDSolveIntroductoryTutorial.html

Advanced Numerical Differential Equation Solving in the Wolfram Language:

https://reference.wolfram.com/language/tutorial/NDSolveOverview.html

Partial Differential Equations (PDEs):

https://reference.wolframcloud.com/language/tutorial/DSolvePartialDifferentialEquations.html

Numerical Solutions of Partial Differential Equations:

http://reference.wolfram.com/language/tutorial/NDSolvePDE.html

Economic, Financial, and Optimization Applications

This chapter explains Mathematica's functionality related to economics and finance. It covers how to access economic information, price financial instruments, such as bonds and derivates, and access both, historical and real-time financial data. It also discusses constrained optimization using mainly examples related to operations research and investment portfolio management. **FindShortestTour**, a new function to solve problems like the Traveling Salesman Problem in a very efficient way, is shown in the context of creating a tour of South American cities. The last section provides a very short introduction to blockchains and cryptocurrencies, including a brief overview of the program's capabilities in these two areas.

11.1 Accessing Economic Information

One of the most useful features available in *Mathematica* is the free-form input, enabling us to directly access the functionality of **Wolfram|Alpha**). Here are some economic-related examples:

- We can use the free-form input to add different currencies using exchanges rates that are at most a few minutes old. For real-time calculations, we would need access to the Wolfram Finance platform.

In[ ]:=

Out[ ]= $60.39

- Type **Japan vs. Germany** using the **Wolfram|Alpha Query** style:

In[ ]:=
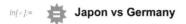

Output not shown

- Several frames (pods) will be displayed, each of them containing information about an aspect of the question. In the right corner of each frame, the symbol ⊞ appears. Click on one of them and you will see a menu with several options. In this example, we chose the **Economic Properties** frame and clicked on the **Subpod content**. The following entry was generated afterward:

In[ ]:= `WolframAlpha["Japon vs Germany",`
`    {{"EconomicProperties:WorldDevelopmentData", 1}, "Content"}]`

Out[ ]=

	Japan	Germany
GDP	$5.058 trillion per year (world rank: 3rd) (2020)	$3.846 trillion per year (world rank: 4th) (2020)
GDP at parity	$5.334 trillion per year (world rank: 4th) (2020)	$4.561 trillion per year (world rank: 5th) (2020)
real GDP	$4.381 trillion per year (world rank: 3rd) (2020)	$3.436 trillion per year (world rank: 4th) (2020)
GDP per capita	$40 193 per year per person (world rank: 34th) (2020)	$46 253 per year per person (world rank: 26th) (2020)
GDP real growth	−4.586% per year (world rank: 133rd) (2020)	−4.57% per year (world rank: 132nd) (2020)
Gini index	0.329 (world rank: 47th) (2013)	0.317 (world rank: 36th) (2018)
consumer price inflation	−0.23% per year (world rank: 182nd) (2021)	+3.14% per year (world rank: 96th) (2021)

- We repeat the same steps for "China vs US" but in this case instead of **Subpod content**, we choose **ComputableData**. The output is shown below. The advantage of this choice is that the output is being generated as a list allowing easy subsequent manipulation.

In[ ]:= **WolframAlpha["China vs US",**
        {{"EconomicProperties:WorldDevelopmentData", 1}, "ComputableData"}]

Out[ ]= $\Big\{$ {, China, United States},

   $\Big\{$GDP, \$1.47227×10^{13} per year , \$2.0953×10^{13} per year $\Big\}$,

   $\Big\{$GDP at parity, \$2.42832×10^{13} per year , \$2.0953×10^{13} per year $\Big\}$,

   $\Big\{$real GDP, \$1.46318×10^{13} per year , \$1.92945×10^{13} per year $\Big\}$,

   $\Big\{$GDP per capita, \$10430. per person per year ,

    \$63210. per person per year $\Big\}$,

   $\Big\{$GDP real growth, 2.348% per year , -3.405% per year $\Big\}$,

   {Gini index, 0.382, 0.415},

   $\Big\{$consumer price inflation, 0.98% per year , 4.7% per year $\Big\}\Big\}$

> In the **WolframAlpha** style you can select the information you are interested in by choosing from the menu that appears after clicking on **+** ▶ **Subpod content** or **ComputableData** in the desired frame. An *input cell* will be generated to get only the desired information.

■ The graph below shows the evolution of the gross domestic product, GDP, of China from 1980 to 2021.

In[ ]:= **DateListPlot[CountryData["China", {{"GDP"}, {1980, 2021}}]]**

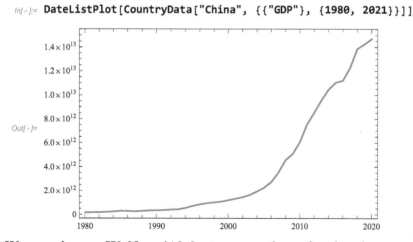

We can also use **Wolfram|Alpha** to access data related to demographics, and make predictions.

■ Here the free-form input is used to import data. In this example, we use it to forecast China's population.

In[ ]:=

Out[ ]:= TimeSeries [⊞ 〰 Time: 01 Jan 1980 to 01 Jan 2022
                              Data points: 43]

- Time series data can be fitted using TimeSeriesModelFit.

In[ ]:= `ts = TimeSeriesModelFit[%]`

Out[ ]:= TimeSeriesModel [⊞ 〰 Family: ARIMA
                                 Order: {2, 3, 0}]

- Now, the evolution of China's population over the next 10 years can be estimated using TimeSeriesForecast.

In[ ]:= `DateListPlot[{ts["TemporalData"], TimeSeriesForecast[ts, {10}]},`
        `PlotLegends → {"Historical", "Forecast"},`
        `PlotLabel → Style["Chinese Population", Bold]]`

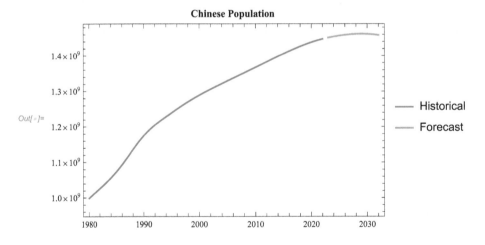

11.2 Financial Information

11.2.1 FinancialData

The `FinancialData` function in many occasions is the best way to access finance-related information. Although in theory the program can display thousands of indicators, in reality the accessible ones are mostly those related to the US markets. In any case, regardless of whether the information may or may not be directly accessible from `FinancialData`, nowadays it would be easy to access it. For example, many bank platforms enable their clients to download data in an Excel format that could then be imported into *Mathematica* and analyzed using any of the multiple functions related to finance:

Those readers who may need reliable access to other markets or real-time financial data would probably be interested in taking a look at the Wolfram Finance Platform, a product from Wolfram Research (the developers of *Mathematica*) that focuses on finance:

http://www.wolfram.com/finance-platform/

FinancialData[*"name","property"*,{*start,end,interval*}] gives the value of the specified property for the financial entity "name" (it could be an index, a stock, a commodity, etc.) for a given period of time (by default it returns the most recent value). We recommend reading the documentation pages to know more about this function's capabilities.

- The symbols ("name") used are the ones available in http://finance.yahoo.com/ since Yahoo Finance supplies the information to the function. In the website, we can find that the symbol for the Standard & Poor's 500 index, that includes the 500 largest public companies in terms of market capitalization that trade in either NYSE or NASDAQ is: ^GSPC (we can also use SP500). We can check it as follows:

In[ ]:= **FinancialData["^GSPC", "Name"]**

Out[ ]= **S&P 500 Index**

- If we want to see the index value (with a few minutes' delay) we can type:

In[ ]:= **FinancialData["SP500"]**

Out[ ]= **3677.95**

- We can also access historical data. In this example, we show the trajectory of the S&P 500 index.

In[ ]:= **DateListPlot[FinancialData["^GSPC",**
         {"January 1 2000", "September 30 2022"}], Joined → True]

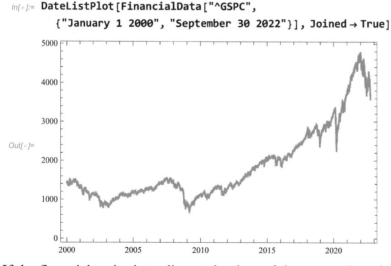

If the financial entity is trading at the time of the query, the price that we get is the one from a few minutes earlier (usually the delay is 15 minutes). The access to information in "real time" requires additional payments to 3rd-party providers such as Bloomberg or Reuters. There are even people willing to pay very large amounts of money to gain an advantage of a few milliseconds when trading:

http://adtmag.com/articles/2011/07/29/why-hft-programmers-earn-top-salaries.aspx.

■ The following function will dynamically update the S&P500 index (as long as the command is executed during trading hours):

```
In[•]:= sp500 := FinancialData["SP500"]
```

```
In[•]:= FinancialData["SP500"]
```

```
Out[•]= 3677.95
```

■ The output changes every two seconds:

```
In[•]:= Dynamic[sp500, UpdateInterval → 2]
```

```
Out[•]= sp500
```

■ We can also represent it in a self-updating graph:

```
data = {};
Dynamic[Last[AppendTo[data, sp500]],
 UpdateInterval → 2, TrackedSymbols → {}]

Dynamic[DateListPlot[data, Joined → True], SynchronousUpdating → False]
```

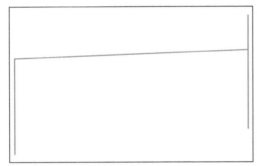

■ As mentioned earlier, if the entity is not available in FinancialData we may still be able to import the information from specialized websites. An interesting one is Nasdaq Data Link (formerly known as Quandl), https://data.nasdaq.com/, that provides access to numerous free and premium economic and financial databases. We can first select the data and download it in our chosen format: Excel, CSV or any other. Then we can use Import to analyze them using *Mathematica*. Alternatively, we can access the data directly using Import or URLFetch.

11.2.2 Visualization

The number of specialized graphs for economic and financial visualization has increased significantly in the latest versions of *Mathematica*.

■ All the functions below end in **Chart**. We have already seen some of them in Chapter 6. In this section, we're going to cover the ones normally used for analyzing stock prices.

```
In[•]:= ? *Chart
```

∨ System`					
BarChart	CandlestickChart	InteractiveTradingChart	PairedBarChart	RectangleChart	TradingChart
BoxWhiskerChart	DistributionChart	KagiChart	PieChart	RenkoChart	
BubbleChart	GeoBubbleChart	LineBreakChart	PointFigureChart	SectorChart	

Choose any of them and click on it to access its documentation.

These graphs complement the capabilities of FinancialData but they can also be used with other functions as shown below:

- Let's visualize the historical closing prices for the Google using different graphs.

In[ ]:= **data = FinancialData["GOOGL", "Close", {{2021, 1, 1}, {2022, 9, 30}}];**

- If we click on any point inside the plot, we'll see the information associated to it.

In[ ]:= **GraphicsColumn[**
{KagiChart[data], PointFigureChart[data], LineBreakChart[data]}]

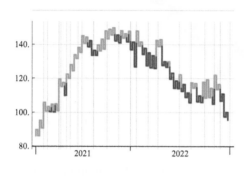

Out[ ]=

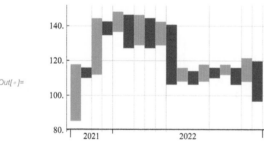

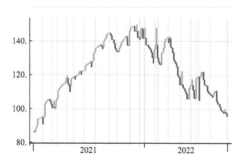

- The next graph uses the function RenkoChart with EventLabels and Placed to add symbols ("→" y "↓") indicating in what position they must be placed.

In[ ]:= `RenkoChart[{"NYSE:GE", {{2020, 01, 1}, {2021, 12, 31}}},`
        `EventLabels → {{2020, 01, 30} → Placed["→", Before],`
        `{2020, 05, 31} → Placed["↓", Above]}]`

Out[ ]=

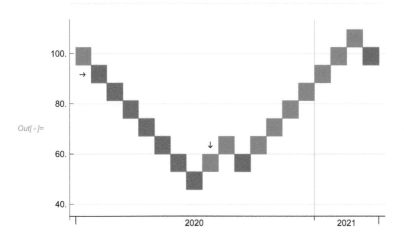

- The graph below displays the trajectory of GE in the NYSE from January 1, 2021 to December 31, 2021. For each trading day, we can see the open, high, low, and close prices.

In[ ]:= `CandlestickChart[{"NYSE:GE", {{2021, 1, 1}, {2021, 12, 31}}}]`

Out[ ]=

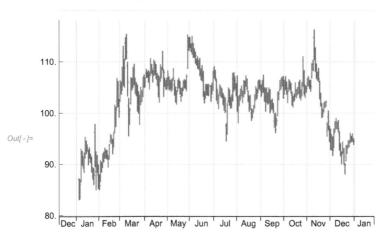

- ChartElementFunction is an option that when added to BarChart or PieChart3D, improves the presentation of the data.

In[ ]:= `GraphicsRow[{BarChart[{{1, 5, 2}, {4, 5, 7}},`
        `ChartElementFunction → "GradientScaleRectangle"],`
        `PieChart3D[{3, 2, 2, 4, 1}, ChartElementFunction → "TorusSector3D",`
        `SectorOrigin → {Automatic, 1}, ChartStyle → "Rainbow"]}, Frame → All]`

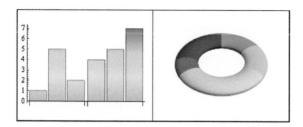

In the world of professional stock market investors, there's a group that uses chart analysis as its main tool for making investment decisions. Its members are known as chartists and their techniques are known as technical analysis. They closely study the past performance of stocks or indices to find trends and support and resistance levels. These people associate certain graph shapes to future markets ups and downs. Although there's no sound mathematical theory justifying their predictions, all economic newspapers have a section dedicated to this type of analysis.

- The powerful function InteractiveTradingChart, includes practically all the functionality that a chartist may require. In this example, we display GE's share price history for a given period. Notice that in the upper area of the graph, we can see a trimester at a time and using the bottom part we can move through the entire period. We can also choose the time interval (days, weeks or months) and even what chart type and indicators to show.

In[ ]:= **InteractiveTradingChart[{"NYSE:GE", {{2020, 01, 1}, {2022, 05, 31}}}]**

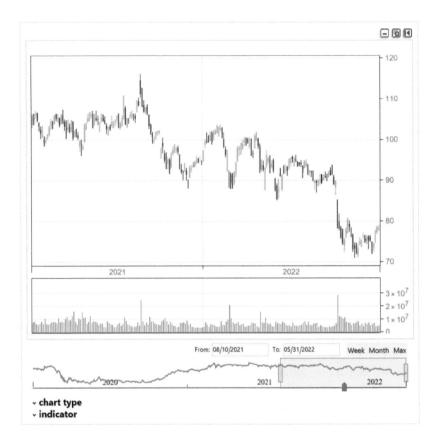

11.2.3 Automatic Adjustments

With *Mathematica* we can automatically fit our financial data to a given distribution (by default the program will use the maximum likelihood method) with the function EstimatedDistribution.

- Let's fit Google's share price to a lognormal distribution after downloading the data.

In[ ]:= `googleStock =`
        `FinancialData["GOOG", {{2015, 1, 1}, {2022, 09, 30}, "Day"}, "Value"];`

In[ ]:= `dist = QuantityMagnitude[`
        `EstimatedDistribution[googleStock, LogNormalDistribution[μ, σ]]]`

Out[ ]= `LogNormalDistribution[4.08435, 0.483913]`

- The fit is poor on both tails. As a matter of fact, all the efforts that have been made trying to predict financial markets have produced unsatisfactory results. These markets exhibit what Mandelbrot (the father of fractal geometry) called "wild randomness" (single observations can impact the total in a very disproportionate way).

In[ ]:= **QuantilePlot[googleStock, dist, Filling → Automatic, FillingStyle → Red]**

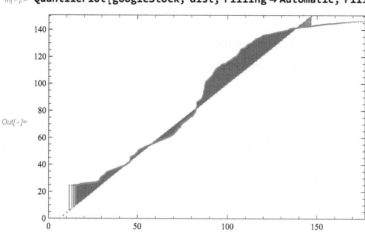

Out[ ]=

11.3 Financial Functions

The functions related to financial calculations can be found in guide/Finance. We will give an overview of some of them in this section.

11.3.1 Time Value Computations

```
ClearAll["Global`*"]
```

The following functions are included under the *Time value Computations* category: TimeValue, EffectiveInterest, Annuity, AnnuityDue and Cashflow. Let's see some examples.

TimeValue["s", "i", "t"] calculates the time value of a security s at time t for an interest specified by i.

- If we invest an amount c during t years at an annual interest rate of i, to find the future value *Mathematica* will apply the following formula:

In[ ]:= **TimeValue[c, i, t]**

Out[ ]= $c\,(i+1)^t$

- How much do we need to invest at 3.5% annual interest rate to get €100,000 in 10 years?

In[ ]:= **Solve[TimeValue[quantity, .035, 10] == 100000, quantity]**

Out[ ]= {{quantity → 70 891.9}}

- We have two options to invest €60,000 for 10 years. Option A offers a fixed annual interest rate of 4% during the 10 years. Option B offers an interest rate of 3% during years 1 and 2, 4 % in years 3 to 5, and 4.5% until the end of the investment period. Which option is the best one?

In[ ]:= **{TimeValue[60 000, 0.04, 10] (*product A*),**
  TimeValue[60000, {{0, 0.03}, {2, 0.04}, {5, 0.045}}, 10] (*product B*)}

Out[ ]= {88 814.7, 89 229.2}

The function EffectiveInterest["r", "q"] returns the effective interest rate corresponding to an interest specification r, compounded at time intervals q. It is very useful to avoid making mistakes when comparing nominal and effective interest rates. `Annuity["p", "t"]` represents an annuity of fixed payments *p* made over *t* periods. Below we show examples using both functions.

- What is the effective rate of an annual nominal interest rate of 4% compounded quarterly?

In[∘]:= `EffectiveInterest[.04, 1/4]`

Out[∘]= 0.040604

- We'd like to deposit €60,000 in an account with an annual interest of 4% compounded quarterly. How much money will we be able to withdraw every quarter so the money lasts for 10 years?

In[∘]:= `Solve[TimeValue[Annuity[{wd, {-60000}}], 10, 1/4],`
    `EffectiveInterest[.04, 1/4], 10] == 0, wd]`

Out[∘]= {{wd → 1827.34}}

- The next set of functions creates a loan amortization table for a given period and interest rate. In this case, we borrow €150,062 and make monthly payments of €1,666 during 10 years at 6% nominal annual rate. The same approach can be used with different numbers:

In[∘]:= `years = 10; monthlypayments = 1666; monthlyinterest = 0.06/12;`

In[∘]:= `loanValue[month_] = TimeValue[`
    `Annuity[monthlypayments, years*12 - month], monthlyinterest, 0];`

In[∘]:= `amortizationList =`
    `Table[{i + 1, loanValue[i], monthlypayments, loanValue[i] *`
        `monthlyinterest, monthlypayments - loanValue[i] *monthlyinterest,`
        `loanValue[i + 1]}, {i, 0, years*12 - 1}] // Round;`

In[∘]:= `amortizationTable[month1_, month2_] :=`
    `TableForm[amortizationList[[month1 ;; month2]], TableHeadings →`
        `{None, {"Month", "Beginning\nPrincipal", "Monthly\nPayment",`
           `"Interest\nPayment", "Principal\nPayment", "Ending\nPrincipal"}}]`

- Let's see the amortization table during the first 12 months. Try to display the entire period of the loan.

In[∘]:= `amortizationTable[1, 12]`

Out[∘]//TableForm=

Month	Beginning Principal	Monthly Payment	Interest Payment	Principal Payment	Ending Principal
1	150 062	1666	750	916	149 147
2	149 147	1666	746	920	148 226
3	148 226	1666	741	925	147 302
4	147 302	1666	737	929	146 372
5	146 372	1666	732	934	145 438
6	145 438	1666	727	939	144 499
7	144 499	1666	722	944	143 556
8	143 556	1666	718	948	142 607
9	142 607	1666	713	953	141 654
10	141 654	1666	708	958	140 697
11	140 697	1666	703	963	139 734
12	139 734	1666	699	967	138 767

- We can check graphically how the remaining principal goes down over time until it becomes 0 at the end of the loan:

```
In[ ]:= ListLinePlot[Table[{i + 1, loanValue[i]}, {i, 0, years * 12 - 1}],
        AxesLabel → {"Month", "Remaining principal"}, ImageSize → 300]
```

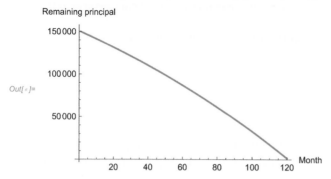

11.3.2 Bonds

```
ClearAll["Global`*"]
```

Bonds are financial assets that we can also analyze with *Mathematica*. When an organization needs capital, instead of borrowing from banks, they can issue bonds. Bonds can be considered a type of loan in which the lenders are the ones purchasing them. Those lenders can usually be anyone. There are different types of bonds. Normally they pay a fixed amount of interest at predetermined time intervals: annually, half-annually, quarterly or even monthly. Some bonds issued for short periods of time (less than one year), such as *Treasury Bills*, don't pay interest explicitly, they are sold at a discount from their par value.

For analyzing bonds, *Mathematica* has the function FinancialBond. It takes the following arguments (some of them are optional):

"FaceValue"	face value, par value
"Coupon"	coupon rate, payment function
"Maturity"	maturity or call/put date
"CouponInterval"	coupon payment interval
"RedemptionValue"	redemption value
"InterestRate"	yield to maturity or yield rate
"Settlement"	settlement date
"DayCountBasis"	day count convention

Let's see some examples.

- The yield to maturity of a €1,000 30-year bond maturing on June 31, 2018, with a coupon of 6% paid quarterly that was purchased on January 6, 2012 for €900 would be:

```
In[ ]:= FindRoot[FinancialBond[{"FaceValue" → 1000, "Coupon" → 0.05,

          "Maturity" → {2018, 6, 31}, "CouponInterval" → 1/4},

          {"InterestRate" → y, "Settlement" → {2012, 1, 6}}] == 900, {y, .1}]
```

```
Out[ ]= {y → 0.069269}
```

- A zero-coupon bond with a redemption value of €1,000 after 10 years is currently sold for €400. Find the implied yield rate compounded semiannually:

```
In[ ]:= FindRoot[FinancialBond[

            {"FaceValue" → 1000, "Coupon" → 0, "Maturity" → 10, "CouponInterval" → 1/2},

            {"InterestRate" → y, "Settlement" → 0}] == 400, {y, .1}]

Out[ ]= {y → 0.0937605}
```

11.3.3 Derivatives

```
In[ ]:= Clear["Global`*"]
```

In Chapter 41 of the book of Genesis, the Egyptian Pharaoh asks Joseph to
interpret a dream he's had about 7 sleek and fat cows followed by 7 ugly and gaunt
ones. Joseph interprets the dream as periods of plenty followed by periods of
famine and gives the Pharaoh the following piece of advice: "... Let Pharaoh
appoint commissioners over the land to take a fifth of the harvest of Egypt during
the seven years of abundance. They should collect all the food of these good years
that are coming and store up the grain under the authority of Pharaoh, to be kept in
the cities for food. This food should be held in reserve for the country, to be used
during the seven years of famine that will come upon Egypt, so that the country
may not be ruined by the famine". Some people consider this story as the earliest
mention of the derivative concept based on the idea of fixing in advance the future
price of a certain item. Closer to our current era, on April 3, 1848, the *Chicago
Board of Trade* was established to create a marketplace for the buying and selling
of several commodities (grains, cattle, hogs, etc.) at fixed prices and following
certain standards. Normally, the buyer would pay in advance a small portion of the
total amount to the seller. The purpose was clear: the buyer could set in advance
how much things were going to cost in the future and the seller could secure a
certain selling price; it could be considered as an insurance against uncertainty.
Since then, this idea has been applied to many other areas: raw materials, shares,
stock indexes, etc.

In finance, a derivative product is one whose price depends on the price of another
product (the underlying asset). We say that the price of the former derives from the
price of the latter. For example, we can create a derivative based on the price of
gold without the need to physically own the gold. In principle, these instruments
should play a beneficial economic role. Very often, however, people associate
them with speculative transactions more similar to gambling than to uncertainty
reduction. There's a kernel of truth is this opinion since investments in these
products can generate either large losses or large gains. This is the reason why
there have always been investors trying to (and sometimes succeeding) influence
the price of the underlying for speculative purposes. Warren Buffett, the most
successful investor of the 20th century and one of world's richest men, even
described them as "weapons of mass destruction" although it's known that some of
his investments take advantage of them, and that if used properly they help to
reduce uncertainty.

When considering derivatives as an alternative to purchasing stocks, the two most
important things to keep in mind are: i) For the same number of shares, derivatives

require a much smaller investment. As a consequence, gains and losses are significantly higher; ii) when the share price goes up, everybody wins (more precisely, the latent value is the same for everybody). With derivatives, when someone wins somebody else loses and the other way around (it's a zero-sum game).

An example of a financial derivative is a stock option, or a well-known variant of it more commonly available to regular investors, a warrant. An option or warrant gives the owner the right to buy (*Call*) or sell (*Put*) the underlying asset at a given price usually determined at the beginning of the contract. Both have prices (premiums) that are normally much smaller than the stock associated to them. For example, we could buy a warrant to purchase Telefónica shares in a year for €15 each and pay €0.5 for it. At expiration, if the shares were trading above €15, let's say €16.5, we would make a profit of €1 per share, after subtracting the €0.5 initially paid for it (it's common to just settle the differences, without actual delivery of the underlying asset). However, if at expiration Telefónica was trading below €15, we would just lose the premium paid.

An informative overview of financial derivatives and their statistical measurement can be found in:

https://www.imf.org/external/pubs/ft/wp/wp9824.pdf.

The list below contains some of the most commonly used terms when describing derivatives transactions:

Underlying asset: the financial instrument on which a derivative's price is based.
Call option. An agreement that gives the buyer the right (but not the obligation) to buy the underlying asset at a certain time and for a certain price from the seller.
Put option. An agreement that gives the buyer the right (but not the obligation) to sell the underlying asset at a certain time and for a certain price to the seller.
Premium. The amount paid for the right to buy or sell the underlying asset.
Market price. The current price at which the underlying asset can be bought or sold.
Strike. The price at which the agreement can be exercised.

The Greek letters delta, gamma, theta, and vega are used to describe the risk of changes in the underlying asset.

Delta: It's the option sensitivity to changes in the underlying asset. It indicates how the price of the option changes when the underlying price goes up or down. For an option at the money (strike price same as market price), delta is usually around 50%. This means that if the price of the underlying asset changes by €2, the price of the option will change by €1.
Gamma: It measures the sensitivity of delta to price changes in the underlying asset. Since delta is not a linear function, gamma is not linear either. Gamma decreases as the uncertainty increases.
Vega: This metric indicates how sensitive the option price is to volatility changes in the underlying asset. It represents the change in the option price if the volatility of the underlying asset changes by 1%. When volatility goes up, the price of the option also goes up. Vega may vary depending on whether the option is at the money, in the money or out of the money.
Theta: A measure of the decrease in the value of an option as a result of the passage of time (easy to remember since both start with t). The longer the time period until exercise, the higher Theta is: There's more time for the underlying asset's price to move in a direction favorable for the option holder.

Mathematica gives us a very wide array of options when using FinancialDerivative. This function returns either the value of the specified financial instrument (e.g., an American call) , or certain specified properties, such as the implied volatility or the Greeks.

To make the valuation examples in this section as realistic as possible we will need certain market data.

Often, to make the purchase of an option cheaper, we can sell another one in such a way that we limit our potential gains but we also reduce the premium paid. This is the idea behind *zero cost* strategies.

An example of this strategy would be as follows: We purchase an option with a strike price equal to the exercise price (an *at the money* option) and sell another one with a higher strike price, let's suppose 18% higher. Then, this value will be our maximum return. Obviously, the higher the potential return, the higher the potential risk and the other way around. This strategy is known as a *Call Spread*.

- In this case we use Telefónica shares trading in the NYSE (the data and functions used in the example come from Sebastián Miranda from Solventis).

In[ ]:= `FinancialData["NYSE:TEF", "Name"]`

Out[ ]= Telefónica

- We're going to need the 12-month EURIBOR rate. This type of interest rate is published by Euribor EBF (http://euribor-rates.eu). Let's import the 10 most recent rates (remember that in Chapter 2 we explained how to extract specific parts of a web page) as follows:

In[ ]:= `Import["http://www.euribor-rates.eu/euribor-rate-12-months.asp", "Data"][[`
`    3, 2, 1]]`

Out[ ]= $\left(\begin{array}{lllll} \{4/14/2022, 0.003\ \%\} & \{4/13/2022, -0.014\ \%\} & \{4/12/2022, 0.005\ \%\} & \{4/11/2022, -0.030\ \%\} & \{4/8/2 \\ \{4/1/2022, -0.086\ \%\} & \{3/1/2022, -0.363\ \%\} & \{2/1/2022, -0.431\ \%\} & \{1/3/2022, -0.499\ \%\} & \{12/1/2 \\ \{1/3/2022, -0.499\ \%\} & \{1/4/2021, -0.502\ \%\} & \{1/2/2020, -0.248\ \%\} & \{1/2/2019, -0.121\ \%\} & \{1/2/2 \end{array}\right)$

- However, unless we're creating a program to automate the process, probably the easiest thing to do is to copy the values directly. Besides, it's unusual for websites to keep their structure for a long time so, by the time you read these lines, the instruction above may not work. We'll store in memory the first rate that we will use for our calculations later on.

In[ ]:= `euribor12m = 0.003 / 100`

Out[ ]= `0.00003`

- For our computations, we need to know several indicators associated to the stock, starting with its price (Figure 11.1). The values returned are time sensitive and depend on the time of execution of the query. We show the commands below so readers can replicate the example.

In[ ]:= `FinancialData["NYSE:TEF", "Price"] (* stock price,`
`    in USD, in NYSE at the time of the query*)`

Out[ ]= `$3.26`

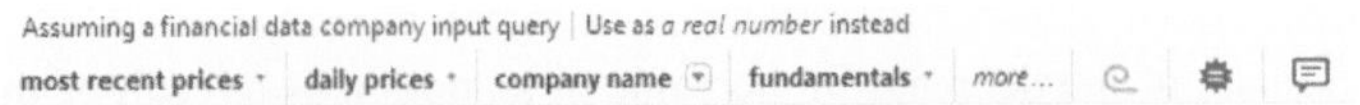

Assuming a financial data company input query | Use as a *real number* instead

most recent prices · | daily prices · | company name ▾ | fundamentals · | *more...* | ⊙ | ✿ | ⊟

Figure 11.1   Predictive interface for financial data.

In[ ]:= `spotPrice = 3.36;`

- Stock volatility during the past 50 days. (Volatility is a measure of the variability of the stock price over a specific time horizon. It's usually the standard deviation of the closing stock price during the last 50 trading sessions, or an alternative chosen time horizon):

In[ ]:= `FinancialData["NYSE:TEF", "Volatility50Day"]`

Out[ ]= `28.1187%`

In[ ]:= `vol = 0.3069;`

- Last year's dividend yield, computed as the ratio between the dividends paid and the current stock price.

In[ ]:= `FinancialData["NYSE:TEF", "DividendYield"]`

Out[ ]= `Missing[NotAvailable]`

In[ ]:= `div = 0.0;`

- Now we can evaluate the price of a call option given its expiration date (in this case, in one year), and using the current stock price as the strike price:

In[ ]:=
```
FinancialDerivative[{"American", "Call"},
    {"StrikePrice" → spotPrice, "Expiration" → 1},
    {"InterestRate" → euribor12m , "Volatility" → vol,
     "CurrentPrice" → spotPrice, "Dividend" → div}]
```

Out[ ]= `0.399738`

- In the next graph, we can see the evolution of the value of our call spread as a function of the underlying asset price and the remaining time. To make it easier to visualize, the solid green line indicates the current price, our strike price; and the dashed red line, the price of the underlying asset at which we maximize our profit. We need to keep in mind that our starting point in the time scale is 1. Remember that our goal is to have a maximum gain of 18%.

In[ ]:=
```
data = {#, 1 - #2, FinancialDerivative[{"American", "Call"},
            {"StrikePrice" → spotPrice, "Expiration" → 1},
            {"InterestRate" → euribor12m , "Volatility" → 0.5,
             "CurrentPrice" → #, "Dividend" → div, "ReferenceTime" → #2}] -
          FinancialDerivative[{"American", "Call"},
            {"StrikePrice" → 1.18 * spotPrice, "Expiration" → 1},
            {"InterestRate" → euribor12m , "Volatility" → 0.5,
             "CurrentPrice" → #, "Dividend" → div, "ReferenceTime" → #2}]} & @@@
       Flatten[Table[{price, referencetime}, {referencetime, 0, 1, 0.05},
         {price, 0.8 spotPrice, 1.4 * spotPrice, 1}], 1];
```

```
In[ ]:= ListPlot3D[data,
    (* Contours *)
    Mesh → {{{spotPrice, {Green, Thick}},
        {1.18 * spotPrice, {Red, Thick, Dashed}}}},
    (* Formats *)
    AxesLabel →
        (Style[#, 12] & /@ {"Underlying Price", "Remaining Time", "Profit"}),
    ColorFunction → ColorData["LightTemperatureMap"],
    ColorFunctionScaling → True]
```

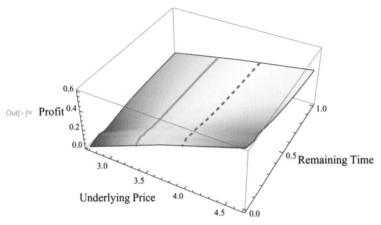

- At expiration, the profile of our profit as a function of the underlying asset price is:

```
In[ ]:= ListPlot[Cases[data, {x_, 0., y_} → {x, y}],
    Joined → True, PlotStyle → Directive[Thickness[0.02], Red],
    AxesLabel → {"Underlying Price", "Profit"}]
```

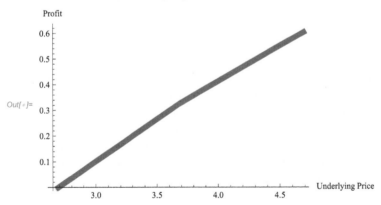

In this example, the warrants that we've used are of the American type, meaning that they can be exercised at any time until the expiration date, although it is not usually optimum to do so.

However, *Mathematica,* in addition to financial derivatives price calculations, can also compute their sensitivities, known as *Greeks.* These sensitivities are the partial derivatives of the option price with respect to different valuation parameters. For example, the *delta* of an option is the partial derivative of its price with respect to the price of the underlying asset. This measure not only gives us an idea of how sensitive the option value is to small changes in the price of the underlying asset but also, for standard options (also known as *vanilla*), it gives us the probability that the option will be exercised. *Gamma,* the second derivative with respect to the underlying asset price, tells us how sensitive the *delta* is to changes to it. This metric is very important for hedging purposes. *Rho* is the partial derivative with respect to the interest rate and in the case of stock options is the least sensitive Greek. *Theta* is the derivative with respect to time to expiration. Initially, since options are not usually bought *in the money i.e.* if a ticker trades at $12, purchasing an option for $10 is not very common, they don't have intrinsic value. They only get value from their time component. The final one, *Vega*, is the partial derivative with respect to the volatility of the underlying asset. This is such an important parameter in option valuation that for certain options, their prices are quoted in terms of the volatility of their respective underlying assets.

- The graph below shows how the Greeks change with respect to the remaining time to expiration:

```
In[ ]:= ticks[min_, max_] := Table[
          {j, ToString[Round[12 - j 12]] <> " Months", {.03, 0}}, {j, min, max, 0.2}]
```

```
In[ ]:= TabView[Rule @@@ Transpose[{{"Delta", "Gamma", "Rho", "Theta", "Vega"},
          ListPlot[#, Ticks → {ticks, Automatic}] & /@ Transpose[Table[{i, #} & /@
            FinancialDerivative[{"American", "Call"}, {"StrikePrice" → 55.00,
              "Expiration" → 1}, {"InterestRate" → 0.1, "Volatility" → 0.5,
              "CurrentPrice" → 50, "Dividend" → 0.05, "ReferenceTime" → i},
              "Greeks"][[All, 2]], {i, 0, 0.95, 0.05}]]}]]
```

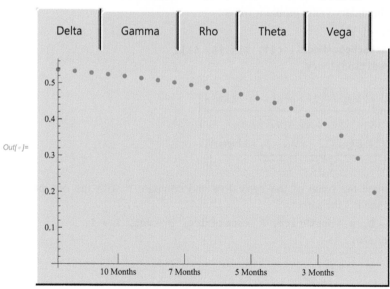

The same Greek letters are used to indicate the risks of portfolio of options due to changes in the price of the underlying assets or market conditions.

Delta: the delta of a portfolio of options with the same underlying assets is the weighted average of the delta of the individual options.

The delta factor = *the change in the price of the option as a result of changes in the price of the underlying assets*. For an at-the-money option, the delta factor is usually close to 50%, so if the underlying price changes by €1, the price of the option will change by €0.5.

11.3.4 The Black–Scholes Equation

In this example, we first compute the price of a derivative using *Mathematica*'s existing function and then compare it with the result obtained by solving the Black–Scholes equation directly

http://www.wolfram.com/language/11/partial-differential-equations/find-the-value-of-a-european-call-option.html

- Let's start by using FinancialDerivative to compute the price of a European vanilla call option with the data from the previous section:

In[ ]:= `spotPrice = 3.26; euribor12m = 0.00003; vol = 0.3069;`

In[ ]:= `FinancialDerivative[{"European", "Call"}, {"StrikePrice" → spotPrice,`
`    "Expiration" → 1}, {"InterestRate" → euribor12m, "Volatility" → vol,`
`    "CurrentPrice" → spotPrice}]`

Out[ ]= `0.397621`

- Now, let's do the same computation using the Black–Scholes equation (To create more visually appealing documents you can convert the input cell to the traditional format: **Cell ▸ Convert to ▸ TraditionalForm**):

In[ ]:= `BlackScholesModel =`

$$\left\{-r\,c(t,\,s) + r\,s\,\frac{\partial c(t,\,s)}{\partial s} + \frac{1}{2}\,s^2\,\sigma^2\,\frac{\partial^2 c(t,\,s)}{\partial s^2} + \frac{\partial c(t,\,s)}{\partial t} = 0,\; c(T,\,s) = \max(s - k,\,0)\right\};$$

- After solving the boundary value problem we get:

In[ ]:= `(dsol = c[t, s] /.`
`    DSolve[BlackScholesModel, c[t, s], {t, s}][`
`    1]) // TraditionalForm`

Out[ ]//TraditionalForm=

$$\frac{1}{2}\,e^{-r\,T}\left(s\,e^{r\,T}\,\mathrm{erfc}\!\left(\frac{2\log(k) + \left(2\,r + \sigma^2\right)(t - T) - 2\log(s)}{2\,\sqrt{2}\,\sigma\,\sqrt{T - t}}\right) - \right.$$
$$\left. k\,e^{r\,t}\,\mathrm{erfc}\!\left(\frac{2\log(k) + \left(2\,r - \sigma^2\right)(t - T) - 2\log(s)}{2\,\sqrt{2}\,\sigma\,\sqrt{T - t}}\right)\right)$$

- Finally, we calculate the price of the derivative and compare it with the one previously obtained. We can see it's the same:

In[ ]:= `dsol /. {t → 0, s → spotPrice, k → spotPrice, σ → vol, T → 1,`
`    r →  euribor12m}`

Out[ ]= `0.397621`

11.3.5. Basel II Capital Adequacy: Internal Ratings-Based (IRB) Approach

The key cornerstone of prudential regulation of banks (and financial institutions in general) is to ensure that each bank holds sufficient equity capital to absorb unexpected losses, that is, the materialization of financial risks, mainly market (price), credit, and operational risks.

The example below, *Basel II Capital Adequacy: Internal Ratings-Based (IRB) Approach,* by Poomjai Nacaskul, shows how to calculate, using the Internal

Ratings-Based approach, the minimum amount of capital that a bank should keep following the standards defined by the Basel II regulatory framework on bank capital adequacy.

http://demonstrations.wolfram.com/BaselIICapitalAdequacyInternalRatingsBasedI RBApproach

- To facilitate the understanding of the example, we also include the code:

```
In[ ]:= (*Setting defaults*)
    DefaultEAD = 1 000 000;
    MinimumPD = 0.0001; MaximumPD = 0.25;
    MinimumLGD = 0.01; DefaultLGD = 0.7; MaximumLGD = 1.0;
    MinimumMaturity = 1 / 250;
    DefaultMaturity = 1;
    MaximumMaturity = 30;
```

```
In[ ]:= (* Create shorthands for Normal CDF & CDF inverse *)
    Phi[x_] := CDF[NormalDistribution[0, 1], x];
    PhiInverse[p_] := InverseCDF[NormalDistribution[0, 1], p];
```

```
In[ ]:= (* Correlation Function, Basel Committee's way of parameterising a 2-
      parameter formula (PD, rho) with 1 parameter (PD). *)
    R[PD_] := 0.12 * (1 - Exp[-50 PD]) / (1 - Exp[-50]) +
        0.24 * (1 - (1 - Exp[-50 PD]) / (1 - Exp[-50]));
    ConditionalPercentageExpectedLoss[PD_, LGD_] := Phi[
        (PhiInverse[PD] + Sqrt[R[PD]] PhiInverse[0.999]) / Sqrt[1 - R[PD]]] * LGD
    UnconditionalPercentageExpectedLoss[PD_, LGD_] := PD * LGD;
```

```
In[ ]:= (* Maturity Adjustment, with a 1-year maturity as a base case *)
    b[PD_] := (0.11852 - 0.05478 * Log[PD]) ^ 2;
    MaturityAdjustment[Maturity_, PD_] :=
        (1 + (Maturity - 2.5) * b[PD]) / (1 - 1.5 * b[PD]);
```

```
In[ ]:= (* Capital Requirement Calculation: note as PD --> 1,
    this quantity --> 0, as credit risk is taken care of via provisioning,
    i.e. E[Loss] = exposure x PD x LGD. *)
    CapitalRequirement[PD_, LGD_, Maturity_] :=
        (ConditionalPercentageExpectedLoss[PD, LGD] -
            UnconditionalPercentageExpectedLoss[PD, LGD]) *
        MaturityAdjustment[Maturity, PD];
```

```
In[ ]:= Manipulate[
      Pane[Column[{Text@Style["Basel II Capital Adequacy", Bold, 16],
          Text@"Regulatory (minimal) capital for financial institutions
              per $ exposure,\nas per Internal Ratings-Based
              Approach (IRB) Risk Weight Function (RWF)\n",
          Grid[{{Plot[CapitalRequirement[PD, LGD, Maturity], {PD, 0.0001, 0.25},
              PlotLabel → "as a function of Probability of Default (PD)\nat
                  a given Maturity (M)", ImageSize → 320],
              Plot3D[CapitalRequirement[PD, LGD, Maturity], {PD, 0.0001, 0.25},
              {Maturity, 1/250, 30}, PlotLabel → "as a function of PD and M\n",
              ImageSize → 280, Mesh → None, ColorFunction →
              Function[{x, y, z}, ColorData["GrayTones"][z]]]}}]},
        Alignment → Center], ImageSize → {610, 350}],
      {{LGD, 0.7, "Loss Given Default (LGD)"},
      0.01, 1, Appearance → "Labeled"},
      {{Maturity, 1, "Maturity M in years (left graph only)"},
      {1/250, 1/50, 1/12, 1/4, 1, 2, 5, 10, 15, 30},
      ControlType → RadioButton}, SaveDefinitions → True]
```

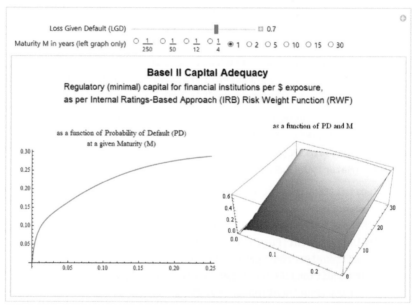

To learn more about financial modeling and risk management using *Mathematica*, Igor Hlivka has created several excellent example notebooks. You can find them in:

http://community.wolfram.com/web/ihcomm.

11.4 Optimization

11.4.1 What is Constrained Optimization?

The optimization problems that we discuss in this section are related to finding an optimum (maximum or minimum, depending on the case) for an objective function of *n* variables, given *r* constraints.

This type of problems is common in several fields such as economics, finance, and engineering as we will show using different examples. You can find an excellent tutorial in the documentation pages: "Introduction to Constrained Optimization in the Wolfram Language" (tutorial/ConstrainedOptimizationIntroduction).

To solve them, *Mathematica* has the following functions (Figure 11.2):

Function	Application	Description
FindMinimum / FindMaximum	Local Numerical optimization	Linear programming methods, non - linear, interior points, and the use of second derivatives
NMinimize / NMaximize	Global Numerical optimization	Linear programming methods, Nelder - Mead, *differential evolution, simulated annealing,* random search
Minimize / Maximize	Exact Global optimization	Linear programming methods, algebraic cylindrical decomposition, Lagrange multipliers and other analytical methods, integer linear programming.
LinearProgramming LinearOptimization	Linear optimization	Simplex, modified simplex, interior point, ...

Figure 11.2 *Mathematica* functions for optimization.

The following tree can help us decide the most appropriate function to find the minimum. (To find the maximum we would use the functions Maximize, NMaximize or FindMaximum instead). Note: Since *Mathematica* 12, LinearProgramming has been superseded by LinearOptimization, but the command can still be used.

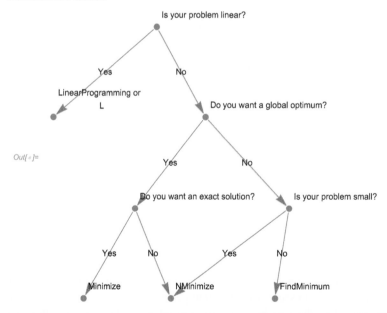

In many cases, the problem can be solved using several functions. With the exception of LinearProgramming and LinearOptimization, the syntax for the rest of the functions is very similar: **f**[{objective function, *constraints*}, {*variables*}].

```
In[ ]:= Clear["`Global`*"]
```

Let's see some examples.

- Given the objective function (ob) with constraints c1 and c2:

In[ ]:= **var = {x, y}; ob = x + y; c1 = 0 ≤ x ≤ 1; c2 = 0 ≤ y ≤ 2;**

- The maximum can be found using any of the previously mentioned commands:

In[ ]:= **{NMaximize[{ob,c1,c2}, var], Maximize[{ob,c1,c2}, var],
FindMaximum[{ob,c1,c2}, var]}**

$$Out[ ]= \begin{pmatrix} 3. & \{x \to 1., y \to 2.\} \\ 3 & \{x \to 1, y \to 2\} \\ 3. & \{x \to 1., y \to 2.\} \end{pmatrix}$$

- We can see that the maximum occurs when $\{x \to 1, y \to 2\}$ and is 3. We can verify it (remember that "/." is used for substitutions).

In[ ]:= **x + y /. {x → 1, y → 2}**

Out[ ]= 3

- The minimum can be computed as follows:

In[ ]:= **Minimize[{ob,c1,c2}, var]**

Out[ ]= $\{0, \{x \to 0, y \to 0\}\}$

- We can interpret the problem graphically by drawing the plane $x + y$ only in the region: $\{0 \le x \le 1, 0 \le y \le 2\}$.

In[ ]:= **Plot3D[x + y, {x, -1, 2}, {y, -1, 2},
RegionFunction → Function[{x, y}, 0 ≤ x < 1 && 0 ≤ y < 2],
AxesLabel → Automatic, Mesh → None]**

Out[ ]=

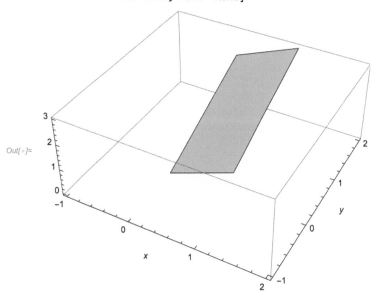

The graph clearly shows that the maximum is in (1,2) and the minimum in (0,0). Use the mouse to move the graph around to see it from different perspectives.

In certain situations, we'd like the variables to take only integer values. To add this extra requirement in *Mathematica,* we use the built-in symbol **Integers**.

- Let's consider the following model in which we want to minimize the objective function (ob) (notice that equality when typing constraints is "=="):

In[ ]:= **Clear["`Global`*"]**

```
In[ ]:= var = {x, y};
        ob = x + 2 y;
        c1 = - 5 x + y == 7;
        c2 = x + y ≥ 26;
        c3 = x ≥ 3;
        c4 = y ≥ 3;

In[ ]:= NMinimize[{ob, c1, c2, c3, c4}, var ∈ Integers]

Out[ ]= {58., {x → 4, y → 27}}
```

Since this is a linear problem, we can also use either LinearOptimization (since *Mathematica* 12) or LinearProgramming. As a matter of fact, these functions are the most appropriate ones for linear problems, especially if the number of variables is large.

```
In[ ]:= LinearOptimization[ob, {c1, c2, c3, c4}, {x ∈ Integers, y ∈ Integers}]

Out[ ]= {x → 4, y → 27}
```

The syntax is: `LinearProgramming[c, m, b]` . This function finds the vector **x** that minimizes the quantity **c.x** subject to the constraints **m.x ≥ b** and **x ≥ 0**. We can limit the values that the variables (some or all of them) can take to just integers with `LinearProgramming[..., Integers]`.

- For comparison purposes, let's solve the same problem using `LinearProgramming`. The syntax is: `LinearProgramming[c, m, b]` . This function finds the vector **x** that minimizes the quantity **c.x** subject to the constraints **m.x ≥ b** and **x ≥ 0**. We can limit the values that the variables (some or all of them) can take to just integers with `LinearProgramming[..., Integers]`.

$x + 2y$	→	$c: \{1, 2\}$		
$-5x + y = 7$	→	$m1: \{-5, 1\}$	$b1: \{7, 0\}$	
$x + y \geq 26$	→	$m2: \{1, 1\}$	$b2: \{26, 1\}$	

Notice that the syntax to indicate the type of constraint is as follows: $\{b_i, 0\}$ if $m_i.x == b_i$; $\{b_i, 1\}$ if $m_i.x \geq b_i$ and $\{b_i, -1\}$ if $m_i.x \leq b_i$.

```
In[ ]:= LinearProgramming[{1, 2}, {{-5, 1}, {1, 1}}, {{7, 0}, {26, 1}},
        {{3, Infinity}, {4, Infinity}}, Integers] // Quiet

Out[ ]= {4, 27}
```

Next we'll show some examples for nonlinear optimization.

```
In[ ]:= Clear["`Global`*"]
```

- In this problem, both, the objective function that we want to minimize and the constraints are nonlinear.

```
In[ ]:= var = {x, y};
        ob = x^2 + (y - 1)^2;
        c1 = x^2 + y^2 ≤ 4;
```

- Let's calculate the global minimum with Minimize:

```
In[ ]:= Minimize[{ob, c1}, var]

Out[ ]= {0, {x → 0, y → 1}}
```

- NMinimize can solve this type of problem faster (although in this example we wouldn't be able notice any difference). We add Chop to remove, when the solution is 0, the spurious values that may appear in the output when using numerical approximations. Execute the command below without it and see what happens.

In[ ]:= **NMinimize[{ob, c1}, var] // Chop**

Out[ ]= $\{-3.45349 \times 10^{-8}, \{x \to 0, y \to 0.999999\}\}$

- If we are looking for a local minimum we can use FindMinimum. Although it's not necessary, this function is much faster if we enter an initial point from which to start the search. It's also a good idea to add Chop.

In[ ]:= **FindMinimum[{ob, c1}, {{x, 0}, {y, 0}}] // Chop**

Out[ ]= $\{-3.3699 \times 10^{-8}, \{x \to 0, y \to 0.999999\}\}$

- We present the previous problem graphically with Plot3D and ContourPlot, very convenient functions when trying to find optimum points visually. (The option Contours specifies how many contours we want to display; if not specified, *Mathematica* will choose the optimum number for the given function).

In[ ]:= **GraphicsRow[{Plot3D[x^2 + (y - 1)^2, {x, -3, 3},**
      {y, -3, 3}, RegionFunction → Function[{x, y}, x^2 + y^2 ≤ 4],
      AxesLabel → Automatic, Mesh → All], ContourPlot[x^2 + (y - 1)^2,
      {x, -3, 3}, {y, -3, 3}, RegionFunction → Function[{x, y}, x^2 + y^2 ≤ 4],
      Contours → 100, FrameLabel → Automatic]}]

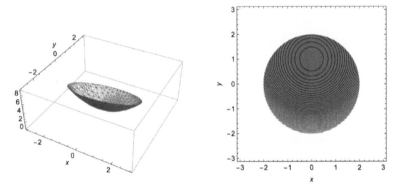

Click inside the output from Plot3D and move the graph around to see the minimum from different angles. In the contour plot, we can see that the minimum corresponds to the coordinates (0, 1). By default, dark colors are associated to small values and light hues are linked to bigger ones. If you place your cursor over the contour limits, you'll see the value that the function takes at that location. (All the points connected by the same line have the same function value.)

- Here is another example of nonlinear optimization, in this case for finding the global minimum (Minimize). The objective function is $\text{Exp}(-xy)$ with the constraint that $x, y \in$ a Circle centered in $\{0, 0\}$ and with a radius $r = 1$. We show the result, both numerically and graphically.

In[ ]:= **m = Minimize[{Exp[-x y], {x, y} ∈ Disk[]}, {x, y}]**

Out[ ]= $\left\{\dfrac{1}{\sqrt{e}}, \left\{x \to -\dfrac{1}{\sqrt{2}}, y \to -\dfrac{1}{\sqrt{2}}\right\}\right\}$

```
In[*]:= g = Plot3D[Exp[-x y], {x, y} ∈ Disk[], Axes → True,
           Boxed → False, PlotStyle → Opacity@0.6, AxesLabel → Automatic];
```

```
In[*]:= Show[g, Graphics3D[{PointSize@Large, Red, Point[{x, y, m[[1]]} /. m[[2]]]}]]
```

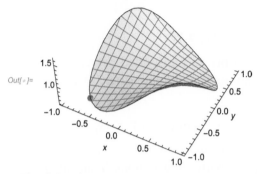

Out[*]=

- The following example shows how the optimization algorithm approaches the optimum (minimum) point for f(x,y) in an iterative way.

```
In[*]:= Clear["Global`*"]
```

```
In[*]:= f[x_, y_] := (x - 1)^2 + 2 (y - x^2)^2
```

```
In[*]:= Module[{pts}, pts = Last[Last[Reap[Sow[{-1.2, 1, f[-1.2, 1]}];
           FindMinimum[f[x, y],
               {{x, -1.2}, {y, 1}}, StepMonitor :> Sow[{x, y, f[x, y]}]]]]];
       Show[Plot3D[f[x, y], {x, -1.4, 1.4}, {y, -0.6, 2}, Mesh → 5,
         PlotStyle → Automatic, PlotLabel → Style[Row[{"Finding the minimum of ",
               f[x, y], " using iteration"}], 12, Bold], AxesLabel → Automatic],
         Graphics3D[{Orange, PointSize[0.025], Point[Rest[Most[pts]]],
           Thick, Line[pts], Darker@Green, PointSize[0.05],
           Point[Last[pts]], Red, Point[First[pts]]}]]]
```

Finding the minimum of $2\left(y - x^2\right)^2 + (x - 1)^2$ using iteration

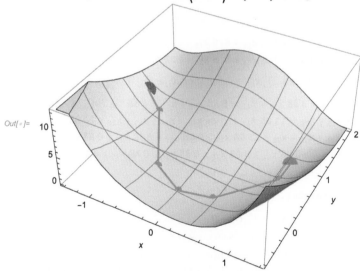

Out[*]=

11.4.2 Application: Supplying Construction Sites

We'd like to optimize the supply of bricks from warehouses to construction sites. There are two brick warehouses $j = \{1,2\}$, where we have in stock 20 and 25 tonnes, respectively. From them we supply 6 sites $i = \{1, 2, 3, 4, 5, 6\}$ with a daily demand in tonnes of: $\{3, 5, 4, 7, 6, 11\}$. We assume that the location (latitude and longitude) of the warehouses and sites is:

```
In[ ]:= warehouses = {{7, 1}, {2, 9}};
```

```
In[ ]:= sites = {{1.25, 1.25}, {8.75, 0.75},
          {0.5, 4.75}, {5.75, 5}, {3, 6.5}, {7.25, 7.75}};
```

We'd like to minimize the total amount of bricks transported per unit of distance, $\sum_{i,j} x_{i,j}\, d_{i,j}$, meeting the demand without exceeding the stock available in the warehouses.

For that purpose, we have to calculate the minimum of $\sum_{i,j} x_{i,j}\, d_{i,j}$, where $x_{i,j}$ corresponds to the quantities transported from the warehouses $i:\{1, 2\}$ to the sites $j:\{1, ..., 6\}$ with $d_{i,j}$ representing the distance between warehouse i and site j.

- To calculate the distances between the warehouses and the sites we use the function `EuclideanDistance` (in the bi-dimensional case, it's the equivalent to the Pythagorean theorem):

```
In[ ]:= EuclideanDistance[{x1, y1}, {x2, y2}]
```

$$Out[ ]= \sqrt{|x1 - x2|^2 + |y1 - y2|^2}$$

- The command below returns triplets, $\{i, j, d_{i,j}\}$, indicating the distance in km between the warehouse i and site j. We use `Round[value, 0.1]` to round the distance to the nearest first decimal.

```
In[ ]:= Table[{i, j, Round[EuclideanDistance[warehouses[[i]], sites[[j]]], 0.1]},
          {i, 1, 2}, {j, 1, 6}]
```

$$Out[ ]= \begin{pmatrix} \{1, 1, 5.8\} & \{1, 2, 1.8\} & \{1, 3, 7.5\} & \{1, 4, 4.2\} & \{1, 5, 6.8\} & \{1, 6, 6.8\} \\ \{2, 1, 7.8\} & \{2, 2, 10.7\} & \{2, 3, 4.5\} & \{2, 4, 5.5\} & \{2, 5, 2.7\} & \{2, 6, 5.4\} \end{pmatrix}$$

- Therefore, the variables and the objective function, $\sum_{i,j} x_{i,j}\, d_{i,j}$, are:

```
In[ ]:= var = {x11, x12, x13, x14, x15, x16, x21, x22, x23, x24, x25, x26};
```

```
In[ ]:= of = 5.8 x11 + 1.8 x12 + 7.5 x13 + 4.2 x14 + 6.8 x15 +
          6.8 x16 + 7.8 x21 + 10.7 x22 + 4.5 x23 + 5.5 x24 + 2.7 x25 + 5.4 x26;
```

- The constraints are: $\sum_i x_{i,j} = \text{demand}_j$, $\sum_j x_{i,j} = \text{stock}_i$ and $x_{i,j} \geq 0$

```
In[ ]:= constraints = {x11 + x21 == 3 && x12 + x22 == 5 && x13 + x23 == 4 && x14 + x24 == 7 &&
          x15 + x25 == 6 && x16 + x26 == 11 && x11 + x12 + x13 + x14 + x15 + x16 ≤ 20 &&
          x21 + x22 + x23 + x24 + x25 + x26 ≤ 25 && x11 ≥ 0 && x12 ≥ 0 && x13 ≥ 0 && x14 ≥ 0,
          x15 ≥ 0 && x16 ≥ 0 && x21 ≥ 0 && x22 ≥ 0 && x23 ≥ 0 && x24 ≥ 0 && x25 ≥ 0 && x26 ≥ 0};
```

- The solution is:

```
In[ ]:= NMinimize[{of, constraints}, var]
```

$$Out[ ]= \{149.4, \{x_{11} \to 3., x_{12} \to 5., x_{13} \to 0., x_{14} \to 7., x_{15} \to 0.,$$
$$x_{16} \to 0., x_{21} \to 0., x_{22} \to 0., x_{23} \to 4., x_{24} \to 0., x_{25} \to 6., x_{26} \to 11.\}\}$$

This result indicates that the optimum solution, 149.4 tonnes × km, is obtained by transporting 3 tonnes from warehouse 1 to site 1, 5 tonnes from warehouse 1 to site 2 and so on.

In "Optimal Transport Scheduling", a demonstration by Yifan Hu, you can see a dynamic visualization of the resolution of this problem:

http://demonstrations.wolfram.com/OptimalTransportScheduling

If in the demonstration you use 20 and 25 tonnes, you'll see that the solution is the same (there's actually a small difference due to rounding).

11.4.3 Application: Portfolio Optimization

A bank commissions a consulting firm to analyze the most profitable way to invest €10,000,000 for two years given the options in Figure 11.3. The table includes the expected rate of return (bi-annual) and associated risk.

Product (i)	Return (%), b_i	Risk (%) r_i
Mortgages	9	3
Mutual funds	12	6
Personal loans	15	8
Commercial loans	8	2
Certificates / Bonds	6	1

Figure 11.3 Investment options.

The capital not invested in any of the products will be placed in government bonds (assumed to be riskless) with a bi-annual rate of return of 3%. The objective of the consulting firm is to allocate the capital to each of the products to meet the following goals:

(a) Maximize the return per € invested.
(b) Keep the possibility of loss to a maximum of 5% of the total amount invested.
(c) Invest at least 20% in commercial loans.
(d) Allocate to mutual funds and personal loans an amount no larger than the one invested in mortgages.

- Variables: x_i is the percentage of capital invested in product i. The amount placed in government bonds can be considered as a new product with an expected return, in percentage, $b_6 = 3$ and risk $r_6 = 0$. Therefore:

In[ ]:= `var = {x₁(*Mortgages*), x₂(*Mutual funds*),`
        `x₃(*Personal loans*), x₄ (*Commercial loans*), x₅`
        `(*Certificates/Bonds*), x₆(*Government debt*)};`

- Objective function to maximize: $\sum_{i=1}^{6} b_i x_i$.

In[ ]:= `of = 9 x₁ + 12 x₂ + 15 x₃ + 8 x₄ + 6 x₅ + 3 x₆;`

- Constraint 1: All the money is invested: $\sum_{i=1}^{6} x_i = 1$.

In[ ]:= `c1 = x₁ + x₂ + x₃ + x₄ + x₅ + x₆ == 1;`

- Constraint 2: The average risk is $R = \sum_{i=1}^{6} r_i x_i / \sum_{i=1}^{6} x_i \le 5$ as $r_6 = 0$ then:

In[ ]:= $$c2 = \frac{3 x_1 + 6 x_2 + 8 x_3 + 2 x_4 + x_5}{x_1 + x_2 + x_3 + x_4 + x_5} \le 5$$

Out[ ]= $$\frac{3 x_1 + 6 x_2 + 8 x_3 + 2 x_4 + x_5}{x_1 + x_2 + x_3 + x_4 + x_5} \le 5$$

- Constraint 3: At least 20% of the capital has to be invested in commercial loans ($x_4 \ge 0.2$).

In[ ]:= `c3 = x₄ ≥ 0.2;`

- Constraint 4: The percentage invested in mutual funds (x_2) and personal loans (x_3) cannot be bigger than the one invested in mortgages (x_1).

In[ ]:= `c4 = x₂ + x₃ ≤ x₁;`

- Constraint 5: No percentage invested can be negative ($x_i \geq 0$).

In[ ]:= `c5 = Map[# ≥ 0 &, var];`

- Using NMaximize, *Mathematica* tells us that we'd get a return of 11.2% investing 40% in mortgages, 40% in personal loans, and 20% in commercial loans.

In[ ]:= `sol = NMaximize[{of, c1, c2, c3, c4, c5}, var]`

Out[ ]= $\{11.2, \{x_1 \to 0.4, x_2 \to 0., x_3 \to 0.4, x_4 \to 0.2, x_5 \to 0., x_6 \to 0.\}\}$

In[ ]:= `"€" <> ToString@AccountingForm[10 000 000 sol[[1]] / 100, DigitBlock → 3]`

Out[ ]= €1,120,000.

11.4.4 Application: Calculating Cod Consumption

In[ ]:= `Clear["Global`*"]`

Many problems in economics involve Cobb–Douglas functions: $F(x, y) = A\, x^a\, y^b$ with $x, y \geq 0$ and A, a, b constants. Let's see one example:

The cod consumption in certain region follows the function $c(p, i) = 1.3\, p^{-1.5}\, i^{0.01}$, where p is the price per kg and i is the daily income in €/day. The previous expression is valid for daily incomes between €20 and €200 per day and the price per kg varies between €4 and €12. What would be the maximum and minimum consumption?

- We can proceed as usual defining the variables, objective function, and constraints:

In[ ]:= `c[p_, i_] = 1.3 p^-1.05 i^0.01;`

In[ ]:= `NMinimize[{c[p, i], 4 ≤ p ≤ 12 && 20 ≤ i ≤ 200}, {p, i}]`

Out[ ]= $\{0.0985856, \{p \to 12., i \to 20.\}\}$

In[ ]:= `NMaximize[{c[p, i], 0.4 ≤ p ≤ 1.2 && 20 ≤ i ≤ 200}, {p, i}]`

Out[ ]= $\{3.58749, \{p \to 0.4, i \to 200.\}\}$

- We can verify the solution visually using Plot3D and ContourPlot. If we place the cursor on the second graph and move it around, we can see the different values of the function $c(r, t)$. This way we can check that the upper-left corner corresponds to the largest value (maximum), and the lower-right corner to the smallest one (minimum). The boundaries of the curve correspond to the constraints.

In[ ]:=
```
GraphicsRow[{Plot3D[c[p, i], {p, 4, 12}, {i, 0, 200},
      RegionFunction → Function[{p, i}, 4 ≤ p ≤ 12 && 20 ≤ i ≤ 200],
      AxesLabel → {"Price", "Income", "c"}, Mesh → None],
    ContourPlot[c[p, i], {p, 4, 12}, {i, 0, 200}, RegionFunction →
        Function[{p, i}, 4 ≤ p ≤ 12 && 20 ≤ i ≤ 200], Mesh → None,
      Contours → 100, FrameLabel → Automatic, PlotLegends → Automatic]},
   Frame → All, AspectRatio → 1/3, ImageSize → {600, 300}]
```

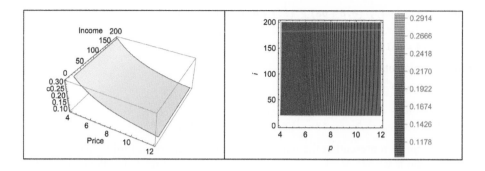

11.4.5 Transportation Optimization

The department store chain Carrefa would like to outsource the supply of milk cartons for its centers in Barcelona, Madrid, Seville, and Valladolid. For that purpose, it receives offers from three suppliers: Pascualo, Simone, and Pulova. The prices per carton are as follows: Pascualo €3.00, Simone €2.80, and Pulova €2.70. The maximum number of cartons that each supplier can send daily is shown in the table below, under the column "Maximum Supply". The daily demand for each center is in the "Demand" row. The rest of the information in Figure 11.4 relates to the transportation costs per carton, in €.

	Maximum Supply	Barcelona	Madrid	Seville	Valladolid
Pascualo	2500	€ 0.5	€ 0.6	€ 0.7	€ 0.5
Simone	1700	€ 0.5	€ 0.6	€ 0.8	€ 0.5
Pulova	1500	€ 0.4	€ 0.5	€ 0.9	€ 0.7
	Demand	800	1100	500	300

Figure 11.4   Transportation optimization input data.

Carrefa would like to know how much to order from each supplier to minimize the total cost.

In[ ]:= **Clear["Global`*"]**

This is the type of problem that can be solved with LinearProgramming or the newest function LinearOptimization. With LinearProgramming, matrix notation is required. For didactic purposes we're going to compare both functions using algebraic notation for LinearOptimization although matrix notation can also be used.

With LinearOptimization the variables can be defined as follows: x_{ij} = number of cartons from supplier i , i:$\{1, 2, 3\}$, with $\{1 = \text{Pascualo}, 2 = \text{Simone}, 3 = \text{Pulova}\}$, to center j: $\{1, 2, 3, 4\}$, with $\{1 = \text{Barcelona}, 2 = \text{Madrid}, 3 = \text{Seville}, 4 = \text{Valladolid}\}$.

In[ ]:= **var = {x_{11}, x_{12}, x_{13}, x_{14}, x_{21}, x_{22}, x_{23}, x_{24}, x_{31}, x_{32}, x_{33}, x_{34}};**

- To define the objective function, we have to group terms by supplier. For example: the total cost of sending x units from Pascualo to Barcelona (x_{11}) would be the transportation cost plus the purchasing price: (0.5 + pa). We use the same approach for calculating the rest of the variables:

In[ ]:= **{pa, sim, pul} = {3, 2.8, 2.7};**

$In[\circ]:=$ **of = { (0.5 + pa), (0.6 + pa), (0.7 + pa), (0.5 + pa),**
(0.5 + sim), (0.6 + sim), (0.8 + sim), (0.5 + sim),
(0.4 + pul), (0.5 + pul), (0.9 + pul), (0.7 + pul) };

- Or in algebraic notation:

$In[\circ]:=$ **of1 = of .var**

$Out[\circ]=$ $3.5\,x_{11} + 3.6\,x_{12} + 3.7\,x_{13} + 3.5\,x_{14} + 3.3\,x_{21} +$
$3.4\,x_{22} + 3.6\,x_{23} + 3.3\,x_{24} + 3.1\,x_{31} + 3.2\,x_{32} + 3.6\,x_{33} + 3.4\,x_{34}$

- The constraint **c1** representing the maximum number of cartons of milk that we can receive from Pascualo is written as (the values of those variables not related to the constraint should be 0), m1 and bi:

$In[\circ]:=$ **c1 = x_{11} + x_{12} + x_{13} + x_{14} ≤ 2500;**

$In[\circ]:=$ **m1 = {1, 1, 1, 1, 0, 0, 0, 0, 0, 0, 0, 0}; b1 = {2500, -1};**

where -1 is equivalent to "≤"; 0 to "=="; and 1 to "≥".

- The constraint **c2** or **m2**, related to the maximum capacity of Simone is:

$In[\circ]:=$ **c2 = x_{21} + x_{22} + x_{23} + x_{24} ≤ 1700;**

$In[\circ]:=$ **m2 = {0, 0, 0, 0, 1, 1, 1, 1, 0, 0, 0, 0}; b2 = {1700, -1};**

- The constraint **c3** or m3, dealing with the supply coming from Pulova:

$In[\circ]:=$ **c3 = x_{31} + x_{32} + x_{33} + x_{34} ≤ 1500;**

$In[\circ]:=$ **m3 = {0, 0, 0, 0, 0, 0, 0, 0, 1, 1, 1, 1}; b3 = {1500, -1};**

- The constraint **c4**, $x_{11} + x_{21} + x_{31} == 800$, indicating the demand from Barcelona:

$In[\circ]:=$ **c4 = x_{11} + x_{21} + x_{31} == 800;**

$In[\circ]:=$ **m4 = {1, 0, 0, 0, 1, 0, 0, 0, 1, 0, 0, 0}; b4 = {800, 0};**

- The constraint **c5**, $x_{12} + x_{22} + x_{32} == 1100$, referring to the demand from Madrid:

$In[\circ]:=$ **m5 = {0, 1, 0, 0, 0, 1, 0, 0, 0, 1, 0, 0}; b5 = {1100, 0};**

$In[\circ]:=$ **c5 = x_{12} + x_{22} + x_{32} == 1100;**

- The constraint **c6**, $x_{13} + x_{23} + x_{33} == 500$, about Seville:

$In[\circ]:=$ **c6 = x_{13} + x_{23} + x_{33} == 500;**

$In[\circ]:=$ **m6 = {0, 0, 1, 0, 0, 0, 1, 0, 0, 0, 1, 0}; b6 = {500, 0};**

- And finally the constraint **c7**, or m3 to make sure that the needs of Valladolid are being satisfied:

$In[\circ]:=$ **c7 = x_{14} + x_{24} + x_{34} == 300;**

$In[\circ]:=$ **m7 = {0, 0, 0, 1, 0, 0, 0, 1, 0, 0, 0, 1}; b7 = {300, 0};**

- Now we can solve the problem (LinearProgramming assumes that all the variables are nonnegative so we don't have to include those constraints).

$In[\circ]:=$ **sol1 = LinearProgramming[of,**
{m1, m2, m3, m4, m5, m6, m7}, {b1, b2, b3, b4, b5, b6, b7}]

$Out[\circ]=$ {0., 0., 0., 0., 400., 0., 500., 300., 400., 1100., 0., 0.}

- Using LinearOptimization the constraint that all the variables are nonnegative should be included.

$In[\circ]:=$ **c8 = Thread[var ≥ 0]**

Out[ ]= $\{x_{11} \geq 0, x_{12} \geq 0, x_{13} \geq 0, x_{14} \geq 0, x_{21} \geq 0, x_{22} \geq 0, x_{23} \geq 0, x_{24} \geq 0, x_{31} \geq 0, x_{32} \geq 0, x_{33} \geq 0, x_{34} \geq 0\}$

In[ ]:= `sol2 = LinearOptimization[of1, {c1, c2, c3, c4, c5, c6, c7, c8}, var]`

Out[ ]= $\{x_{11} \to 0., x_{12} \to 0., x_{13} \to 0., x_{14} \to 0., x_{21} \to 0., x_{22} \to 400.,$
$\quad x_{23} \to 500., x_{24} \to 300., x_{31} \to 800., x_{32} \to 700., x_{33} \to 0., x_{34} \to 0.\}$

- Both commands return the optimal value for each variable. If we'd like to calculate the minimum total transportation cost, we just need to multiply the objective function by the solution (we use "." for matrix multiplication). Although sol1 and sol2 are different , the optimization values are the same; this happens when the problem has more than one solution.

In[ ]:= `of.sol1`

Out[ ]= 8870.

In[ ]:= `of1 /. sol2`

Out[ ]= 8870.

11.5 The Shortest Path Problem

11.5.1 The Traveling Salesman Problem

There are problems in mathematics that look very specialized and easy to solve at first sight, but in reality they are very complicated and their resolution can have applications in a wide variety of fields. One of the most famous examples is the *Traveling Salesman Problem* (TSP). This problem can be defined as follows: A traveler would like to visit N cities starting and finishing in the same arbitrarily chosen one, minimizing the total distance traveled and passing through each city only once. If you think about it, there are many real-world problems that are actually equivalent to the TSP: route optimization (transportation, logistics), optimum network layout (trains, roads, electricity), even the connections inside microprocessors to minimize calculation time.

A graphical demonstration of the problem created by Jon McLoone can be found in:

http://demonstrations.wolfram.com/TravelingSalesmanProblem/

Mathematically, the problem consists of finding a permutation $P = \{c_0, c_1, ..., c_{n-1}\}$ such that $d_P = \sum_{i=1}^{n-1} d[c_i, c_{i+1}]$ would be a minimum, with d_{ij} representing the distance from city i to city j.

The seemingly simplest solution would be to enumerate all the possible routes, calculate the distance traveled for each route and select the shortest one. The problem with this approach is that computing time increases dramatically when the number of cities gets bigger. For example, a computer would take 3 seconds to solve the problem if there were 10 cities, a bit more than half a minute for 11 cities, and almost 80,000 years! to solve the problem for 20 cities.

Conclusion: This solving method is not feasible. Therefore, several alternative algorithms have been developed that use a variety of strategies to simplify the problem (something similar to computer chess programs that instead of calculating all possible moves, use different criteria to choose the most appropriate ones).

For this type of problem, *Mathematica* has the function:

`FindShortestTour[{e1, e2, ...}]`. This function takes a list of coordinates: {*e1*, *e2*, ...} and attempts to find the optimum ordering so that the total distance traveled is minimized and all the points have been visited just once. The output returns both, the total distance and the order of traveling.

Example: Given a set of points p = {1, 2, 3, 4, 5, 6} each with coordinates{x,y}:{{4, 3},{1, 1}, {2, 3},{3, -5},{-1, 2},{3, 4}}, find the shortest path that will visit all the points only once.

In[ ]:= `d = {{4, 3}, {1, 1}, {2, 3}, {3, -5}, {-1, 2}, {3, 4}};`

In[ ]:= `{dist, order} = FindShortestTour[d]`

Out[ ]= $\{2\sqrt{2} + \sqrt{5} + 3\sqrt{10} + \sqrt{65}, \{1, 4, 2, 5, 3, 6, 1\}\}$

- We sort the points following the order from the output above and represent them graphically:

In[ ]:= `d[[order]]`

Out[ ]=
$$\begin{pmatrix} 4 & 3 \\ 3 & -5 \\ 1 & 1 \\ -1 & 2 \\ 2 & 3 \\ 3 & 4 \\ 4 & 3 \end{pmatrix}$$

In[ ]:= `Graphics[Line[%]]`

Out[ ]=

11.5.2 A Tour Around South American Cities

The next example is about organizing a trip to several cities in South America. In what order should we visit them to minimize the total distance traveled?

- The cities we're visiting are the following (we indicate both, the cities and their respective regions and countries to avoid ambiguity):

```
In[ ]:= cities={Entity["City", {"Asuncion", "Asuncion", "Paraguay"}],
       Entity["City", {"Bogota", "DistritoCapital", "Colombia"}],
       Entity["City", {"RioDeJaneiro", "RioDeJaneiro", "Brazil"}],
       Entity["City", {"BuenosAires", "BuenosAires", "Argentina"}],
       Entity["City", {"Caracas", "DistritoCapital", "Venezuela"}],
       Entity["City", {"LaPaz", "LaPaz", "Bolivia"}],
       Entity["City", {"Lima", "Lima", "Peru"}],
       Entity["City", {"Montevideo", "Montevideo", "Uruguay"}],
       Entity["City", {"Quito", "Pichincha", "Ecuador"}],
       Entity["City", {"Santiago", "Metropolitana", "Chile"}]};
```

- Now we can calculate the best visiting order:

```
In[ ]:= order = Last[FindShortestTour[GeoPosition[cities]]]
```

```
Out[ ]= {1, 8, 4, 10, 6, 7, 9, 2, 5, 3, 1}
```

- Finally, we can display the route on a map:

```
In[ ]:= GeoListPlot[cities[[order]], Joined → True]
```

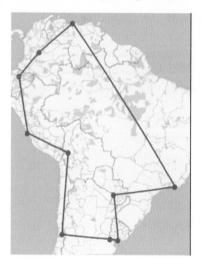

11.5.3 Moon Tourism

It's the year 2035, and we are planning to organize sightseeing tours around the Moon. We want to find the shortest route for visiting all the Apollo landing sites.

- "Apollo landings" is one of the available entity classes that have been included in the function MannedSpaceMissionData.

```
In[ ]:= MannedSpaceMissionData["Classes"]
```

Out[ ]= { manned space missions , Apollo landings , Apollo crewed mission ,

Chinese crewed mission , Earth orbit crewed mission , lunar crewed mission ,

manned space mission , Mercury Atlas crewed mission , Mercury Redstone crewed mission ,

pre-flight test crewed mission , project Gemini crewed mission ,

project Mercury crewed mission , Russian crewed mission ,

Shenzhou crewed mission , Soyuz crewed mission , SpaceX Dragon crewed mission ,

SpaceX crewed mission , STS crewed mission , sub-orbital flight crewed mission ,

United States crewed mission , Voskhod crewed mission , Vostok crewed mission }

```
In[ ]:= MannedSpaceMissionData[ ▦ Apollo manned mission  MANNED SPACE MISSIONS ]
```

Out[ ]= { Apollo 1 , Apollo 7 , Apollo 8 , Apollo 9 , Apollo 10 , Apollo 11 ,

Apollo 12 , Apollo 13 , Apollo 14 , Apollo 15 , Apollo 16 , Apollo 17 }

```
In[ ]:= apollolandings =
        MannedSpaceMissionData[ ▦ Apollo landings  MANNED SPACE MISSIONS ,
        {"Position"}, "EntityAssociation"]
```

Out[ ]= ⟨| Apollo 11 → {GeoPosition[{0.6741, 23.47}, Moon]},

Apollo 12 → {GeoPosition[{-3.012, -23.42}, Moon]},

Apollo 14 → {GeoPosition[{-3.645, -17.47}, Moon]},

Apollo 15 → {GeoPosition[{26.13, 3.634}, Moon]},

Apollo 16 → {GeoPosition[{-8.973, 15.50}, Moon]},

Apollo 17 → {GeoPosition[{20.19, 30.77}, Moon]} |⟩

- The names of the manned landing missions are:

```
In[ ]:= missionnames = Keys[apollolandings]
```

Out[ ]= { Apollo 11 , Apollo 12 , Apollo 14 , Apollo 15 , Apollo 16 , Apollo 17 }

- Notice that Apollo 13 is missing. Let's find out why:

```
In[ ]:= MannedSpaceMissionData[ Apollo 13  MANNED SPACE MISSION , "Description"]
```

Out[ ]= {intended to be the third manned mission to land on the moon,
          mid−mission onboard explosion forced landing to be aborted, returned safely to Earth}

- Create a map showing the locations of the landings:

```
In[ ]:= GeoListPlot[missionnames, GeoRange → All,
        GeoProjection → "Orthographic", GeoLabels → True, LabelStyle → White]
```

- To extract the coordinates of those landings:

In[ ]:= **landingcoords = Values[apollolandings] // Flatten**

Out[ ]= {GeoPosition[{0.6741, 23.47}, Moon], GeoPosition[{−3.012, −23.42}, Moon],
GeoPosition[{−3.645, −17.47}, Moon], GeoPosition[{26.13, 3.634}, Moon],
GeoPosition[{−8.973, 15.50}, Moon], GeoPosition[{20.19, 30.77}, Moon]}

- Next, we find out that the shortest route starts where Apollo 11 landed:

In[ ]:= **order = Last[FindShortestTour[landingcoords]]**

Out[ ]= {1, 6, 4, 2, 3, 5, 1}

- Finally, we create a plot showing the path of the tour:

In[ ]:= **GeoListPlot[landingcoords⟦order⟧, Joined → True, GeoLabels → Automatic]**

11.6 Optimum Flows

In this example, we want to maximize the total number of freight cars that can be sent from Vancouver to Winnipeg given the following network characteristics:

The code generating graph g is not shown.

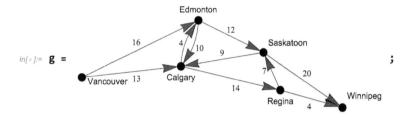

In[ ]:= **g =**   **;**

- The previous illustration has been created using the Graph function. For simplifying purposes, let's just focus on 3 cities: { (1) Vancouver, (2) Calgary, (3) Edmonton}. The flow directions (i→j) and capacities (EdgeCapacity) are: {From (1) Vancouver to (2) Calgary, 13, from (1) Vancouver to (3) Edmonton, 4, from (3) Edmonton to (2) Calgary, 10 and from (2) Edmonton to (3) Calgary, 4}. We display that information using DirectedEdge and with the help of VertexLabels and Placed position the labels in the desired locations. The style is defined using Style[#, "Style"]&. You can use the documentation to learn more about the syntax for each function.

```
In[ ]:= g1 = Graph[{
        "Vancouver", "Calgary", "Edmonton"},
       {{{  1, 2}, {1, 3}, {2, 3}, {3, 2}}, Null}, { EdgeCapacity → {13, 16, 4, 10},
        EdgeLabels → { DirectedEdge["Vancouver", "Calgary"] → Placed[13, {0.5, {2, 1}}],
          DirectedEdge["Vancouver", "Edmonton"] → Placed[16, {0.5, {-1.5, 2}}] ,
          DirectedEdge["Edmonton", "Calgary"] → Placed[10, {0.5, {-0.2, 1.5}}],
          DirectedEdge["Calgary", "Edmonton"] → Placed[4, {0.5, {-0.7, 1.5}}] },
        GraphStyle → "BasicBlack",
       VertexLabels → { "Edmonton" → Placed[" (3)Ed.", Above, Style[#, Red] & ],
         "Vancouver" → Placed[" (1)Vanc.", After, Style[#, Blue] & ],
         "Calgary" → Placed[" (2)Cal.", Above, Style[#, FontFamily → "Helvetica"] & ]},
        VertexSize → {Small}}, ImagePadding → All, ImageSize → Small]
```

Out[ ]=

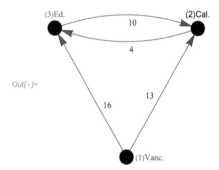

- In the next example, we add VertexCoordinates to specify the coordinates for the center of the vertices. We also modify several options to make it easier to overlay the graph on a map of Canada later.

```
Graph[{
    "Vancouver", "Calgary", "Edmonton"},
  {{{1, 2}, {1, 3}, {2, 3}, {3, 2}}, Null}, { EdgeCapacity → {13, 16, 4, 10},
   EdgeLabels → { DirectedEdge["Edmonton", "Calgary"] → Placed[10, {0.5, {-0.2, 1}}],
     DirectedEdge["Vancouver", "Edmonton"] → Placed[16, {0.5, {1, 0}}],
     DirectedEdge["Vancouver", "Calgary"] → Placed[13, {0.6, {1, 1.5}}]]},
   GraphStyle → "BasicBlack", ImagePadding → {{0, 40}, {0, 10}},
   VertexCoordinates → {{-0.253, 0.112}, {-0.1419, 0.1239}, {-0.122, 0.1735}},
   VertexLabels → { "Edmonton" →
      Placed["Edmonton", Above, Style[#, FontFamily → "Helvetica"] &],
     "Vancouver" →   Placed["Vancouver",
       {{1.5, -1}, {0, 0}},    Style[#, FontFamily → "Helvetica"] &],
     "Calgary" →  Placed["Calgary", Below, Style[#, FontFamily → "Helvetica"] &]},
   VertexSize → {Small}}]
```

Out[ ]=

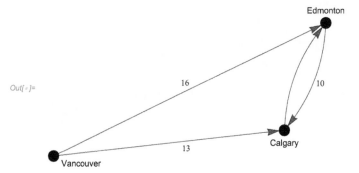

- To find the optimum flow we use the function FindMaximumFlow.

In[ ]:= **F = FindMaximumFlow[g, "Vancouver", "Winnipeg", "OptimumFlowData"]**

Out[ ]= OptimumFlowData[⊞ ◇ Flow value: 23]

In[ ]:= **F["FlowValue"]**

Out[ ]= 23

- We can see the optimum flow per edge as follows:

In[ ]:= **Text@Grid[Prepend[{#, F[#]} & /@ F["EdgeList"],**
        {Style["Railroad", Bold], Style["Edge Flow", Bold]}]]

Railroad	Edge Flow
Vancouver ⇝ Calgary	7
Vancouver ⇝ Edmonton	16
Calgary ⇝ Regina	11
Edmonton ⇝ Calgary	4
Edmonton ⇝ Saskatoon	12
Regina ⇝ Saskatoon	7
Regina ⇝ Winnipeg	4
Saskatoon ⇝ Winnipeg	19

Out[ ]=

- Here we display the result on top of the map of Canada, downloaded using **CountryData** (for further details see Chapter 5).

In[ ]:= `Panel[Show[{CountryData["Canada", "Shape"], F["FlowGraph"]},`
       `PlotRange → {{-0.31, 0.083}, {0.04, 0.25}}]]`

Out[ ]=
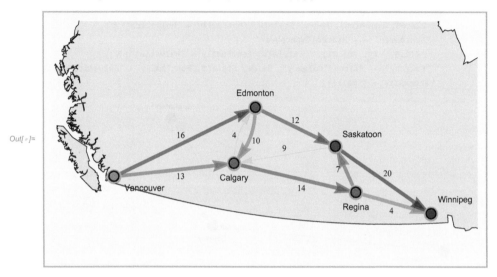

11.7 BlockChain

Let's assume that I'd like to transfer $500 from one of my bank accounts to another one. This kind of transaction doesn't actually require the movement of dollar bills from one bank to another. It just needs a simple change in the banks' electronic balances. Similar transactions happen every day affecting millions of accounts without involving any physical money, just modification of electronic bits.

Until recently, this type of financial activity was mostly done using a network of banks. However, in 2008 someone (or maybe a group of people) under the pseudonym of Satoshi Nakamoto, sent a message to a cryptography forum containing a 9-pages document describing a method for creating a digital coin that could be used as a unit of account and medium of exchange. The following year, the ideas behind that paper were implemented using a virtual coin named bitcoin: https://bitcoin.org/bitcoin.pdf .

Although Nakamoto was talking about a concrete coin, he was really describing a revolutionary technology named blockchain with a multitude of potential applications.

This technology eliminates intermediaries by decentralizing the network. Users, instead of banks or other centralized organizations, take control, and it's them that keep track and verify all the transactions being recorded on the blockchain. One of the most recent applications is the recording of the virtual ownership of digital files through something called non-fungible tokens or NFTs.

Blockchains are basically decentralized digital ledgers in which the entries (blocks) are interconnected and encrypted to protect the privacy and safety of transactions. They are, in other words, distributed and secured databases that can contain any kind of records, not just economic ones. These chains of blocks have a very important requirement: There have to be users (nodes) in charge of verifying and approving the transactions so that the blocks containing them can be added to the ledger.

Although a detailed explanation of blockchains is beyond the scope of this book, it's important to mention that to design one, we would need to consider first a list of basic requirements known as the "five pillars of blockchain":

1. Permissioned/Private. Writing records is exclusive to members; third parties can be granted read access, with the general public excluded. The permissions architecture goes beyond "access = everything" and allows third-party access to specific raw data, as deemed appropriate, for interoperability and application requirements.

2. Decentralized/P2P. Allowing for equal control over the shared database among all permissioned participants and of equal importance; distributing the number of full copies of the ledger to maximize the probability that there will always be a complete record in existence and available for those with permissions to access.

3. Immutability & Data Integrity. Records are guaranteed to be cryptographically secure, with no possibility for bad actors to threaten data integrity.

4. Scalability. The ability to secure trillions of transactions or records without compromising the networks synchronization, security, accessibility, or data integrity.

5. Security. Support for data encryption and the management and enforcement of complex permission settings for participants and third parties.

Mathematica includes functionality to design and manage blockchains and there are many videos covering these topics at:

https://www.wolfram.com/broadcast/s?sx=blockchain

The functions available related to blockchains can be used for:

Configuration ($BlockchainBase)

Reading (BlockchainData, BlockchainBlockData, BlockchainTransactionData and BlockchainAddressData)

Writing purposes (BlockchainTransaction, BlockchainTransactionSign and BlockchainTransactionSubmit):

Here are some examples:

- We can first specify the default blockchain to be used for blockchain computations with $BlockchainBase:

In[ ]:= **$BlockchainBase = "Bitcoin"**

Out[ ]= Bitcoin

- Now we can read the information about the specified default blockchain using BlockchainData:

In[ ]:= **BlockchainData[] // Dataset**

Type	Bitcoin
Name	BTC.main
Core	Bitcoin
Blocks	607 858
LatestBlockHash	0000000000000000000044f9678b5ad59d667b71a25d53e22f9d9c7586e0e516c
MinimumFee	12 601 sat

- The function below shows how to create a Pay-to-Public-Key-Hash (P2PKH) transaction, the most common form of transaction on the Bitcoin network using BlockchainTransaction:

```
In[ ]:= BlockchainTransaction[<|"Inputs" → {<|"TransactionID" →
          "9bef194418607318673caaf8f45d52fb60e8c8586e6ce8414c377a279ca14399",
          "Index" → 0|>}, "Outputs" → {<|"Amount" → $0.0002` , "Address" →
          "munDTMqa9V9Uhi3P21FpkY8UfYzvQqpmoQ"|>, <|"Amount" → $0.16223` ,
          "Address" → "mo9QWLSJ1g1ENrTkhK9SSyw7cYJfJLU8QH"|>},
       "BlockchainBase" → {"Bitcoin", "Testnet"}|>]
```

Out[ ]= BlockchainTransaction[blockchain base: {Bitcoin, Testnet} signed: False fee: 7000 sat]

As of May 2022, *Mathematica* supports three virtual coins: Bitcoin, Ethereum, and Ark (https://ark.io) . This last coin could be very useful if you were interested in developing blockchain applications since there is a written agreement between ARK Ecosystem and Wolfram Blockchain Labs (WBL) that integrates the ARK public API into the Wolfram Language.

WBL: https://www.wolframblockchainlabs.com/

Notice the appearance of the word "Hash" in both, one of the properties of the BlockchainData output and in the name of the blockchain transaction. This term refers to a type of mathematical function that converts an input of arbitrary length into an output of fixed length consisting of a series of unique characters. This function is crucial to validate transactions, safely store passwords, and sign electronic documents.

11.8 Additional Resources

The Wolfram Finance Platform: http://www.wolfram.com/finance-platform
Mathematica's finance related functions:
http://reference.wolfram.com/mathematica/guide/Finance.html
Constrained optimization tutorial: tutorial/ConstrainedOptimizationOverview
Financial, Economics and more: https://www.wolfram.com/wolfram-u/courses/catalog/?topic=finance/ and https://www.wolfram.com/solutions/

12

Faster, Further

This chapter discusses how to make computationally intensive tasks more efficient by distributing the calculations among several cores using parallel computing. In addition, if our computer has a NVidia graphics card, we also have the option of using the card processors through the Compute Unified Device Architecture (CUDA) programming interface, integrated in Mathematica, for extra processing power. The chapter also describes complementary tools such as Workbench, an excellent Integrated Development Environment (IDE) for large Mathematica-based project. Finally, some indications to connexion with other programs and devices are given.

12.1 Parallel Computing

Computer programs are usually run sequentially: A problem is broken down into a series of instructions performed one by one in order. Only one instruction is carried out at any time, with its output being passed to the next one until the entire program has been executed.

Parallel computing consists of breaking down programs into discrete components which can be executed simultaneously. In *Mathematica,* this is done by distributing the individual program parts among multiple kernels. This is possible when the computer we're using has several processors, if we distribute the computation among several machines or a combination of both. (B. Blais, *Introduction to Parallel Computing,* Lawrence Livermore National Laboratory:

https://hpc.llnl.gov/documentation/tutorials/introduction-parallel-computing-tutorial

To do parallel computing with several processors, one of them must act as the master distributing and synchronizing the operations with the rest of the processors (located in the same computer and/or in other computers connected to the grid). This task implies an extra amount of time (communication, synchronization, etc.) that must be taken into account. The objective is to reduce the total computation time compared to how long it would take to perform the calculation sequentially. There are many scientific and technical fields where parallel computing is very useful or even the only realistic way to perform certain types of calculations. For example: in meteorology, there used to be predictive models that took longer to evaluate than the actual prediction window.

In the past, parallel programming was a niche area only available for expensive hardware. However, today even PCs have more than one processor. If you have an NVidia graphics card, you can also use CUDA (included in *Mathematica*) for parallel computing. This computation power gets even bigger if we connect computers to a grid. You can even participate in collaborative projects: One of the best examples of this approach was the well-known SETI program, another example of a collaborative participation project is the study of protein folding: https://fold.it/.

In summary, *Mathematica* has the capability to perform calculations in parallel right out of the box. In addition, it's possible to set up a grid to distribute computations to several computers using either gridMathematica or Wolfram Lightweight Grid Manager

gridMathematica: https://www.wolfram.com/gridmathematica

WLGM: https://www.wolfram.com/lightweight-grid-manager

12.2 Parallel Programming

12.2.1 How to Use More than One Kernel

Usually, computer programs are executed sequentially in a single CPU. Although not exactly the same, we can say that in *Mathematica* a sequential program will be executed in a single kernel, while a program that includes parallel calculations will be distributed among all the available kernels and its instructions executed simultaneously.

- Let's start with a list of numbers that we'd like to factorize (transform them into a product of prime numbers).

```
In[ ]:= numbers = {342895, 423822, 1546,
          168972, 123555, 23440, 1378, 957, 348300, 2987};
```

Factorization plays a crucial role in cryptography. As a matter of fact, the transmission of encrypted information is often done using numbers obtained from multiplying two prime numbers. This technique is the most commonly used one when doing secure commercial transactions over the Internet.

- For factorization purposes, *Mathematica* has the `FactorInteger` function that when used along with `Map` (or `/@`) can be applied to more than one number at once:

In[ ]:= `list = Map[FactorInteger, numbers]`

Out[ ]= {{{5, 1}, {7, 1}, {97, 1}, {101, 1}}, {{2, 1}, {3, 1}, {7, 1}, {10091, 1}},
{{2, 1}, {773, 1}}, {{2, 2}, {3, 1}, {14081, 1}},
{{3, 1}, {5, 1}, {8237, 1}}, {{2, 4}, {5, 1}, {293, 1}},
{{2, 1}, {13, 1}, {53, 1}}, {{3, 1}, {11, 1}, {29, 1}},
{{2, 2}, {3, 4}, {5, 2}, {43, 1}}, {{29, 1}, {103, 1}}}

- The previous result is formatted as a list. Next, we display a table where we can see each number and its corresponding prime factors.

In[ ]:= `Grid[Transpose[{numbers, CenterDot @@ (Superscript @@@ #) & /@ list}],`
`Frame → All]`

Out[ ]=

342 895	$5^1 \cdot 7^1 \cdot 97^1 \cdot 101^1$
423 822	$2^1 \cdot 3^1 \cdot 7^1 \cdot 10\,091^1$
1546	$2^1 \cdot 773^1$
168 972	$2^2 \cdot 3^1 \cdot 14\,081^1$
123 555	$3^1 \cdot 5^1 \cdot 8237^1$
23 440	$2^4 \cdot 5^1 \cdot 293^1$
1378	$2^1 \cdot 13^1 \cdot 53^1$
957	$3^1 \cdot 11^1 \cdot 29^1$
348 300	$2^2 \cdot 3^4 \cdot 5^2 \cdot 43^1$
2987	$29^1 \cdot 103^1$

Operationally, the kernel or logical core of *Mathematica* (not to be mistaken with the physical core of the computer) factorized each number sequentially. That is, it first factorized 342,895, then 423,822 and so on. In this case, the factorization time was very small and so wouldn't justify the use of parallel computations. However, when the factorization involves big numbers, the calculation time increases dramatically so we'd be very interested in using methods to reduce it, such as parallel computing.

For parallel computing purposes, *Mathematica* has certain functions that are added to the usual ones to parallelize calculations, but before we can take advantage of them, we need to set up *Mathematica* properly.

- The standard installation of *Mathematica* includes the possibility of using more than one kernel in the host machine with the actual number depending on the license. We can access the computer configuration by going to: **Edit ▸ Preferences ▸ Parallel**. We will see a screen similar to the one in Figure 12.1 (for further details you can watch the video http://www.wolfram.com/broadcast/screencasts/parallelcomputing).

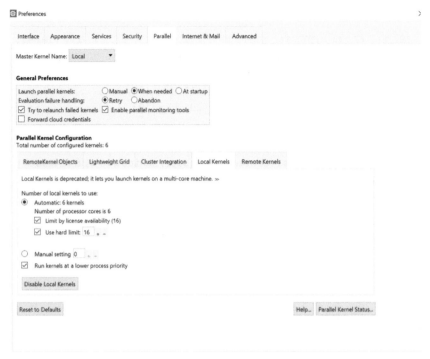

Figure 12.1   Parallel settings in *Mathematica*.

In this case, the number of kernels that will be launched automatically is 4, the same as the number of computer cores. Notice that the kernels are set by default to low priority, giving preference to other computer processes.

- To launch the available kernels we click on the **Parallel Kernel Status...** button. We can also do the same by going the menu bar and choosing: **Evaluation ▸ Parallel Kernel Status**. In both cases, a window similar to the one in Figure 12.2 will appear:

Figure 12.2   Parallel kernel status.

- An alternative way to launch the kernels directly is to write down the following command (for this to work, the kernels should not have been launched previously using any of the methods described above):

In[ ]:= **LaunchKernels[]**

If you launch more cores than the ones available in the computer, you'll get a $Failed message for each core not found (this can happen, for example, if you have configured your machine for WLGM, to which we will refer later, but the application has not been activated yet).

- We can always find out our available kernels using the following command:

In[ ]:= **Kernels[]**

To execute a program in parallel, in addition to having more than one kernel available, the instructions have to explicitly state that their execution should be done in parallel. Usually this is done using **Parallelize[***exp***]**, with *exp* being the regular *Mathematica* command. **Parallelize** will launch the kernels first, if they haven't been launched already using any of the methods described in the previous paragraphs.

- You can also use the commands that directly enable parallel evaluation:

In[ ]:= **?Parallel***

∨ System`								
Parallel`. Array	Parallel`. Comb`. ine	Parallel`. epipe`. d	Paralleli`. zation	Parallel`. Map	Parallel`. ogra`. m	Parallel`. Subm`. it	Parallel`. Table	
Parallel`. AxisP`. lot	Parallel`. Do	Parallel`. Evalu`. ate	Paralleli`. ze	Parallel`. Needs	Parallel`. Produ`. ct	Parallel`. Sum	Parallel`. Try	

- To factorize the numbers previously defined, we put the command we used earlier inside **Parallelize** (Instead of **Parallelize[**Map[*f,expr*]**]** we can also use the equivalent expression ParallelMap[*f,expr*]).

In[ ]:= **Parallelize[Map[FactorInteger, numbers]]**

Out[ ]= {{{5, 1}, {7, 1}, {97, 1}, {101, 1}}, {{2, 1}, {3, 1}, {7, 1}, {10091, 1}},
{{2, 1}, {773, 1}}, {{2, 2}, {3, 1}, {14081, 1}},
{{3, 1}, {5, 1}, {8237, 1}}, {{2, 4}, {5, 1}, {293, 1}},
{{2, 1}, {13, 1}, {53, 1}}, {{3, 1}, {11, 1}, {29, 1}},
{{2, 2}, {3, 4}, {5, 2}, {43, 1}}, {{29, 1}, {103, 1}}}

In this case, we have used all the available kernels to simultaneously factorize the numbers.

We can try to use this command with other functions, although as mentioned before, the execution may not be faster. That will depend on many factors, such as the type of calculation, the number of kernels available, etc. If the computation time is short (a few seconds or less) we will not notice any difference or the parallel calculation may even take longer.

- In the example below, we show the computation time with and without parallelization. When measuring computation times for comparison purposes, it's recommended to start a new session (you can use Quit [] to exit the current one) since, as we have seen in previous chapters, *Mathematica* will automatically reuse previous calculations reducing the execution time significantly. We use AbsoluteTiming instead of Timing because AbsoluteTiming takes into account the entire calculation time while Timing only considers the CPU time. The ";" tells *Mathematica* not to print the output of the command since we're only interested in its computation time.

$In[\circ]:=$ $\sum_{i=1}^{10\,000} i\,!;$ // AbsoluteTiming

$Out[\circ]=$ {0.928053, Null}

$In[\circ]:=$ Parallelize$\left[\sum_{i=1}^{10\,000} i\,!\right];$ // AbsoluteTiming

$Out[\circ]=$ {0.415421, Null}

- Sometimes is not possible to parallelize a computation:

$In[\circ]:=$ Parallelize$\left[\int \frac{1+x^2}{x}\,dx\right]$

$\cdots$ Parallelize: $\int \frac{1+x^2}{x}\,dx$ cannot be parallelized; proceeding with sequential evaluation.

$Out[\circ]=$ $\frac{x^2}{2} + Log[x]$

12.2.2 Defining Functions

When doing parallel programming, it's very important to remember that if we define a function and we'd like it to be used by all the kernels, we have to explicitly state it, as shown in the next example.

- We define a function that generates a random number between 1 and 10.

$In[\circ]:=$ rand := RandomInteger[10]

- Next, we use DistributeDefinitions to parallelize the previous function so it can be used by all the available kernels.

$In[\circ]:=$ DistributeDefinitions[rand];

- Finally, we execute the function in all the different kernels.

$In[\circ]:=$ ParallelEvaluate[rand]

$Out[\circ]=$ {10, 7, 0, 6, 6, 6}

- Once a function has been parallelized, it will remain so even after closing the kernels. In this example, we close the kernels and open them again.

In[ ]:= `CloseKernels[]; LaunchKernels[];`

- The definition still works with the new kernels:

In[ ]:= `ParallelEvaluate[rand]`

Out[ ]= {6, 2, 6, 4, 9, 2}

- Commands in *Mathematica* that start with `Parallel`, such as `ParallelTable` (equivalent to `Parallelize[Table[...]]`), will automatically parallelize the definitions, without the need to use `DistributeDefinitions`, as the following example shows:

In[ ]:= `f1[n_] := 2^n`

In[ ]:= `ParallelTable[f1[k], {k, 1, 6}]`

Out[ ]= {2, 4, 8, 16, 32, 64}

- An alternative way to execute the previous command is by using `With` to give values to the local variables:

In[ ]:= `WaitAll[Table[With[{i = i}, ParallelSubmit[2^i]], {i, 6}]]`

Out[ ]= {2, 4, 8, 16, 32, 64}

- Sometimes, we may be interested in performing the calculation in a specific kernel. In this example, we choose kernel 2 to execute the computations:

In[ ]:= `CloseKernels[]; LaunchKernels[];`

In[ ]:= `kernel2 = Kernels[][[2]]`

Out[ ]= KernelObject[⊞ ◉ Name: Local kernel
KernelID: 14]

Remember that if we don't parallelize variables, we will find unexpected behavior.

- In this example, the equality **a === 2** (remember that **SameQ** (===) checks whether two expressions are identical) returns false even though **a = 2**, because **a** has not been defined for all the kernels:

In[ ]:= `a = 2;`
`ParallelEvaluate[a === 2, kernel2]`

Out[ ]= True

- We can fix the problem using `With`:

In[ ]:= `With[{a = 2}, ParallelEvaluate[a === 2, kernel2] ]`

Out[ ]= True

In[ ]:= `CloseKernels[];`

12.2.3 ParallelTry

In[ ]:= `LaunchKernels[];`

With `ParallelTry` we can evaluate a command in different kernels and get the result from the kernel that executes it first. This can be useful, for example, when comparing different computation methods.

- Here we compare different methods for evaluating the minimum of a function. The output shows the fastest one:

In[ ]:= `AbsoluteTiming[ParallelTry[`

$$\{\#, \text{NMinimize}[\{e^{Sin[50 x]} - Sin[10 (x + y)] + \frac{1}{4} (x^2 + y^2), y \geq 0 \&\& x \geq 0\},$$

$$\{x, y\}, \text{Method} \rightarrow \#]\} \&, \{\text{"DifferentialEvolution"},$$

`"NelderMead", "RandomSearch", "SimulatedAnnealing"}]]`

Out[ ]= {0.141646, {NelderMead, {-0.628918, {x → 0.0942305, y → 0.0625364}}}}

12.2.4 ParallelCombine

`ParallelCombine` distributes part of a computation among all the available kernels combining the partial results. It's normally used with list elements. Its syntax is: **`ParallelCombine`**[*f*, {e_1, ..., e_n}] that applies *f* to the elements of a list and groups the partial results.

- The function below calculates the first 9 primes. It's the same as **`Prime`**[{1, ..., 9}] except that the calculation in this case is done in parallel:

In[ ]:= `ParallelCombine[Prime, {1, 2, 3, 4, 5, 6, 7, 8, 9}]`

Out[ ]= {2, 3, 5, 7, 11, 13, 17, 19, 23}

12.2.5 ParallelEvaluate

The function `ParallelEvaluate` evaluates an expression in different kernels and returns the result from each one of them. By default it uses all the available kernels but we can specify which ones to use. This function is commonly employed in calculations involving Monte Carlo simulation. The next example shows how to calculate π through simulation (an illustrative but quite inefficient method).

- First, we generate *m* pairs of real numbers {*x*, *y*} between 0 and 1. Then we find out which ones fall inside the circle with radius *r* = 1 with **`Norm`**[x y] < 1 equivalent to $x^2 + y^2 < 1$. Finally, we multiply by 4 since we're only generating positive {*x*, *y*}, simulating just 1/4 of the circle:

In[ ]:= `With[{n = 10^5}, ParallelEvaluate[`
`    4. Count[RandomReal[{0, 1}, {n, 2}], xy_ /; Norm[xy] < 1] / n]]`

Out[ ]= {3.13396, 3.14492, 3.13684, 3.13684, 3.14632, 3.14472}

- The result is an estimate from each kernel. We calculate the mean to get a final value:

In[ ]:= `Mean[%]`

Out[ ]= 3.1406

12.2.6 ParallelMap

`ParallelMap` is another commonly used function that, as previously mentioned, is equivalent to `Map` for sequential computations.

- We are going to solve again the previous example of approximating π. We generate pairs of points $\{x, y\}$ with values between -1/2 and 1/2. Next we select those points inside a circle with radius 1/2 centered at (0,0) using again **Norm**[r] $< r$. We then proceed to calculate π using the output given by the command below:

In[∘]:= `Solve[(pi r^2) /1 == interiorPoints / total, pi]`

Out[∘]:= $\left\{\left\{\text{pi} \rightarrow \dfrac{\text{interiorPoints}}{r^2 \, \text{total}}\right\}\right\}$

In[∘]:= `r = 1/2;`

In[∘]:= `DistributeDefinitions[r]`

Out[∘]:= `{r}`

In[∘]:= `pts = ParallelTable[RandomReal[{-r, r}, {25 000, 2}], {Length[Kernels[]]}];`

In[∘]:= `interiorpts = ParallelMap[Select[#, (Norm[#] < r &)] &, pts];`

- By combining the results, we get the following approximation:

In[∘]:= `DistributeDefinitions[pts, interiorpts];`
`N[Mean[ParallelTable[Length[interiorpts[[i]]] / (Length[pts[[i]]] * r^2),`
`{i, Length[Kernels[]]}]]]`

Out[∘]:= `3.14755`

12.2.7 Finding the Roots of an Equation

We'd like to find the solutions to the equation $\cos(t^2) = e^{-t} + 0.3$ with `FindRoot`, a function that uses numerical algorithms requiring an initial guess.

- The graph below displays the multiple solutions (actually an infinite number of them) that correspond to the intersection points of both curves:

In[∘]:= `Plot[{Cos[t^2], Exp[-t] + .3}, {t, 0, 10}]`

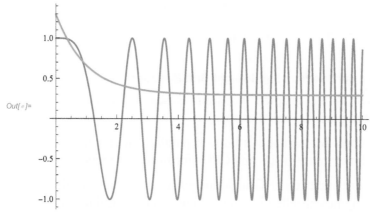

- If we only give the function one initial guess, we will usually get a solution close to it:

In[∘]:= `FindRoot[{Cos[t^2] == Exp[-t] + .3}, {t, 3}]`

Out[∘]:= `{t → 2.26454}`

We are going to find the solution with and without parallelization. We just want to compare the computation times so we'll not show the results.

- Using multiple initial guesses, we can find all the equation roots for a given domain. We have used Map since what we're really doing is solving the equation for different starting points:

```
In[ ]:= initval = Table[x, {x, 0, 4 Pi, .1}];
```

```
In[ ]:= vals = Map[FindRoot[{Cos[t^2] == Exp[-t] + .3}, {t, #}] &, initval]; //
        AbsoluteTiming
```

```
Out[ ]= {0.0277301, Null}
```

- The command below solves the problem replacing Map with ParallelMap.

```
In[ ]:= pvals = ParallelMap[FindRoot[{Cos[t^2] == Exp[-t] + .3}, {t, #}] &,
        initval]; // AbsoluteTiming
```

```
Out[ ]= {0.0536711, Null}
```

In this case, the difference is negligible since, as mentioned before, there are many factors that may influence the computation time.

- We visually represent the solutions:

```
In[ ]:= Show[Plot[{Cos[t^2], Exp[-t] + .3}, {t, 0, 12}], Graphics[
        {Red, Point[MapThread[{t /. #1, #2} &, {pvals, Cos[t^2 /. pvals]}]]}]]
```

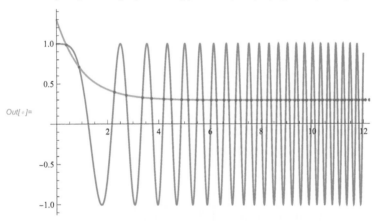

12.2.8 GridMathematica and Wolfram Lightweight Grid Manager

As mentioned earlier in the chapter, Wolfram Research offers two additional programs for performing parallel calculations.

GridMathematica extends Mathematica's built-in parallelization capabilities to run more tasks in parallel over more GPUs and CPUs for faster computations.

Alternatively, Wolfram Lightweight Grid Manager makes it easy to discover and connect to Wolfram engines installed in other computers in a local network.

12.3 The Mandelbrot Set

12.3.1 Defining the Mandelbrot Set

The Mandelbrot set is arguably the most famous fractal set. It can be created by recursion as follows :

$$z_0 = C$$

$$z_{n+1} = z_n^2 + C$$

where C is any complex number.

http://mathworld.wolfram.com/MandelbrotSet.html

If we take C such that z_{n+1} doesn't tend to infinity, the points generated by this recursion belong to the Mandelbrot set.

- The previous series can be easily computed in *Mathematica* with `FixedPointList`.

In[ ]:= `FixedPointList[#^2 + c &, c, 5]`

Out[ ]= $\left\{ c, \ c + c^2, \ c + \left(c + c^2\right)^2, \ c + \left(c + \left(c + c^2\right)^2\right)^2, \right.$

$\left. c + \left(c + \left(c + \left(c + c^2\right)^2\right)^2\right)^2, \ c + \left(c + \left(c + \left(c + \left(c + c^2\right)^2\right)^2\right)^2\right)^2 \right\}$

- With $C = 1$, the series diverges (in the example we display only the first 6 terms). It tends to infinity in absolute terms, so it doesn't belong to it.

In[ ]:= `FixedPointList[#^2 + 1 &, 0, 5]`

Out[ ]= `{0, 1, 2, 5, 26, 677}`

- However if $C = -1$, the series is bounded and therefore belongs to the set.

In[ ]:= `FixedPointList[#^2 - 1 &, 0, 5]`

Out[ ]= `{0, -1, 0, -1, 0, -1}`

12.3.2 Visualizing the Mandelbrot Set

Often, the Mandelbrot set is represented using the escape-time coloring algorithm based on applying different colors depending on the number of iterations required to prove if a series belongs to the set. Darker colors (in our example bright red) indicate that only a few iterations are needed while lighter ones mean that many iterations were required. There's always a limit for the maximum number of iterations allowed. Since *Mathematica* 10, the build-up function `MandelbrotSetPlot` is available, however, we will build a function step by step to represent the Mandelbrot set.

- `FixedPointList` with the option `SameTest` enables us to set a logical test to stop the iteration as soon as it returns true starting with the first two values of the series. The command below returns the number of iterations needed to get an absolute value greater than 2 (the limit traditionally used) and if after 50 iterations, the absolute value still doesn't meet the condition, it returns the number of iterations +1. The higher the number of iterations, the higher the certainty that the series belongs to the Mandelbrot set:

In[ ]:= `Length[`
        `FixedPointList[#^2 - 0.2 + 0.1 I &, 0, 50, SameTest → (Abs[#] > 2 &)]]`

Out[ ]= `51`

- We define a function that implements the previous approach for a given complex number. In this case we use a maximum of 20 iterations:

In[ ]:= `mf1[z_] :=`
        `Length[FixedPointList[#^2 + z &, z, 20, SameTest → (Abs[#] > 2 &)]]`

- We apply it to find out what points in the complex plane $x + i\,y$ belong to it. We add `AbsoluteTiming` to compare the computation times using different methods.

In[ ]:= `pts1 = Table[mf1[x + I y], {x, -2, 1, 0.002}, {y, -1, 1, 0.002}]; //`
         `AbsoluteTiming`

Out[ ]= {22.5427, Null}

- We use `ArrayPlot` to visualize the previous points.

In[ ]:= `ArrayPlot[pts1, ColorFunction → Hue]`

Out[ ]=

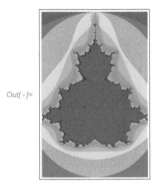

In[ ]:=

12.3.3 Faster

In this section, we show several techniques, apart from parallelization, to reduce computation time.

These techniques require the presence of a C compiler in the computer. (CCompilerDriver/tutorial/Overview). For 32-bit systems, an option is to install Visual C++ Express, a Microsoft application that can be downloaded for free from:

http://www.microsoft.com/visualstudio

In case of 64-bit systems, one option is to install the 64-bit version of MinGW . In our case, we have downloaded mingw-w64-bin_i686-mingw_20111220 and copied it to the folder C:\mingw.

Mingw-w64: http://mingw-w64.sourceforge.net

- To run the C compiler we need to load the following package first:

In[ ]:= `Needs["CCompilerDriver`"]`

- The command below will display all the C compilers available in your computer:

In[ ]:= `CCompilers[]`

Out[ ]= { }

- If the function above returns an empty list it may be that either you don't have any compiler installed or, as in our case, you have installed MinGW in folder: C:/MinGW. To let *Mathematica* know that you have the open-source software in your computer, execute the following commands:

In[ ]:= `Needs["CCompilerDriver`GenericCCompiler`"]`

```
In[ ]:= $CCompiler =
          {"Compiler" → GenericCCompiler, "CompilerInstallation" → "C:/MinGW",
           "CompilerName" → "x86_64-w64-mingw32-gcc.exe"};
```

Now we can launch the available kernels as explained before. To use kernels over a grid we need to make sure that we have access to them through WLGM.

- Here we repeat the calculation from the previous section but in parallel:

```
In[ ]:= mf2[z_] :=
          Length[FixedPointList[#^2 + z &, z, 20, SameTest → (Abs[#] > 2 &)]]

In[ ]:= DistributeDefinitions[mf2];

In[ ]:= pts2 = ParallelTable[mf2[x + I y],
          {x, -2, 1, 0.002}, {y, -1, 1, 0.002}]; // AbsoluteTiming

Out[ ]= {5.33571, Null}
```

- We can still significantly reduce the computation time with the command `Compile` (not limited to parallel calculations). We use it to indicate that we're going to evaluate a function only for numeric inputs. This way the program saves time by not performing the checks associated to symbolic calculations and can apply its own routines optimized for numerical calculations.

```
In[ ]:= mf3 = Compile[{{c, _Complex}},
          Length[FixedPointList[#^2 + c &, c, 50, SameTest → (Abs[#2] > 2.0 &)]]];

In[ ]:= DistributeDefinitions[mf3];
          pts3 = ParallelTable[mf3[x + y ⅈ],
          {x, -2, 1, 0.002}, {y, -1, 1, 0.002}]; // AbsoluteTiming

Out[ ]= {1.05738, Null}
```

- We can perform the computation even faster by taking advantage of some of the options of `Compile`. Using this function we can convert the *Mathematica*-defined function into C code that is compiled and dynamically linked:

```
In[ ]:= cfun = Compile[ {{x}}, x Sin[x^2],
          CompilationTarget → "C", RuntimeAttributes → {Listable}];
          cfun[{3, 4}]

Out[ ]= {1.23636, -1.15161}
```

- Now, we can define the fastest function so far:

```
In[ ]:= mf4 = Compile[{{c, _Complex}}, Module[{num = 1}, FixedPoint[ (num++;
          #1^2 + c) &, 0, 99, SameTest → (Re[#1]^2 + Im[#1]^2 ≥ 4 &)];
          num], CompilationTarget → "C",
          RuntimeAttributes → {Listable}, Parallelization → True];

In[ ]:= pts4 = mf4[Table[x + ⅈ y, {x, -2, 1, 0.002}, {y, -1, 1, 0.002}]]; //
          AbsoluteTiming

Out[ ]= {1.46059, Null}
```

- We display the results from the three functions using different color templates and compare them with the built-in function `MandelbrotSetPlot`:

In[ ]:= `GraphicsRow[{ArrayPlot[pts2, ColorFunction → "GreenPinkTones"],`
`        ArrayPlot[pts3, ColorFunction → "BlueGreenYellow"],`
`        ArrayPlot[pts4 , ColorFunction → "Rainbow"]}]`

In[ ]:= `GraphicsRow[{MandelbrotSetPlot[ColorFunction → "GreenPinkTones"],`
`        MandelbrotSetPlot[ColorFunction → "BlueGreenYellow"],`
`        MandelbrotSetPlot[ColorFunction → "Rainbow"]}]`

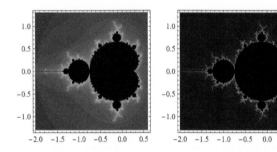 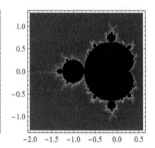

If you are a programmer you can even generate C code that can be converted into an executable file (CCodeGenerator/tutorial/Overview) or you can call external programs (guide/CallingExternalPrograms).

- To generate C code we need to load the CCodeGenerator package. In the example below, we load it after clearing all the variables from memory:

In[ ]:= `Clear["Global`*"]`

In[ ]:= `Needs["CCodeGenerator`"]`

- The next command generates optimized C code for a given expression:

In[ ]:= `s = ExportString[Compile[x, Sin[x^2]], "C"];`

- We use `Panel` in the following example to show the output of the previous instruction inside a frame displaying the generated code along with all the required headings and libraries necessary to compile an executable file.

```
In[ ]:= Panel[z[StringReplace[s, "\r\n" → "\n"],
        {Full, 300}, Scrollbars → {Automatic, True}],
      BaseStyle → {FontFamily → "Courier", FontSize → 8}]
```

```
z[#include "math.h"

#include "WolframRTL.h"

static WolframCompileLibrary_Functions funStructCompile;

static mbool initialize = 1;

#include "m-4a3e4639-9687-41a0-8159-98ba6a7d8dc2.h"

DLLEXPORT int Initialize_m-4a3e4639-9687-41a0-8159-98ba6a7d8dc2(WolframLibraryData libData)
{
if( initialize)
{
funStructCompile = libData->compileLibraryFunctions;
initialize = 0;
}
return 0;
}

DLLEXPORT void Uninitialize_m-4a3e4639-9687-41a0-8159-98ba6a7d8dc2(WolframLibraryData libData)
{
if( !initialize)
{
initialize = 1;
}
}

DLLEXPORT int
    m-4a3e4639-9687-41a0-8159-98ba6a7d8dc2(WolframLibraryData libData, mreal A1, mreal *Res)
{
mreal R0_0;
mreal R0_1;
mreal R0_2;
R0_0 = A1;
R0_1 = R0_0 * R0_0;
R0_2 = sin(R0_1);
*Res = R0_2;
funStructCompile->WolframLibraryData_cleanUp(libData, 1);
return 0;
}

, {Full, 300}, Scrollbars → {Automatic, True}]
```

Out[ ]= is shown to the left of the panel.

12.4 Comparing Organisms Genetically

Let's see how parallel computing is used in the real world. In this case, we are trying to determine the genetic proximity of organisms using a string metric known as the Levenshtein distance.

The Levenshtein distance is used in information theory to measure the similarity between two sequences. We can define it as the minimum number of single character edits necessary to convert one sequence into the other one.

For example: To go from kitten to sitting, we proceed as follows:

kitten → sitten (we substitute 'k' with 's')
sitten → sittin (we substitute 'e' with 'i')
sittin → sitting (we insert 'g' at the end).

This means that the Levenshtein distance is 3 since we have made 3 edits to turn kitten into sitting.

- This metric can be calculated in *Mathematica* with `EditDistance`:

In[ ]:= `EditDistance["kitten", "sitting"]`

Out[ ]= 3

Let's apply this metric to compare the genomes of three organisms using the files located in the Data directory that, along with other examples in this book, can be downloaded from http://diarium.usal.es/guillermo.

- We first set the working directory to the one where the files are located (Data). This way we can use Import directly without the need to state the entire path:

In[ ]:= `SetDirectory[FileNameJoin[{NotebookDirectory[], "Data"}]]`

Out[ ]= `G:\Mi unidad\Mathematica\MBN13\Data`

In[ ]:= `picrogenes = Import["picrophilus_torridus_gene.fna"];`
  `acidogenes = Import["thermoplasma_acidophilum_gene.fna"];`
  `volcaniumgenes = Import["thermoplasma_volcanium_gene.fna"];`

In the previous example, the computation time was very short, but, when the number of characters to compare increases, the time required for the calculations increases drastically.

- Let's see the partial contents of one of the files containing the genome of the picrogenes. The command below shows some of their DNA bases (A,C, G, T):

In[ ]:= `picrogenes`

- These three organisms are very simple and their genome size (total number of genes) is, respectively:

In[ ]:= `Length /@ {picrogenes, acidogenes, volcaniumgenes}`

Out[ ]= `{1598, 1575, 1610}`

- Let's load the imported data to analyze them using parallelization:

In[ ]:= `With[{ picrogenes = picrogenes, acidogenes = acidogenes,`
  `volcaniumgenes = volcaniumgenes}, ParallelEvaluate[picro = picrogenes;`
  `acido = acidogenes; volcanium = volcaniumgenes;];];`

- The command below shows the data for one gene that will be used as an input for the first kernel:

In[ ]:= `ParallelEvaluate[picro⟦1⟧, Kernels[]⟦1⟧]`

Out[ ]= ATGAATTCGGATAACTTTCAATCCTATTTAGGTTATTATTTAACCTGGATTTCCTTGGCCATATTTGCCGAGA⸫
TACTATACTTTCAATCCTATTTAGGTTATTATTTAACATATATCGGATTTCAAACCCTTATAGATAGACT⸫
AGACTTTCAATCCTATTTAGGTTATTATTTAACGCCAGAGGCATTTCTCCGTGCGTGGTCCTTTAATGAC⸫
TTTCAATCCTATTTAGGTTATTATTTAACACAACAGGCAAAAACAGAATTAAGACTGGCCATTGAGCTTT⸫
CAATCCTATTTAGGTTATTATTTAACCCGTTTCTAATAAACAATATAATAAATGATTTTTAA

For the next example, we want to compare the first eight genes of *thermoplasma acidophilum* with those of *thermoplasma volcanium*. However, before doing it, let's discuss a simpler case first. In the example below we compare the initial 4 bases of the first 2 genes of volcaniumgenes, with their respective counterparts in *thermoplasma acidophilum*.

- We start by extracting the relevant bases and genes from both organisms:

In[ ]:= `{vo1, vo2, aci1, aci2} =`
    `{StringTake[volcaniumgenes⟦1⟧, 4], StringTake[volcaniumgenes⟦2⟧, 4],`
    `StringTake[acidogenes⟦1⟧, 4], StringTake[acidogenes⟦2⟧, 4] }`

Out[ ]= `{ATGA, ATGA, ATGT, ATGG}`

Notice how the first two bases of *thermoplasma volcanium* are actually the same.

- To make the comparison we'll use a combination of the `EditDistance` and `Min` functions. However, in the command below we use **f1** and **f2** first to gain a better understanding of how the actual instruction works. We show the logic in four steps:

In[ ]:= `f2[Map[f1[x, #] &, {aci1, aci2}]]`

Out[ ]= `f2[{f1[x, ATGT], f1[x, ATGG]}]`

In[ ]:= `Map[Function[{x}, f2[Map[f1[x, #] &, {aci1, aci2}]]], {vo1, vo2}]`

Out[ ]= `{f2[{f1[ATGA, ATGT], f1[ATGA, ATGG]}],`
    `f2[{f1[ATGA, ATGT], f1[ATGA, ATGG]}]}`

- Here **f1** and **f2** are replaced with `EditDistance` and `Min`, respectively:

In[ ]:= `Map[Function[{x}, f2[Map[EditDistance[x, #] &, {aci1, aci2}]]],`
    `{vo1, vo2}]`

Out[ ]= `{f2[{1, 1}], f2[{1, 1}]}`

In[ ]:= `Map[Function[{x}, Min[Map[EditDistance[x, #] &, {aci1, aci2}]]],`
    `{vo1, vo2}]`

Out[ ]= `{1, 1}`

- Finally we apply the previous pattern using `ParallelMap` to compare the first eight genes of *thermoplasma acidophilum* with those of *thermoplasma volcanium*:

In[ ]:= `acidovolcanium =`
    `ParallelMap[Function[{x}, Min[Map[EditDistance[x, #] &, acido]]],`
    `volcaniumgenes⟦1 ;; 8⟧]; // AbsoluteTiming`

Out[ ]= `{20.9208, Null}`

Depending on the computer being used, the calculation time may take up to several minutes.

In[ ]:= **acidovolcanium**

Out[ ]= {304, 700, 265, 227, 497, 243, 122, 104}

- We proceed the same way to compare the genes between *picrophilus torridus* and *thermoplasma volcanium*:

In[ ]:= **picrovolcanium =**
    ParallelMap[Function[{x}, Min[Map[EditDistance[x, #] & , picro]]],
    volcaniumgenes〚1 ;; 8〛]; // AbsoluteTiming

Out[ ]= {22.301, Null}

In[ ]:= **picrovolcanium**

Out[ ]= {399, 968, 442, 353, 550, 240, 180, 124}

- The graph below shows the genetic proximity of the first eight genes in both cases: *thermoplasma acidophilum* vs. *thermoplasma volcanium* and *picrophilus torridus* vs. *thermoplasma volcanium*:

In[ ]:= **ListLinePlot[{acidovolcanium, picrovolcanium},**
    PlotLegends → {"acido vs. volcanium", "picro vs. volcanium"}]

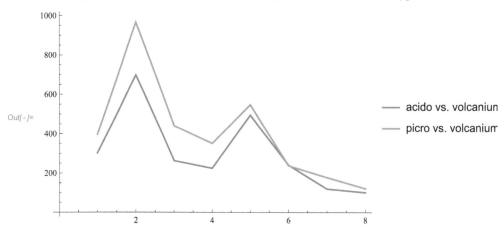

- Comparing *thermoplasma acidophilum* vs. *thermoplasma volcanium* and *picrophilus torridus* vs. *thermoplasma volcanium* in terms of their mean distances we get:

In[ ]:= **{Mean[acidovolcanium], Mean[picrovolcanium]} // N**

Out[ ]= {307.75, 407.}

- A more efficient alternative calculation method using WaitAll is:

In[ ]:= **(acidovolcanium = Map[Function[{gene}, ParallelSubmit[**
    Min[EditDistance[gene, #] & /@ acido]]], volcaniumgenes〚1 ;; 8〛];
  picrovolcanium = Map[Function[{gene}, ParallelSubmit[
    Min[EditDistance[gene, #] & /@ picro]]], volcaniumgenes〚1 ;; 8〛];
  acidovolcanium;
  picrovolcanium;
  acidovolcaniumresult = WaitAll[acidovolcanium];
  picrovolcaniumresult = WaitAll[picrovolcanium];) // AbsoluteTiming

Out[ ]= {23.511, Null}

The calculation time is shorter than the sum of the times of the previous two computations.

- The results are identical to the previously obtained ones:

In[ ]:= `Mean[acidovolcaniumresult] // N`

Out[ ]= `307.75`

In[ ]:= `Mean[picrovolcaniumresult] // N`

Out[ ]= `407.`

12.5 Software Development with Wolfram Workbench

If you are going to build large software systems in *Mathematica* (tutorial/BuildingLargeSoftwareSystemsInTheWolframLanguage) you may be interested in using Wolfram Workbench (http://www.wolfram.com/products/workbench), an Integrated Development Environment (IDE) based on software from the Eclipse Foundation (http://www.eclipse.org). Workbench can help you edit, test and debug code. It can also generate help files than can be integrated within the *Mathematica* documentation. It's highly recommended if you'd like to develop a high-quality enterprise application.

A large project will require two things: Code (consisting normally of several packages) and extensive documentation. Workbench will help you get both.

Wolfram Workbench requires the following software:

1. *Mathematica*

2. Java Development Kit (JDK) 8 or later

3. Eclipse 4.6 (Neon) or later

Note that the Wolfram Workbench plugin is compatible with the various standard Eclipse IDEs.

Wolfram recommends having the latest version of these programs installed before setting up the Wolfram Workbench plugin.

- To set up Wolfram Workbench, install the plugin inside Eclipse and configure it to point to your *Mathematica* installation as it is shown in the following: http://support.wolfram.com/kb/27221 instructions.

- Ones the Wolfram Workbench plugin is installed, to verify it, choose **Help** ▶ **About Eclipse** and identify the Wolfram Workbench logo toward the bottom of the window.

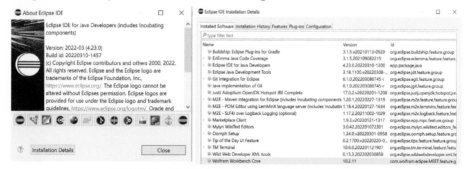

Figure 12.3   The Wolfram Workbench plugin is now installed.

- Configuring the Wolfram Workbench Plugin: Open the Eclipse Preferences (Windows/Linux: **Window ▶ Preferences**, Mac: **Eclipse ▶ Preferences**) and select the Wolfram category on left-hand side of the Preferences window.

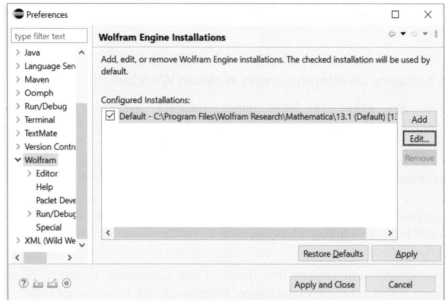

Figure 12.4   Workbench Configuration Page.

- To create a new application project, go to **File ▶ New ▶ Project ▶ Application Project** and type the name of your project.

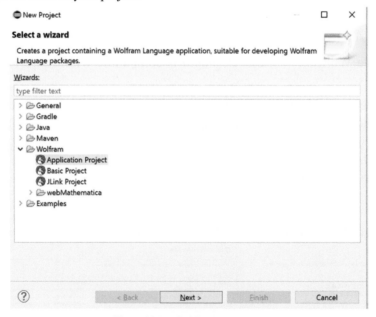

Figure 12.5   Creating a new project.

- After clicking on **Next**, we can assign a name to the application (in this example: Project1).

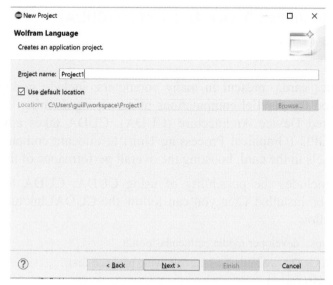

Figure 12.6 Creating an application project.

- Click on **Finish**. You'll see in the Package Explorer tab the projects that you have. Select Project1 and press on PacletInfo.m. Figure 12.6 will appear with basic details about your project: Author, project version, *Mathematica* version, and a brief description.

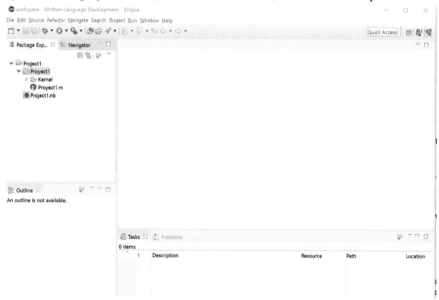

Figure 12.7 Filling in the application details for the new project.

To learn more you can always visit: http://www.wolfram.com/workbench/

At the bottom of the webpage (see Figure 12.7) you will find a link to many useful videos. Additionally, you can also go through examples of completed projects by downloading some of them. It will help you to build your own projects and their associated documentation.

12.6 Compute Unified Device Architecture (CUDA)

■ We start a new session

In[ ]:= **Quit[]**

NVidia graphics cards, present in many computers, include multiple processors that can be used for parallel computations using a programming language named Compute Unified Device Architecture (CUDA). CUDA takes advantage of the power of the GPU (Graphical Processing Unit) to allocate computer instructions among the kernels in the card, boosting the overall performance of the system.

Mathematica includes the possibility of using CUDA. CUDA toolkit for your system should be installed.Then you can follow the CUDALink/tutorial/Overview procedure as follow.

CUDA toolkit: https://developer.nvidia.com/cuda-toolkit

■ To program in CUDA we must load the following package first:

In[ ]:= **Needs["CUDALink`"]**

In[ ]:= **CUDAQ[]**

Out[ ]= True

> If CUDAQ returns False, then *CUDALink* will not work. For more information, read "*CUDALink* Setup".

■ For *CUDALink* to work, you need to have an up-to-date driver. You can check this for your system by running CUDADriverVersion.

In[ ]:= **CUDAInformation[]**

Now we can use the CUDA functions. For further information, please see the tutorial: CUDALink/tutorial/Overview.

12.7 Connexion with Other Programs and Devices

12.7.1 Connexion with Other Programs

Another tool for developers is the Wolfram Engine for Developers. It has the same core engine as *Mathematica* but with a different interface and different licensing. While the complete program is used primarily for interactive computing with the Wolfram Notebook interface, the Wolfram Engine for Developers is intended to be called by other programs or web servers, using a variety of program communication interfaces. This tool is licensed for pre-production use in developing software and unlike *Mathematica*, it is not licensed for generating outputs for commercial or organizational use.

According to the Wolfram website: "The Wolfram Engine is the heart of Wolfram products such as Mathematica, Wolfram|One and Wolfram|Alpha".

"Wolfram Engine Community Edition is freely available for pre-production software development": https://www.wolfram.com/engine/

If you are a Julia, NodeJS, Python, R, or Ruby user, you can perform computations using theses languages within *Mathematica*. Please refer to "ref/ExternalEvaluate" and "workflow/ConfigurePythonForExternalEvaluate" in the Documentation Center for further details.

■ For instance, if you have installed Python, use > at the beginning of a input cell and after selecting it as the language, you can start writing Python code (Figure 12.8):

Type > to get a Python code cell that uses ExternalEvaluate to evaluate:

```
import math
[math.sqrt(i) for i in range(10)]
```

Out[5]= {0., 1., 1.41421, 1.73205, 2., 2.23607, 2.44949, 2.64575, 2.82843, 3.}

Figure 12.8   Using Python within *Mathematica*.

12.7.2 External Devices

Mathematica's connectivity capabilities have been greatly enhanced with the availability of a Wolfram Language framework for connecting external devices.

"The Wolfram Device Framework creates symbolic objects that represent external devices, streamlines interaction with devices, and facilitates the authoring of device drivers. A "device" in the framework can represent both an actual device, such as a temperature sensor, or encapsulate a port, such as a serial port"

The tutorial Developing Device Drivers: https://reference.wolfram.com/language/tutorial/DevelopingDeviceDrivers.html explains the internals of the framework for advanced users and developers of device drivers.

 For details of the interaction with devices see: "Using Connected Devices":

https://reference.wolfram.com/language/guide/UsingConnectedDevices.html

As example, if you happen to be an amateur astronomer, you may enjoy the following blog entry:

http://blog.wolfram.com/2014/12/29/serial-interface-control-of-astronomical-telescopes/

It covers examples of how to use the Wolfram Language to control telescopes using the LX200 or Celestron NexStar protocols, very popular among astronomy fans.

12.7.3 WolframAlpha application, APIs and Widgets

Another area that you may be interested in, is the development of applications and widgets for mobile devices such as smartphones and tablets that make using WolframAlpha easier. You can find tools to develop them in:

http://products.wolframalpha.com/developers/

An API function with one parameter named x applied to an Association :

```
In[ ]:= func = APIFunction[{"x" → "Integer"}, #x + 3 &];
```

In[ ]:=   **func[<|"x" → 5|>]**

Out[ ]= 8

https://reference.wolfram.com/language/ref/APIFunction.html

12.8 Additional Resources

The Wolfram Language and the Wolfram Resources are continuously updating. A summary of the new functionality included in the latest *Mathematica* versions can be found in: the "New Features" section of the Documentation Center (**Help ▶ Wolfram Documentation**) or, alternatively, online:
https://reference.wolfram.com/language/guide/RecentlyAddedFeatures.html

To access the following resources, if they refer to the help files, write their locations in a notebook (e.g., CUDALink/guide/CUDALink), select them and press <F1>. In the case of external links, copy the web addresses in a browser:

Parallelization tutorial:
reference.wolfram.com/mathematica/ParallelTools/tutorial/Overview.html.
Wolfram Research products product to parallelization: gridMathematica and Wolfram Lightweight Grid Manager (WLGM).
CUDALink: CUDALink/guide/CUDALink

Information about Workbench: http://www.wolfram.com/products/workbench

Web applications can be developed with webMathematica:
http://www.wolfram.com/products/webmathematica

Wolfram Engine: https://www.wolfram.com/engine/

Index

Printed and bound by CPI Group (UK) Ltd, Croydon, CR0 4YY

17/10/2024

01775666-0004